The Method of Weighted Residuals and Variational Principles

Books in the Classics in Applied Mathematics series are monographs and textbooks declared out of print by their original publishers, though they are of continued importance and interest to the mathematical community. SIAM publishes this series to ensure that the information presented in these texts is not lost to today's students and researchers.

Classics in Applied Mathematics

C. C. Lin and L. A. Segel, *Mathematics Applied to Deterministic Problems in the Natural Sciences*

Johan G. F. Belinfante and Bernard Kolman, *A Survey of Lie Groups and Lie Algebras with Applications and Computational Methods*

James M. Ortega, *Numerical Analysis: A Second Course*

Anthony V. Fiacco and Garth P. McCormick, *Nonlinear Programming: Sequential Unconstrained Minimization Techniques*

F. H. Clarke, *Optimization and Nonsmooth Analysis*

George F. Carrier and Carl E. Pearson, *Ordinary Differential Equations*

Leo Breiman, *Probability*

R. Bellman and G. M. Wing, *An Introduction to Invariant Imbedding*

Abraham Berman and Robert J. Plemmons, *Nonnegative Matrices in the Mathematical Sciences*

Olvi L. Mangasarian, *Nonlinear Programming*

*Carl Friedrich Gauss, *Theory of the Combination of Observations Least Subject to Errors: Part One, Part Two, Supplement*. Translated by G. W. Stewart

U. M. Ascher, R. M. M. Mattheij, and R. D. Russell, *Numerical Solution of Boundary Value Problems for Ordinary Differential Equations*

K. E. Brenan, S. L. Campbell, and L. R. Petzold, *Numerical Solution of Initial-Value Problems in Differential-Algebraic Equations*

Charles L. Lawson and Richard J. Hanson, *Solving Least Squares Problems*

J. E. Dennis, Jr. and Robert B. Schnabel, *Numerical Methods for Unconstrained Optimization and Nonlinear Equations*

Richard E. Barlow and Frank Proschan, *Mathematical Theory of Reliability*

Cornelius Lanczos, *Linear Differential Operators*

Richard Bellman, *Introduction to Matrix Analysis, Second Edition*

Beresford N. Parlett, *The Symmetric Eigenvalue Problem*

Richard Haberman, *Mathematical Models: Mechanical Vibrations, Population Dynamics, and Traffic Flow*

Peter W. M. John, *Statistical Design and Analysis of Experiments*

Tamer Başar and Geert Jan Olsder, *Dynamic Noncooperative Game Theory, Second Edition*

Emanuel Parzen, *Stochastic Processes*

Petar Kokotović, Hassan K. Khalil, and John O'Reilly, *Singular Perturbation Methods in Control: Analysis and Design*

*First time in print.

Classics in Applied Mathematics (continued)

Jean Dickinson Gibbons, Ingram Olkin, and Milton Sobel, *Selecting and Ordering Populations: A New Statistical Methodology*

James A. Murdock, *Perturbations: Theory and Methods*

Ivar Ekeland and Roger Témam, *Convex Analysis and Variational Problems*

Ivar Stakgold, *Boundary Value Problems of Mathematical Physics, Volumes I and II*

J. M. Ortega and W. C. Rheinboldt, *Iterative Solution of Nonlinear Equations in Several Variables*

David Kinderlehrer and Guido Stampacchia, *An Introduction to Variational Inequalities and Their Applications*

F. Natterer, *The Mathematics of Computerized Tomography*

Avinash C. Kak and Malcolm Slaney, *Principles of Computerized Tomographic Imaging*

R. Wong, *Asymptotic Approximations of Integrals*

O. Axelsson and V. A. Barker, *Finite Element Solution of Boundary Value Problems: Theory and Computation*

David R. Brillinger, *Time Series: Data Analysis and Theory*

Joel N. Franklin, *Methods of Mathematical Economics: Linear and Nonlinear Programming, Fixed-Point Theorems*

Philip Hartman, *Ordinary Differential Equations, Second Edition*

Michael D. Intriligator, *Mathematical Optimization and Economic Theory*

Philippe G. Ciarlet, *The Finite Element Method for Elliptic Problems*

Jane K. Cullum and Ralph A. Willoughby, *Lanczos Algorithms for Large Symmetric Eigenvalue Computations, Vol. I: Theory*

M. Vidyasagar, *Nonlinear Systems Analysis, Second Edition*

Robert Mattheij and Jaap Molenaar, *Ordinary Differential Equations in Theory and Practice*

Shanti S. Gupta and S. Panchapakesan, *Multiple Decision Procedures: Theory and Methodology of Selecting and Ranking Populations*

Eugene L. Allgower and Kurt Georg, *Introduction to Numerical Continuation Methods*

Leah Edelstein-Keshet, *Mathematical Models in Biology*

Heinz-Otto Kreiss and Jens Lorenz, *Initial-Boundary Value Problems and the Navier-Stokes Equations*

J. L. Hodges, Jr. and E. L. Lehmann, *Basic Concepts of Probability and Statistics, Second Edition*

George F. Carrier, Max Krook, and Carl E. Pearson, *Functions of a Complex Variable: Theory and Technique*

Friedrich Pukelsheim, *Optimal Design of Experiments*

Israel Gohberg, Peter Lancaster, and Leiba Rodman, *Invariant Subspaces of Matrices with Applications*

Lee A. Segel with G. H. Handelman, *Mathematics Applied to Continuum Mechanics*

Rajendra Bhatia, *Perturbation Bounds for Matrix Eigenvalues*

Barry C. Arnold, N. Balakrishnan, and H. N. Nagaraja, *A First Course in Order Statistics*

Charles A. Desoer and M. Vidyasagar, *Feedback Systems: Input-Output Properties*

Stephen L. Campbell and Carl D. Meyer, *Generalized Inverses of Linear Transformations*

Alexander Morgan, *Solving Polynomial Systems Using Continuation for Engineering and Scientific Problems*

I. Gohberg, P. Lancaster, and L. Rodman, *Matrix Polynomials*

Galen R. Shorack and Jon A. Wellner, *Empirical Processes with Applications to Statistics*

Richard W. Cottle, Jong-Shi Pang, and Richard E. Stone, *The Linear Complementarity Problem*

Rabi N. Bhattacharya and Edward C. Waymire, *Stochastic Processes with Applications*

Robert J. Adler, *The Geometry of Random Fields*

Mordecai Avriel, Walter E. Diewert, Siegfried Schaible, and Israel Zang, *Generalized Concavity*

Rabi N. Bhattacharya and R. Ranga Rao, *Normal Approximation and Asymptotic Expansions*

Françoise Chatelin, *Spectral Approximation of Linear Operators*

(continued)

Classics in Applied Mathematics (continued)

Yousef Saad, *Numerical Methods for Large Eigenvalue Problems, Revised Edition*

Achi Brandt and Oren E. Livne, *Multigrid Techniques: 1984 Guide with Applications to Fluid Dynamics, Revised Edition*

Bernd Fischer, *Polynomial Based Iteration Methods for Symmetric Linear Systems*

Pierre Grisvard, *Elliptic Problems in Nonsmooth Domains*

E. J. Hannan and Manfred Deistler, *The Statistical Theory of Linear Systems*

Françoise Chatelin, *Eigenvalues of Matrices, Revised Edition*

David Colton and Rainer Kress, *Integral Equation Methods in Scattering Theory*

Bruce A. Finlayson, *The Method of Weighted Residuals and Variational Principles*

The Method of Weighted Residuals and Variational Principles

Bruce A. Finlayson
University of Washington
Seattle, Washington

Society for Industrial and Applied Mathematics
Philadelphia

This SIAM edition is a republication of the work first published by Academic Press in 1972.

10 9 8 7 6 5 4 3 2 1

Author photo used with permission from University of Washington.
Figure 3.1 used courtesy of Heywood-Temple Industrial Publications, Ltd.
Tables 4.4 and 4.5 reproduced by kind permission of The Royal Aeronautical Society's *Aeronautical Journal.*
Figure 5.5-8 and Tables 5.7 and 5.8 used with permission from Elsevier.
Figure 8.1 and Table 4.2 used with permission from John Wiley and Sons.
Tables 9.3 and 9.4 used with permission from The Institution of Chemical Engineers.
Figure 9.1 used with permission from Cambridge University Press.
Table 11.4 used with permission from ACM.

Library of Congress Cataloging-in-Publication Data
Finlayson, Bruce A.
The method of weighted residuals and variational principles / Bruce A. Finlayson, University of Washington, Seattle, Washington.
pages cm. -- (Classics in applied mathematics ; 73)
Originally published: New York : Academic Press, 1972.
Includes bibliographical references and index.
ISBN 978-1-611973-23-5
1. Differential equations--Numerical solutions. 2. Approximation theory. I. Title.
QA371.F53 2014
515'.35--dc23
2013036574

Contents

PART II—VARIATIONAL PRINCIPLES

Chapter 10 **On the Search for Variational Principles**

Chapter 11 **Convergence and Error Bounds**

Preface to the Classics Edition

This book was originally published in 1972, when computers were just beginning to make an impact on the solution of differential equations. The first part of it treats differential equations governing transport problems of flow, heat, and mass, using approximate methods and a series of functions but only a few terms. The second part describes variational principles (or not) for those problems and uses variational principles to derive error bounds that can be calculated for some problems. This was important because the series used in the approximate methods for nonlinear problems could not be extended to a large number of terms without the use of a computer. Since then computers have advanced considerably and much more sophisticated numerical techniques are available, but the mathematical principles described herein are still valid and still useful. Thus, I am very pleased to have it included in the SIAM Classics series. Areas such as optimization, spectral methods, and error bounds using mesh refinement have advanced considerably, and suggestions for current information are given below.

In my own historical journey, in 1972 it was important to have simple approximate solutions that could be calculated easily to show the basic behavior of the solutions to differential equations using reduced models. The Galerkin, collocation, least squares, and integral methods were all subsumed under the phrase *method of weighted residuals* [5]. But the approximate nature of the solution made it important to test the solutions with mathematically rigorous error bounds. Several of my graduate students did this in their theses, and their results are included in the book. At the time, I used the term *variational methods* exclusively for problems with a functional that was to be made stationary or an extremum, and chapters in the book show how to find such a function, if it exists. But, in the intervening time, the term *variational* has been applied to any method that makes the residual orthogonal to a set of complete functions, whether a stationary variational principle exists or not. At the time the book was written, the finite element method had been developed extensively in civil engineering for structural problems, especially linear ones (which usually had variational principles), and it was just beginning to be used to solve differential equations arising in heat transfer and fluid mechanics. Thus, the treatment here of the finite element method is elementary and concise but clear. An important point is the role of natural boundary conditions, since they are derived from a variational principle. Currently, it is clear to the author that some people who apply the finite element method don't understand the relation between the natural boundary condition and the variational principle. Thus, they expect a natural boundary condition to be satisfied exactly, when its role is really to integrate the impact of the surroundings on the solution domain. The treatment of variational principles here explores that topic mathematically. In addition, the adjoint variational problem is introduced for almost any nonlinear problem, and this sometimes helps in establishing error bounds for an approximate solution. The last chapter provides a number of theorems on error bounds. The error bound is often found in terms of the residual, which is the differential equation with the approximate solution inserted. Those theorems are still valid and can be used in modern computer packages to provide rigorous error bounds. Many other theorems are available now, of course (see the books listed below), but often only for linear problems.

My 1980 book, "Nonlinear Analysis in Chemical Engineering" [7], is heavily oriented towards numerical methods that can be applied to nonlinear problems. Yet it builds on the first one by expanding the treatment of the orthogonal collocation method that was being used widely within chemical engineering. Outside of chemical engineering, the polynomials are called Chebyshev, Legendre, or Jacobi polynomials, and spectral methods are used, as referenced below. The 1980 book also introduced the ideas for integration in time of stiff differential equations, as developed primarily by Gear [11], that made it possible to solve complicated nonlinear problems without spending great effort figuring out what time step size would be suitable. The finite element method was now in full application to fluid flow, heat transfer, and mass transfer, so that these methods were expanded greatly. But errors were assessed not by mathematical error bounds but by doing mesh refinement and seeing whether the solution was affected. Of course, there weren't error bounds for all problems, but the material in Chapter 11 here is still valid and can still be applied. Some finite element software, such as Comsol Multiphysics (see [9]), does allow the user to calculate the residual after a solution is found and has an option for mesh refinement using an estimate of the global error.

My 1992 book, "Numerical Methods for Problems with Moving Fronts" [8], treated numerical methods for problems with steep gradients that are moving, which is one type of problem not in the "Nonlinear Analysis" book. In some applications, the approximate methods in this book can be combined analytically with the numerical solution to ease the numerical problem of high convection.

My 2006 book, "Introduction to Chemical Engineering Computing" [9], was written with the realization that many differential equations are solved today by packages written by others. The user's role then is to validate the solution rather than program the computer to solve the problem. Details can be left to websites, such as the author's description of the numerical methods in

http://faculty.washington.edu/finlayso/ebook.

In addition, hp-methods involve reducing the mesh size (h) and increasing the degree of the polynomials (p) [3].

Today, solution on different meshes, each more refined than the last, are often the only guide to the error in the solution. Here again, though, packages like Comsol Multiphysics (see [9]) are useful because they make it possible to easily calculate the residual, and it is possible to see the residual decrease as the approximation improves. If one of the theorems in Chapter 11 holds, the error bound is established. The website www.ChemEComp.com has additional information about the Introduction to Chemical Engineering book. Since error bounds are important (and often neglected), some new problems derived from Chapter 11 of this Classics edition are available from SIAM (www.siam.org/books/cl73) and www.ChemEComp.com/MWR. Some problems have solutions, and some do not. Instructors may ask SIAM (textbooks@siam.org) for the key to the problems without a key.

Optimization is not discussed in the 1972 book, but the orthogonal collocation method led to a fast method for dynamic optimization, as described by Biegler [1]. The speed-up resulted because the solution was expanded in terms of orthogonal polynomials but the collocation equations didn't need to be solved at each step in the optimization, using successive quadratic programming. The number of iterations was sometimes less than five percent of a standard method. Later, Biegler [2] reviewed simultaneous strategies for dynamic optimization in which the state and control profiles in time are discretized using collocation on finite elements. While this leads to large-scale nonlinear program problems, efficient strategies to solve them have been found. They are now called the direct transcription method or the simultaneous collocation method [20]. Applications include collision avoidance for aircraft and trajectories for satellites, among others

[2]. These references are cited because the work on orthogonal collocation in the 1972 and 1980 books led to an important part of the algorithm. In addition, SIAM has published a variety of books on optimization [4, 6, 13, 14, 17, 18].

Another area that has advanced considerably since this book was written is in spectral methods. Lanczos [15, 16] is referred to for Chebyshev (orthogonal) polynomials, which basically became know as the orthogonal collocation method in chemical engineering. But Chevyshev polynomials have been used extensively and connect with Fourier spectral methods for solving differential equations. See especially "Chebyshev Polynomials in Numerical Analysis" [10] by Fox and Parker, "Numerical Analysis of Spectral Methods" [12] by Gottlieb and Orszag (including advection problems and mesh refinement), "Spectral Methods in MATLAB" [21] by Trefethen, "Spectral Methods: Evolution to Complex Geometries and Application to Fluid Dynamics" [3] by Canuto et al., and "Chebyshev Polynomials" by Mason and Handscomb [19] (including error analysis).

While the main thrust of modern computing seems to be to solve the entire problem in all its detail using the computer, sometimes part of the solution is essentially known, such as what happens in a thin region with heat transfer perpendicular to the thin region; it is really unnecessary to use a finite element mesh inside the thin region for certain parameters. Then it is possible to use the simple, approximate methods described below in conjunction with the computer programs, leading to a less complex model that includes more understanding of the problem. Paraphrasing Einstein, make the model as complicated as needed, but no more complicated. As a simple example, shown here and in my later books, for many problems a small time solution can be obtained analytically, and this can be combined with the full numerical problem to provide a solution that is accurate at all times and does not oscillate at small times; frequently the finite difference method and finite element method, by themselves, oscillate from node to node in the first few time steps, and the analytical approximations described here can be used to define a problem without this difficulty.

I encourage modern readers to use the full knowledge of mathematics to validate the best solution possible, and I hope this SIAM Classics book will contribute to that progress. The book was a highlight of my early career, and I've been thankful that Academic Press originally published it. It has been wonderful to work with Sara Murphy and Gina Rinelli at SIAM while this book was being produced for the Classics edition.

Bruce A. Finlayson
July, 2013

References

[1] Biegler, L. T. (1984), Solution of Dynamic Optimization Problems by Successive Quadratic Programming and Orthogonal Collocation, *Comp. Chem. Eng.* **8**, 243–248.

[2] Biegler, L. T. (2007), An Overeview of Simultaneous Strategies for Dynamic Optimization, *Chem. Eng. Proc.* **46**, 1043–1053.

[3] Canuto, C., Hussaini, M. Y., Quarteroni, A., and Zang, T. A. (2007), "Spectral Methods: Evolution to Complex Geometries and Application to Fluid Dynamics," Springer-Verlag, New York.

[4] Conn, A. R., Scheinberg, K., and Vicente, L. N. (2009), "Introduction to Derivative-Free Optimization," SIAM, Philadelphia.

[5] Crandal, S. H. (1956), "Engineering Analysis," McGraw-Hill, New York.

[6] Dennis, J. E., Jr. and Schnabel, R. B. (1996), "Numerical Methods for Unconstrained Optimization and Nonlinear Equations," SIAM, Philadelphia.

[7] Finlayson, B. A. (1980), "Nonlinear Analysis in Chemical Engineering," McGraw-Hill, New York.

[8] Finlayson, B. A. (1992), "Numerical Methods for Problems with Moving Fronts," Ravenna Park, Seattle.

[9] Finlayson, B. A. (2006), 2nd ed. (2012), "Introduction to Chemical Engineering Computing," John Wiley & Sons, New York. See www.ChemEComp.com.

[10] Fox, L. and Parker, I. B. (1972), "Chebyshev Polynomials in Numerical Analysis," 2nd ed., Oxford University Press, London.

[11] Gear, C. W. (1971), "Numerical Initial-Value Problems in Ordinary Differential Equations," Prentice-Hall, Englewood Cliffs, NJ.

[12] Gottlieb, D. and Orszag, S. A. (1977), "Numerical Analysis of Spectral Methods," SIAM, Philadelphia.

[13] Griva, I., Nash, S. G., and Sofer, A. (2009), "Linear and Nonlinear Optimization," 2nd ed., SIAM, Philadelphia.

[14] Gunzburger, M. D. (2002), "Perspectives in Flow Control and Optimization," SIAM, Philadelphia.

[15] Lanczos, C. (1938), Trigonometric Interpolation of Empirical and Analytical Functions, *J. Math. Phys.* **17**, 123–199.

[16] Lanczos, C. (1956),"Applied Analysis," Prentice-Hall, Englewood Cliffs, NJ.

[17] Levy, A. B. (2009), "The Basics of Practical Optimization," SIAM, Philadelphia.

[18] Locatelli, M. and Schoen, F. (2013), "Global Optimization: Theory, Algorithms, and Applications," SIAM, Philadelphia.

[19] Mason, J. C. and Handscomb, D. C. (2003), "Chebyshev Polynomials," Chapman & Hall/CRC, Boca Raton, FL.

[20] Nie, Y., Biegler, L. T., and Villa, C. M. (2013), Reactor Modeling and Recipe Optimization of Polyether Polyol Processes: Polypropylene Glycol, *AIChE J.* **59**, 2515–2529.

[21] Trefethen, L. N. (2000), "Spectral Methods in MATLAB," SIAM, Philadelphia.

Preface

This is a book for people who want to solve problems formulated as differential equations in science and engineering. The subject area is limited to fluid mechanics, heat and mass transfer. While making no pretense at completely covering these subjects and their relationship to variational principles and approximate methods, the book is intended to give the novice an introduction to the subject, and lead him through the difficult research problems being treated in the current literature. The first four chapters give a relatively simple treatment of many classical problems in the field. The literature is full of simple, one-term approximations, but the method of weighted residuals (MWR) can be used to obtain answers of any desired accuracy, and there are several methods specifically adapted to the computer. In many test cases MWR compares favorably to finite difference computations in that the MWR results are either more accurate or require less computation time to generate or both. Chapter 4 discusses the developments by Professor D. E. Abbott and his students at Purdue University on laminar boundary layer flows. Orthogonal collocation is illustrated in Chapter 5. This method was advanced in 1967 by Professor W. E. Stewart at the University of Wisconsin and J. V. Villadsen at Danmarks Tekniske Højskole. It drastically reduces the drudgery of setting up the problem, and, when applicable, is highly recommended. Chapter 6 studies the Galerkin method as applied to convective instability problems, where it proves effective and accurate. Chapters 5 and 7 relate MWR to finite element methods, which is a promising technique, especially for linear problems with irregular boundaries.

The ideas behind the method of weighted residuals are relatively simple and are easily applied. Variational principles are only slightly more complicated, but are often irrelevant to applications in engineering. Consequently this material takes a more appropriate back seat. Variational principles in fluid mechanics, heat and mass transfer are presented in Chapters 7–9. Chapter 10 is a short summary of my opinions on the attempt to derive "variational" principles based on a principle of minimum (or maximum) rate of entropy production. The final chapter gives a summary of results on convergence and error bounds, subjects which are rapidly expanding. While numerical convergence often suffices, the theorems in Chapter 11 give the conditions necessary to ensure convergence in difficult nonlinear problems.

The book is intended to be comprehensible to a graduate student with some knowledge of and interest in mathematics through differential equations. I hope it will find use as a reference book for graduate courses discussing approximate and numerical solutions, as well as a convenient reference for those working in the field. Problem sets are included to aid in self-study and course work.

One of the pleasurable aspects of writing a book, I have discovered, is that while organizing the material new results are suggested. These are scattered throughout the book so that I highlight them here. In the variational method the eigenvalue is stationary to small errors in the approximate eigenfunctions. I show that this is also true of the Galerkin method (Chapter 6). While not a new result, Section 7.8 illustrates two very simple and practical methods of obtaining lower bounds on eigenvalues. A variational principle is given for a collection of fluid drops suspended in another fluid in slow flow—extending the previous results for Newtonian fluids in both phases to allow non-Newtonian fluids in both phases. The nonexistence of a variational principle for the steady-state Navier–Stokes equations is shown in Section 8.6 using Fréchet differentials, making more concise the detailed arguments given in 1930 by C. B. Millikan. Indeed the Fréchet derivative proves useful in organizing the many variational principles, or lack of them, in heat and mass transfer (see Chapter 9). The Fréchet derivative is also used to define a variational principle for any differential equation and its "adjoint"—even nonlinear equations—although there appears to be no advantage to do so. One of the difficulties with MWR or variational methods is that the error is often not known. Some of the methods discussed above are specifically designed to allow the results to be carried to numerical convergence. Even so, it is of interest to have some rigorous definition of the error. The residual —the equation which is solved only approximately—provides such a criteria. Material in Section 11.6, largely developed by one of my students, Noble Ferguson, gives error bounds on the solution in terms of error bounds on the residual. Thus the error can be assessed when the exact solution is unknown. The residual also proves a useful guide in cases where the appropriate theorems have not yet been proved.

The review of the literature is complete to October, 1970.

Acknowledgments

Several teachers have been influential in stimulating my interest in the application of mathematics. My high school physics teacher, Mr. Williams, helped me learn calculus at a time when that subject was not taught in many high schools, especially not in a small town in Oklahoma. Professor Vern Denny, then at Rice University, was influential in both stimulating my interest in mathematical techniques as well as encouraging me to continue my graduate education. Professor L. E. Scriven, my advisor at the University of Minnesota, encouraged my investigations of approximate methods and was especially helpful at increasing the generality and widening the scope of my results. Some of the material, and many of the viewpoints, in this book were developed while a graduate student under his direction. The National Science Foundation deserves thanks because while at Minnesota I was a NSF Graduate Fellow, and some of the recent results were derived under a NSF Grant at the University of Washington. I also owe thanks to the University of Washington, where research is considered part of my duties, and I am given the time to do it, as well as to the faculty of the Department of Chemical Engineering, who have flexible arrangements for deciding which courses I teach. As a result some of the material has been presented in the classroom and the presentation improved for the experience.

I thank especially Norm Sather, who has read the manuscript and made several constructive suggestions. My wife, Pat, has been a special help, both to me and in preparation of the manuscript. I owe special thanks to her and our children, who have suffered through long hours of nelect. The effort to write a book is great. I hope the reader finds the result worthwhile.

Part

I

THE METHOD OF WEIGHTED RESIDUALS

Chapter 1

Introduction

Approximate solutions of differential equations satisfy only part of the conditions of the problem: for example the differential equation may be satisfied only at a few positions, rather than at each point. The approximate solution is expanded in a set of known functions with arbitrary parameters. This book covers two ways to determine the parameters: the Method of Weighted Residuals and the Variational Method. In the method of weighted residuals one works directly with the differential equation and boundary conditions whereas in the variational method one uses a functional related to the differential equation and boundary conditions. In both methods there are two strategies. (1) A first approximation may be sufficient; its validity is assessed using our intuition and experience. Furthermore, insight is often gained from the analytical solution. (2) A sequence of approximations can be calculated to converge to the solution. In the second strategy the calculations must be amenable to a computer; successive approximations must be calculated without any reformulation or intervention by the analyst. Both strategies are discussed, in order of ascending difficulty. The first few chapters treat low-order approximations for a wide variety of heat transfer and flow problems. Then computer-oriented methods are discussed, with detailed applications to nonlinear chemical engineering and convective instability problems.

Variational principles are described for fluid mechanics, heat and mass transfer. Finally, theorems on convergence and error bounds are summarized.

The method of weighted residuals (often abbreviated MWR) actually encompasses several methods (collocation, Galerkin, integral, etc.) and provides a framework to compare, contrast, and elucidate the features of individual methods. Variational methods are not applicable to all problems, and thus suffer a lack of generality. Sometimes they provide powerful results, such as upper and lower bounds on quantities of interest. MWR is easy to apply, whereas variational methods require more mathematical manipulation. For this reason the first six chapters contains a complete treatment of MWR. This is followed by a discussion of variational principles and their application.

Several conventions are used. An ordinary derivative is denoted by $u' \equiv du(x)/dx$, whereas a partial derivative is denoted by $u_x = \partial u(x, y)/\partial x$. The symbol u_x may also denote the x component of the velocity vector $\mathbf{u}$, but the usage is clear from the context. Equations are numbered successively in each chapter. Summation signs are sometimes omitted, in which case summation over repeated indices is understood, i.e., the Einstein convention is used. Both vector and tensor notation is used, and boldface symbols denote vectors. Theorems on convergence and error bounds are given in Chapter 11. They are cross referenced throughout the book as Theorem 11.16, for example, for Theorem 16 in Chapter 11.

The next section summarizes the equations governing problems in Chapters 2–4, so that the reader can relate the problems to each other. Then the method of weighted residuals is reviewed from a historical perspective.

1.1 Basic Equations and Their Classification

Heat transfer is governed by the energy balance for an incompressible fluid in local equilibrium (Bird *et al.*, 1960):

$$\rho C_v \left(\frac{\partial T}{\partial t} + \mathbf{u} \cdot \nabla T \right) = -\nabla \cdot \mathbf{q} + \Phi, \tag{1.1}$$

where ρ is the density, C_v is the heat capacity, T is the temperature, t is time, $\mathbf{u}$ is the fluid velocity, $\mathbf{q}$ is the heat flux, and Φ is the viscous dissipation. The symbol ∇ is the gradient operator, $\nabla \cdot$ is the divergence operator, and ∇^2 is the Laplacian operator. In Cartesian coordinates typical uses are

$$\nabla T = \mathbf{e}_x \frac{\partial T}{\partial x} + \mathbf{e}_y \frac{\partial T}{\partial y} + \mathbf{e}_z \frac{\partial T}{\partial z}, \tag{1.2a}$$

$$\nabla \cdot \mathbf{q} = \frac{\partial q_x}{\partial x} + \frac{\partial q_y}{\partial y} + \frac{\partial q_z}{\partial z}, \tag{1.2b}$$

$$\nabla^2 T = \frac{\partial^2 T}{\partial x^2} + \frac{\partial^2 T}{\partial y^2} + \frac{\partial^2 T}{\partial z^2}. \tag{1.3}$$

The equation of change (1.1) is valid for any material. When modeling heat transfer we must relate the heat flux to the temperature field, and this constitutive relation is a property of the material. The most common assumption is Fourier's law,

$$\mathbf{q} = -k\,\nabla T, \tag{1.4}$$

but other choices are possible. For instance, a nonlinear relation may be necessary or the material may be anisotropic. If Fourier's law is used and viscous dissipation is neglected we obtain the transport equation,

$$\rho C_v \left(\frac{\partial T}{\partial t} + \mathbf{u} \cdot \nabla T\right) = \nabla \cdot (k\,\nabla T). \tag{1.5}$$

Similar to Eq. (1.5) but applied to diffusion in a dilute binary system with constant density and a rate of reaction R is the equation (Bird *et al.*, 1960)

$$\frac{\partial c}{\partial t} + \mathbf{u} \cdot \nabla c = \nabla \cdot (\mathscr{D}\,\nabla c) + R, \tag{1.6}$$

where c is the concentration and $\mathscr{D}$ is the diffusivity.

Special cases of these equations are needed below. In the steady state all time derivatives are zero. For constant diffusivity and no chemical reaction,

$$\mathbf{u} \cdot \nabla c = \mathscr{D}\,\nabla^2 c. \tag{1.7}$$

The concentration must also satisfy certain boundary conditions, leading to a boundary value problem. When the flow is along a cylindrical duct, the velocity $\mathbf{u} = (0, 0, u_z)$ is known, and axial diffusion is negligible compared to convection in the z direction. Equation (1.6) reduces to

$$u_z(x, y)\frac{\partial c}{\partial z} = \mathscr{D}\left(\frac{\partial^2 c}{\partial x^2} + \frac{\partial^2 c}{\partial y^2}\right). \tag{1.8}$$

When the velocity is zero, the simplification of Eq. (1.7) for heat transfer is

$$\nabla \cdot (k\,\nabla T) = 0, \tag{1.9a}$$

$$\nabla^2 T = 0. \tag{1.9b}$$

Equation (1.9b) holds when the thermal conductivity is constant. An equivalent mass transfer problem involves the rate of reaction as well, giving

$$\mathscr{D}\,\nabla^2 c + R = 0. \tag{1.10}$$

The one-dimensional unsteady-state diffusion equation is

$$\frac{\partial c}{\partial t} = \frac{\partial}{\partial x}\left(\mathscr{D}\frac{\partial c}{\partial x}\right), \tag{1.11a}$$

$$c_t = \mathscr{D}c_{xx}, \tag{1.11b}$$

where (1.11b) applies when the diffusivity is constant. The concentration field evolves from some initial distribution as time proceeds. These problems are called initial-value problems. Equation (1.8) is also an initial-value problem, in which the variable z replaces time.

The technique of separation of variables applied to a special case of (1.8) leads to an eigenvalue problem,

$$y'' + \lambda(1 - x^2)y = 0. \tag{1.12}$$

Usually this equation (and the accompanying boundary conditions) can be satisfied only for particular values of the parameter λ, called the eigenvalues.

The basic equation in fluid mechanics is the balance of linear momentum, called the Cauchy momentum equation (Aris, 1962),

$$\rho\left(\frac{\partial \mathbf{u}}{\partial t} + \mathbf{u}\cdot\nabla\mathbf{u}\right) = \rho\hat{\mathbf{F}} + \nabla\cdot\mathbf{T}. \tag{1.13}$$

Here $\mathbf{u}$ is the vector velocity, $\hat{\mathbf{F}}$ is the body force per unit mass, and $\mathbf{T}$ is the stress tensor. Throughout the book both dyadic and tensor notation are used. The Cartesian components are related by

$$\mathbf{T} = \mathbf{e}_i\mathbf{e}_jT_{ij}, \qquad T_{ij} = \mathbf{e}_i\cdot\mathbf{T}\cdot\mathbf{e}_j, \tag{1.14}$$

where $\mathbf{e}_i$ and $\mathbf{e}_j$ are unit basis vectors. The equation of continuity is

$$\frac{\partial \rho}{\partial t} + \mathbf{u}\cdot\nabla\rho = -\rho\,\nabla\cdot\mathbf{u}. \tag{1.15}$$

These equations must be augmented by a constitutive relation for the stress tensor. For example, a Newtonian fluid is one satisfying

$$\begin{aligned} T_{ji} &= -(p + \lambda u_{k,k})\delta_{ij} + \mu(u_{i,j} + u_{j,i}), \\ \mathbf{T} &= -(p + \lambda\,\nabla\cdot\mathbf{u})\mathbf{U} + \mu[\nabla\mathbf{u} + (\nabla\mathbf{u})^T], \end{aligned} \tag{1.16}$$

where p is the pressure, μ is the viscosity, λ is the bulk viscosity, $\delta_{ij} = 1$ if $i = j$ or 0 if $i \neq j$, and , denotes differentiation. [See Aris (1962) for additional information on tensor notation.] More general relations for non-Newtonian fluids are considered in Chapter 8. If a fluid is incompressible its density cannot change, and Eq. (1.15) reduces to

$$\nabla\cdot\mathbf{u} = 0. \tag{1.17}$$

An incompressible, Newtonian fluid with constant viscosity is governed by the Navier–Stokes equation, with ν the kinematic viscosity $\nu = \mu/\rho$:

$$\frac{\partial \mathbf{u}}{\partial t} + \mathbf{u} \cdot \nabla \mathbf{u} = \hat{\mathbf{F}} - \frac{1}{\rho} \nabla p + \nu \nabla^2 \mathbf{u}. \tag{1.18}$$

Consider two special cases of the Navier–Stokes equations. In the first case a fluid is flowing in steady laminar, rectilinear flow down a cylinder (with arbitrary cross section), whose generators are in the z direction. The velocity does not change with distance down the cylinder and can be represented as $\mathbf{u} = (0, 0, u(x, y))$. The Navier–Stokes equation then simplifies as follows. The first term vanishes in steady state, the second term is zero for rectilinear flow (the velocity derivatives are only in the x and y directions, and the velocity is only in the z direction), the pressure drop is a given constant, $(\nabla p)/\mu = -\mathbf{e}_z \lambda$, and it is assumed that there are no body forces (or else they are absorbed into the pressure gradient):

$$u_{xx} + u_{yy} = -\lambda. \tag{1.19}$$

This equation resembles Eqs. (1.9) and (1.10) for heat and mass transfer, and similar methods are applicable to solve it. The other special case arises in boundary layer theory. We do not derive the equations here (see Schlichting, 1960), but record them for two dimensions, $\mathbf{u} = (u(x, y), v(x, y), 0)$:

$$\begin{aligned} u_x + v_y &= 0, \\ u u_x + v u_y &= -\frac{1}{\rho} p_x + \nu u_{yy}, \\ U \frac{dU}{dx} &= -\frac{1}{\rho} p_x. \end{aligned} \tag{1.20}$$

The first two equations are applicable near the presence of a solid body, around which the fluid is flowing, and the outer-stream velocity, $U(x)$, is a given function of x, the distance down the body. The pressure distribution is thus specified as

$$u u_x + v u_y = U U' + \nu u_{yy}. \tag{1.21}$$

1.2 Method of Weighted Residuals

The method of weighted residuals (henceforth abbreviated as MWR) is a general method of obtaining solutions to the above equations of change. The unknown solution is expanded in a set of trial functions, which are specified, but with adjustable constants (or functions), which are chosen to give the best solution to the differential equation. The first approximation gives useful qualitative answers (perhaps within 10–20%), but higher approximations can be calculated (usually on a computer) to give as precise an answer as desired. MWR is illustrated by application to heat conduction in a solid.

The elliptic, boundary-value problem is Eq. (1.9b)

$$\nabla^2 T = T_{xx} + T_{yy} = 0 \quad \text{in} \quad V(x, y) \tag{1.22a}$$

$$T = T_0 \quad \text{on} \quad S, \text{ the boundary of } V. \tag{1.22b}$$

A trial solution is taken in the form

$$T = T_0 + \sum_{i=1}^{N} c_i T_i, \tag{1.23}$$

where the functions T_i are specified to satisfy the homogeneous boundary condition $T_i = 0$. Then the trial function (1.23) satisfies the boundary conditions for all choices of the constant c_i. This trial function is substituted into the differential equation to form the residual

$$R(c_i, x, y) = \nabla^2 T_0 + \sum_{i=1}^{N} c_i \nabla^2 T_i. \tag{1.24}$$

If the trial function were the exact solution the residual would be zero. In MWR the constants c_i are chosen in such a way that the residual is forced to be zero in an average sense. The weighted integrals of the residual are set to zero:

$$(w_j, R) = 0, \qquad j = 1, 2, \ldots, N, \tag{1.25}$$

where the inner product (u, w) is defined by

$$(u, w) = \int_V uw \, dV. \tag{1.26}$$

If $(u, w) = 0$ the functions are orthogonal. Combining (1.24) and (1.25) gives

$$\sum_{i=1}^{N} c_i (w_j, \nabla^2 T_i) = -(w_j, \nabla^2 T_0), \tag{1.27}$$

or simply

$$\sum_{i=1}^{N} B_{ji} c_i = d_j, \tag{1.28}$$

where $B_{ji} = (w_j, \nabla^2 T_i)$, $d_j = -(w_j, \nabla^2 T_0)$. Since the functions T_0 and T_i are known and w_j is specified below, B_{ji} and d_j can be calculated. The matrix B_{ji} is inverted to obtain the c_i, which thus gives the approximate solution in Eq. (1.23).

The weighting functions can be chosen in many ways and each choice corresponds to a different criterion of MWR. We could divide the domain V into N smaller subdomains, V_j, and choose

$$w_j = \begin{cases} 1, & \mathbf{x} \text{ in } V_j, \\ 0, & \mathbf{x} \text{ not in } V_j. \end{cases} \tag{1.29}$$

The differential equation, integrated over the subdomain, is then zero, hence the name subdomain method. As N increases, the differential equation is satisfied on the average in smaller and smaller subdomains, and presumably approaches zero everywhere. This method was first advanced in 1923 by the Dutch engineers, Biezeno and Koch, for solution to problems arising in the stability of beams, rods, and plates (Biezeno and Koch, 1923; Biezeno, 1923, 1924; Biezeno and Grammel, 1955). A modern extension to the field of boundary layer flow is known as the method of integral relations (see Section 4.3).

In the collocation method, the weighting functions are chosen to be the displaced Dirac delta function

$$w_j = \delta(x - x_j), \tag{1.30}$$

which has the property that

$$\int_V w_j R \, dV = R \bigg|_{\mathbf{x}_j}. \tag{1.31}$$

Thus the residual is zero at N specified collocation points $\mathbf{x}_j$. As N increases the residual is zero at more and more points and presumably approaches zero everywhere. The method was applied to solve differential equations by Slater (1934, electronic energy bands in metals) and Barta (1937, torsion of a square prism). It was later developed as a general method for solving ordinary differential equations. Frazer *et al.* (1937) used many different trial functions, and placed the collocation points arbitrarily. Lanczos (1938) expanded the solution in terms of Tchebychev polynomials, and used the roots to a Tchebychev polynomial as the collocation points. The work by Frazer *et al.* was more widely noticed; since for low-order approximations the results often depend greatly on the choice of collocation points, the method has not been widely used despite its easy application. Lanczos' method has been revived as the orthogonal collocation method (see Chapter 5) which has proved to be very suitable for nonlinear problems arising in chemical engineering.

The least squares method was originated by Gauss in 1795 for least squares estimation. The work was not published until 1809 however, leading to a controversy with Legendre, who published the same ideas in 1806 (see, e.g., Hall, 1970; Sorenson, 1970). More recently it has been applied as a method of solving differential equations by Picone (1928). The weighting function is $\partial R/\partial c_j$ so that

$$I(c_i) = \int_V R^2(c_i, \mathbf{x})\, dV \tag{1.32}$$

is minimized with respect to the constants c_i. This method often leads to cumbersome equations, but it has been applied to complicated problems arising in nuclear reactor engineering (Becker, 1964). The mean square residual (1.32) has theoretical significance since error bounds can be derived in terms of it (see Section 11.6). Thus, minimization of (1.32) gives the best possible bounds for the error.

One of the best known approximate methods was developed by the Russian engineer Galerkin (1915).† In this method the weighting functions are chosen to be the trial functions, $w_i = T_i$. The trial functions must be chosen as members of a complete set of functions. A set of functions $\{w_i\}$ is complete if any function of a given class can be expanded in terms of the set, $f = \Sigma\, a_i w_i$. Then the series of Eq. (1.23) is inherently capable of representing the exact solution, provided enough terms are used. A continuous function is zero if it is orthogonal to every member of a complete set. Thus the Galerkin method forces the residual to be zero by making it orthogonal to each member of a complete set of functions (in the limit as $N \to \infty$). The Galerkin method is closely related to variational methods (see Chapter 7) and is highly developed for eigenvalue problems (Chapter 6). It was widely advocated in a number of papers by Duncan (1937, 1938a, b, c, 1939) as well as by Kantorovich and Krylov (1958).

The method of moments was developed by Yamada (1947, 1948, 1950) and Fujita (1951) for application to laminar boundary layer problems and nonlinear transient diffusion. For the ordinary differential equations governing these phenomena the weighting functions are $1, x, x^2, x^3, \ldots$. Thus successively higher moments of the residual are required to be zero. For the first approximation, the method of moments is identical to the subdomain method and is usually called the integral method. The integral method was first developed in 1921 by two German engineers, von Kármán (1921) and

† Bubnov (1913) apparently applied a similar method (see Mikhlin, 1964, Chapter 6 reference list) and Russian authors often refer to it as the Bubnov–Galerkin method. Galerkin had a rather interesting start to his career. While he was imprisoned by the Tsar in 1906 for his political views he published his first technical paper on the buckling of bars, columns, and systems of bars. For a summary of his career see Anonymous (1941).

Pohlhausen (1921), for application to boundary layer problems. It is just a first approximation and successive approximations are discussed in Section 4.3. The integral method is also widely used in transient heat transfer studies (Chapter 3).

If the weighting functions are a set of complete functions, not necessarily the trial functions, the method is just a general MWR. Krawchuk (1932, 1936) chose the weighting functions $w_i = L'T_i$, where L' is a differential operator of the same order as in the problem, but otherwise arbitrary.

These several criteria (each developed in a different country) were unified by Crandall (1956) as the Method of Weighted Residuals. Collatz (1960) called them error distribution principles.† Clymer and Braun (1963) reviewed their application to structural problems, Finlayson and Scriven (1966) reviewed their application to the equations of change, and Vichnevetsky (1969) reviewed their application to initial value problems. Summaries of their application in nuclear engineering are in Becker (1964), Kaplan (1966), Stacey (1967), and Fuller and Meneley (1970). The choice of criterion is not crucial, since all comparisons of different methods indicate that similar results are achieved, especially for higher approximations. See comparisons listed in Finlayson and Scriven (1966), Pomraning (1966) for nuclear engineering problems, and Shuleshko (1961) for boundary methods. One possible exception is for problems with variational principles (see Chapters 6 and 7). Then the Galerkin method is preferred because of its equivalence to the variational method, which gives more accurate values of an eigenvalue, for example. However, in most cases the criterion can be chosen based on convenience.

In interior methods the trial functions satisfy the boundary conditions, but not the differential equation. In mixed methods neither the differential equation nor the boundary conditions are satisfied (Shuleshko, 1959). In boundary methods the trial functions satisfy the differential equations but not the boundary conditions. The trial functions are specialized to the problem, but therein lies their power: many conditions of the problem can be automatically satisfied. Low-order approximations may depend on the choice of trial function but high-order ones usually do not. More recent applications are designed for solution by computer, where MWR competes directly—and usually favorably—with finite difference methods. Countless possibilities exist, as is illustrated by examples in Section 2.6 and the remainder of the book.

† The similarity of the methods was recognized early by Biezeno (1923–1924; see Biezeno and Grammel, 1955, p. 180) and Courant (1924). Crandall and Collatz each introduced a different name for the unified methods.

Error bounds are desirable and sometimes possible (see Chapter 11). One possible *criterion* for a good solution is a small mean square residual (1.32). This was apparently first suggested by Yang (1962). In some cases this criterion can be turned into a rigorous bound on the solution

$$\int (T - \bar{T})^2 \, dV \leq K \int R(\bar{T})^2 \, dV, \tag{1.33}$$

where T and $\bar{T}$ are the exact and approximate solutions and K is a constant (see Section 11.6).

REFERENCES

Anonymous. (1941). On the Seventieth Anniversary of the Birth of B. T. Galerkin, *Prikl. Mat. Mech.* **5**, 337 (in English) (Anniversary issue).

Aris, R. (1962). "Vectors, Tensors, and the Basic Equations of Fluid Mechanics." Prentice-Hall, Englewood Cliffs, New Jersey.

Barta, J. (1937). Über die Näherungsweise Lösung einiger Zweidimensionaler Elastizitätsaufgaben, *Z. Angew. Math. Mech.* **17**, 184–185.

Becker, M. (1964). "The Principles and Applications of Variational Methods." MIT Press, Cambridge, Massachusetts.

Biezeno, C. B. (1923–1924). Over een Vereenvoudiging en over een Uitbreiding van de Methode van Ritz, *Christiaan Huygens* **3**, 69.

Biezeno, C. B. (1924). Graphical and Numerical Methods for Solving Stress Problems, *Proc. Int. Congr. Appl. Mech., 1st Delft*, pp. 3–17.

Biezeno, C. B., and Grammel, R. (1955). "Engineering Dynamics," Vol. I, Theory of Elasticity. Blackie, Glasgow and London.

Biezeno, C. B., and Koch, J. J. (1923). Over een Nieuwe Methode ter Berekening van Vlokke Platen met Toepassing op Enkele voor de Techniek Belangrijke Belastingsgevallen, *Ing. Grav.* **38**, 25–36.

Bird, R. B., Stewart, W. E., and Lightfoot, E. N. (1960). "Transport Phenomena." Wiley, New York.

Bubnov, I. G. (1913). Report on the Works of Prof. Timoshenko Which Were Awarded the Zhuranskii Prize, *Symp. Inst. Commun. Eng.*, (Sborn. inta inzh. putei soobshch.) No. 81, All Union Special Planning Office (SPB). [This work is mentioned by Mikhlin (1964). (See the Chapter 6 reference list.)]

Clymer, A. B., and Braun, K. N. (1963). Assumed Mode Methods for Structural Analysis and Simulation, *AIAA Simulation for Aerospace Flight Conf.*, pp. 244–260. AIAA, New York.

Collatz, L. (1960). "The Numerical Treatment of Differential Equations." Springer-Verlag, Berlin.

Courant, R. (1924). *Proc. Int. Congr. Appl. Mech., 1st, Delft*, p. 17.

Crandall, S. H. (1956). "Engineering Analysis." McGraw-Hill, New York.

Duncan, W. J. (1937). Galerkin's Method in Mechanics and Differential Equations, Great Britain Aero Res. Counc. London. Rep. and Memo. No. 1798. Reprinted in *Great Britain Air Ministry Aero. Res. Comm. Tech. Rep.* **1**, 484–516.

Duncan, W. J. (1938a). The Principles of Galerkin's Method, Great Britain Aero. Res. Counc. London. Rep. and Memo. No. 1848. Reprinted in *Great Britain Air Ministry Aero. Res. Comm. Tech. Rep.* **2**, 589–612.

Duncan, W. J. (1938b). Note on Galerkin's Method for the Treatment of Problems Concerning Elastic Bodies, *Phil. Mag.* [7] **25**, 628–633.

Duncan, W. J. (1938c). Application of the Galerkin Method to the Torsion and Flexure of Cylinders and Prisms, *Phil. Mag.* [7] **25**, 636–649.

Duncan, W. J., and Lindsay, D. D. (1939). Methods for Calculating the Frequencies of Overtones, Great Britain Aero Res. Comm., Rep. and Memo. No. 1888.

Finlayson, B. A., and Scriven, L. E. (1966). The Method of Weighted Residuals—A Review, *Appl. Mech. Rev.* **19**, 735–748.

Frazer, R. A., Jones, W. P., and Skan, S. W. (1937). Approximations to Functions and to the Solutions of Differential Equations, Great Britain Aero. Res. Counc. London. Rep. and Memo. No. 1799. Reprinted in *Great Britain Air Ministry Aero. Res. Comm. Tech. Rep.* **1**, 517–549.

Fuller, E. L., Meneley, D. A., and Hetrick, D. L. (1970). Weighted-Residual Methods in Space-Dependent Reactor Dynamics, *Nucl. Sci. Eng.* **40**, 206–223.

Fujita, H. (1951). *Mem. Coll. Agr. Kyoto Imp. Univ.* **59**, 31. [See *Chem. Abstracts* **46**, 3371f (1952).]

Galerkin, B. G. (1915). Rods and Plates. Series in Some Problems of Elastic Equilibrium of Rods and Plates. *Vestn. Inzh. Tech.* (*USSR*) **19**, 897–908. Translation 63–18924, Clearinghouse, Fed. Sci. Tech. Info., Springfield, Virginia.

Hall, T. (1970). "Carl Friedrich Gauss." MIT Press, Cambridge, Massachusetts.

Kantorovich, L. V., Krylov, V. I. (1958). "Approximate Methods of Higher Analysis." Wiley (Interscience), New York.

Kaplan, S. (1966). Synthesis Methods in Reactor Analysis, *Advan. Nucl. Sci. Technol.*, **3**, 233–266.

Krawchuk, M. F. (1932). Application of the Method of Moments to the Solution of Linear Differential and Integral Equations (in Ukranian), *Kiev. Soobshch. Akad. Nauk USSR* **1**, 168. [Referred to by Shuleshko (1959).]

Krawchuk, M. (1936) "Applications of the Method of Moments to the Solution of Linear Differential and Integral Equations," Vol. 2. Ukranian Acad. of Sci. Press. [Referred to by Shuleshko (1959).]

Lanczos, C. (1938). Trigonometric Interpolation of Empirical and Analytical Functions, *J. Math. Phys.* **17**, 123–199.

Picone, M. (1928). Sul Metodo delle Minime Potenze Ponderate e sul Metodo di Ritz per il Calcolo Approssimato nei Problemi della Fisica-Matematica, *Rend. Circ. Mat. Palermo* **52**, 225–253.

Pohlhausen, K. (1921). The Approximate Integration of the Differential Equation for the Laminar Boundary Layer, *Z. Angew. Math. Mech.*, **1**, 252–268, translated by R. C. Anderson, AD 645 784, Clearinghouse Fed. Sci. Tech. Info., Springfield, Virginia.

Pomraning, G. C. (1966). A Numerical Study of the Method of Weighted Residuals, *Nucl. Sci. Eng.* **24**, 291–301.

Schlichting, H. (1960). "Boundary Layer Theory." McGraw-Hill, New York.

Shuleshko, P. (1959). A New Method of Solving Boundary-Value Problems of Mathematical Physics; Generalizations of Previous Methods of Solving Boundary-Value Problems of Mathematical Physics. *Aust. J. Appl. Sci.* **10**, 1–7, 8–16.

Shuleshko, P. (1961). A Method of Integration over the Boundary for Solving Boundary-Value Problems, *Aust. J. Appl. Sci.* **12**, 393–406.

Slater, J. C. (1934). Electronic Energy Bands in Metals, *Phys. Rev.* **45**, 794–801.

Sorenson, H. W. (1970). Least-Squares Estimation: From Gauss to Kalman, *IEEE Spectrum* **7**, No. 7, 63–68.

Stacey, W. M., Jr. (1967). "Modal Approximations." MIT Press, Cambridge, Massachusetts.

Vichnevetsky, R. (1969). Use of Functional Approximation Methods in the Computer Solution of Initial-Value Partial Differential Equations Problems, *IEEE Trans. Comp.* **C-18**, 499–512.

von Kármán, Th. (1921). Über Laminare und Turbulente Reibung, *Z. Angew. Math. Mech.* **1**, 233–252; (1946). NACA Tech. Memo. 1092.

Yamada, H. (1947). *Rep. Res. Inst. Fluid Eng. Kyushu Univ.* **3**, 29.

Yamada, H. (1948). An Approximate Method of Integration of Laminar Boundary-Layer Equations, I (in Japanese), *Rep. Res. Inst. Fluid Eng.* **4**, 27–42; (1950). *Appl. Mech. Rev.* **3**, No. 762.

Yamada, H. (1950). A Method of Approximate Integration of the Laminar Boundary-Layer Equation (in Japanese), *Rep. Res. Inst. Fluid Eng.* **6**, 87–98; (1952). *Appl. Mech. Rev.* **5**, No. 488.

Yang, K. T. (1962). On an Improved Karman–Pohlausen's Integral Procedure and a Related Error Criterion, *Proc. U.S. Nat. Congr. Appl. Mech., 4th, Berkeley, California*, pp. 1419–1429.

Chapter 2

Boundary-Value Problems in Heat and Mass Transfer

The method of weighted residuals is applied to several simple examples governing steady-state heat and mass transfer. In this way the mechanics of the method can be illustrated without undue complexity. We treat heat conduction in one dimension with a temperature-dependent thermal conductivity—leading to a nonlinear ordinary differential equation—and follow this with a two-dimensional example—leading to a partial differential equation. Boundary methods are illustrated and lead to error bounds for the approximate solution. The general steady-state heat conduction problem is treated for three-dimensional, nonlinear cases with a variety of boundary conditions in arbitrary domains. Mass transfer from a fluid or solid sphere to a fluid in Stokes flow is considered next to illustrate the judicious use of perturbation methods to test the accuracy of MWR approximate solutions. In the applications of MWR there are three important steps: choice of a trial function, choice of a criterion, and calculation of successive approximations. We shall see in some of these examples that the choice of a criterion is not too crucial, especially for higher approximations. The choice of trial functions is more important, and the various possibilities are discussed in the final section.

2.1 One-Dimensional Heat Conduction

Consider steady-state heat conduction across a slab with a temperature-dependent thermal conductivity. Each face of the slab is maintained at a different temperature, T_0 and T_1, and we wish to calculate the temperature distribution throughout the slab as well as the heat flux. The differential equation (1.9a) and boundary conditions are

$$\frac{d}{dx'}\left[k(T)\frac{dT}{dx'}\right] = 0, \qquad T(0) = T_0, \quad T(d) = T_1. \tag{2.1}$$

The temperature dependence of the thermal conductivity is taken in the linear form

$$k = k_0 + \alpha(T - T_0), \tag{2.2}$$

where k_0 and α are constants. The equations are made dimensionless by choosing a standard for each variable. For distance the logical choice is the thickness of the slab, d. For the temperature we have several choices, T_0, T_1, $T_1 - T_0$, etc. If possible it is desirable to subtract a constant term, so that we consider $T - T_0$. This has the dimensions of temperature, so we consider its value relative to $T_1 - T_0$. The nondimensional variable is then chosen $\theta = (T - T_0)/(T_1 - T_0)$. We substitute this into (2.1), define $x = x'/d$, substitute for x', and divide the result by the constant k_0/d^2 to obtain the nondimensional form of the problem, with $a = \alpha(T_1 - T_0)/k_0$, $\theta' = d\theta/dx$:

$$[(1 + a\theta)\theta']' = 0, \qquad \theta(0) = 0, \quad \theta(1) = 1. \tag{2.3}$$

In other problems we merely start with the nondimensional form of the equations and skip these steps.

We first choose a trial solution in the form of a polynomial in x because of its simplicity:

$$\theta_N = \sum_{i=0}^{N+1} c_i x^i. \tag{2.4}$$

This function can be made to satisfy the boundary conditions by requiring

$$c_0 = 0, \qquad \sum_{i=1}^{N+1} c_i = 1. \tag{2.5}$$

Equation (2.4) can then be written as

$$\theta_N = x + \sum_{j=1}^{N} A_j(x^{j+1} - x). \tag{2.6}$$

Notice that the first trial function x, satisfies the boundary conditions of the problem and each of the expansion functions satisfies the homogeneous boundary conditions (zero temperature at the walls). Thus we started with a series representation of the solution and, by application of the boundary conditions, were able to deduce reasonable trial functions.

The trial solution is substituted into Eq. (2.3) to form the residual

$$R(x, \theta_N) = (1 + a\theta_N)\theta_N'' + a(\theta_N')^2. \tag{2.7}$$

The weighted residual becomes

$$\int_0^1 w_k R(x, \theta_N)\, dx = 0, \qquad k = 1, 2, \ldots, N. \tag{2.8}$$

We next apply several criteria for comparison. For the first approximation

$$\theta_1 = x + A_1(x^2 - x), \qquad \theta_1' = 1 + A_1(2x - 1), \quad \theta_1'' = 2A_1. \tag{2.9}$$

Apply the collocation method using the collocation point $x = \frac{1}{2}$ because it is the midpoint of the interval. Another point could be chosen, but low-order approximations are likely to be better if the collocation points are distributed somewhat evenly throughout the region. Then the weighted residual (or the residual evaluated at the point $x = \frac{1}{2}$) becomes

$$[1 + \tfrac{1}{2}a(1 - \tfrac{1}{2}A_1)]2A_1 + a = 0, \tag{2.10}$$

which determines A_1. We choose to calculate numerical results only for the case $a = 1$, in which case

$$-\tfrac{1}{2}A_1^2 + 3A_1 + 1 = 0 \qquad \text{or} \qquad A_1 = -0.317. \tag{2.11}$$

The other solution to the quadratic is rejected as being physically unrealistic, since it gives the heat flux in the wrong direction at $x = 1$. The approximate solution is then

$$\theta_1 = x - 0.317(x^2 - x). \tag{2.12}$$

Compare this solution to that obtained with the subdomain method. For the first approximation the domain is taken from zero to one, whereas for higher approximations the interval $0 \to 1$ could be split into segments of equal length. The weighted residual is thus

$$\int_0^1 R(x, \theta_1)\, dx = 0 \tag{2.13}$$

and the differential equation is satisfied on the average. For $a = 1$ this is

$$\int_0^1 [1 + x + A_1(x^2 - x)]2A_1\, dx + \int_0^1 [1 + A_1(2x - 1)]^2\, dx = 0, \tag{2.14}$$

which yields the solution $A_1 = -0.333$. The approximate solution is then

$$\theta_1 = x - 0.333(x^2 - x), \tag{2.15}$$

which differs only slightly from the collocation solution. The method of moments gives the same solution, (2.15), for the first approximation.

In the Galerkin method the weighting function is $w_1 = x(1 - x)$. Then

$$\int_0^1 x(1 - x)R(x, \theta_1)\, dx = 0 \tag{2.16}$$

yields the approximate solution

$$\theta_1 = x - 0.326(x^2 - x). \tag{2.17}$$

How good are the results? For the first approximation precise answers are not expected, but comparisons to the exact solution $\theta = -1 + (1 + 3x)^{1/2}$ in Table 2.1 indicate the solution is accurate to 8 %. Of course, when there is no

TABLE 2.1

TEMPERATURE FOR ONE-DIMENSIONAL NONLINEAR HEAT CONDUCTION

			Method of moments		
x	Collocation Eq. (2.12)	Galerkin Eq. (2.17)	Eq. (2.15)	Eq. (2.19)	Exact solution
0.10	0.129	0.129	0.130	0.143	0.140
0.25	0.309	0.311	0.313	0.332	0.323
0.50	0.579	0.582	0.583	0.594	0.581
0.75	0.809	0.811	0.813	0.809	0.803
0.90	0.929	0.929	0.930	0.925	0.924

exact solution with which to compare, we must compute successive approximations. Convergence of these lends assurance that we have an accurate solution.

We compute the second approximation using the method of moments

$$\int_0^1 R(x, \theta_2)\, dx = 0, \qquad \int_0^1 xR(x, \theta_2)\, dx = 0. \tag{2.18}$$

The algebra now becomes more lengthy (due to the nonlinear nature of the problem) and for more difficult problems the solution for A_i could be found on a computer. The approximation satisfying (2.18) is

$$\theta_2 = \tfrac{3}{2}x - \tfrac{3}{4}x^2 + \tfrac{1}{4}x^3. \tag{2.19}$$

This solution differs only slightly from the first approximation (see Table 2.1) so that we stop with the second approximation. It is within 3% of the exact solution.

In the first approximation the flux at the boundary is

$$
\begin{aligned}
(1 + a\theta_1)\frac{d\theta_1}{dx} &= 1 - A_1 = 1.333 \qquad \text{at} \quad x = 0, \\
(1 + a\theta_1)\frac{d\theta_1}{dx} &= 2(1 + A_1) = 1.333 \qquad \text{at} \quad x = 1.
\end{aligned}
\tag{2.20}
$$

The fluxes at each boundary do not have to agree since the solution is only approximate; they are identical only for the moments solution (2.15). The true answer is 1.5, so that the error is 11%. For the second approximation the flux is 1.5, which is the exact result. Thus we obtain macroscopic or average information which is more accurate than the detailed solution for temperature. First approximations are particularly useful when this macroscopic information is the only part of the solution that is desired. Another feature of the results is that the different criteria gave about the same results, particularly for the higher approximations. The criterion chosen does not seem to be too crucial. Thus the decision can be made on a basis of convenience or other factors.

When very precise results are required, higher approximations must be calculated. These are easily done using the orthogonal collocation method presented in Chapter 5. As N increases we expect the residual to become smaller. This is illustrated in Fig. 2.1, which shows the residual for three orthogonal collocation solutions. For $N = 6$ the residual is very small. This suggests that the residual can be used as an error criteria, and this is rigorously established for many problems (see Sections 2.3, 4.1, 11.6). In problems for which the theorem has not yet been proved, the size of the residual gives a qualitative guide to the accuracy of the approximation. Comparing the residuals in Fig. 2.1, we would conclude that the approximate solution corresponding to the case $N = 6$ would be the best solution. In fact, that solution corresponds to the exact solution to five decimal places. Furthermore, error bounds can often be derived in terms of the residual, so that the error bounds are improved as the residual is decreased. For all problems we can use the residual as a guide to the success of the approximation, and in many problems this guide can be made precise, with a numerical value for the error. Establishing error bounds is a more difficult subject than finding the solution, so that we defer error bounds to Chapter 11.

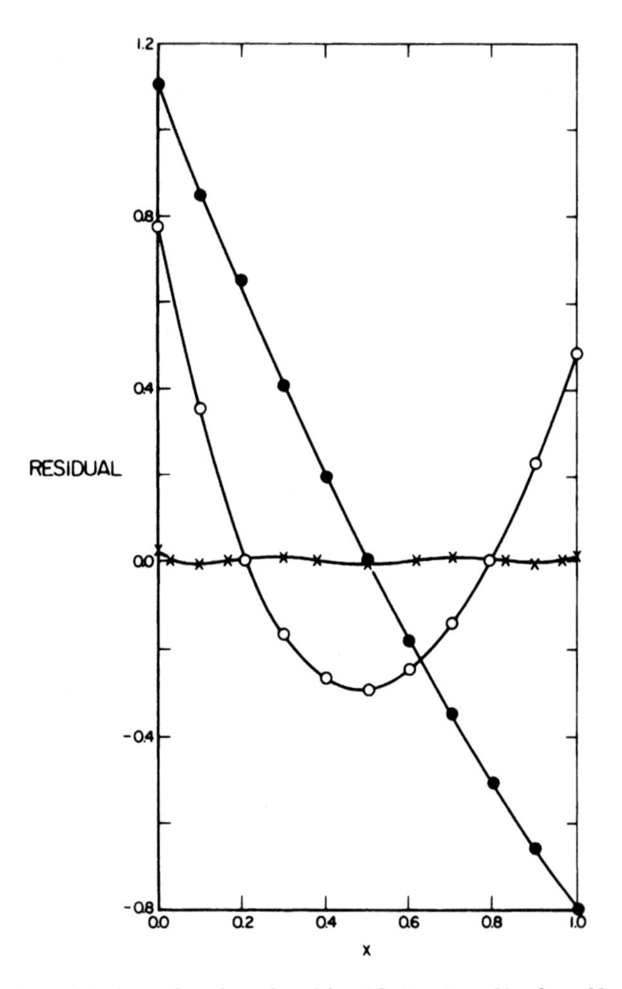

Fig. 2.1. Residual as a function of position (● $N = 1$, ○ $N = 2$, × $N = 6$).

2.2 Reduction to Ordinary Differential Equations

MWR can be used to reduce the dimensionality of a problem. If the solution depends on x and y, for example, trial functions can be taken with the y dependence specified. The partial differential equation (with respect to x and y) is then reduced to an ordinary differential equation (with respect to x), which determines the approximate solution. This approach was first tried by Kantorovich (1933, 1942) and Poritsky (1938). More recently, an adaptation was proposed by Kerr (1968) to eliminate the dependence on the initial choice of trial functions. In Kerr's scheme, the y dependence is assumed, for example, and MWR is used to find the x dependence. This x dependence is then used as a trial function, with unspecified y dependence, which is determined using MWR. The procedure is repeated until further iterations make no change (or until the calculations become unwieldy).

Let us illustrate the reduction to ordinary differential equations by application to heat conduction in a slab which extends from zero to infinity in the y direction and from zero to one in the x direction. On both sides and at infinity the temperature is T_0 while on the face $y = 0$ it takes the value $T_0 + T_1 x(1 - x)$. The dimensionless temperature is taken as $\theta = (T - T_0)/T_1$ and the problem is summarized:

$$\frac{\partial}{\partial x}\left(k\frac{\partial \theta}{\partial x}\right) + \frac{\partial}{\partial y}\left(k\frac{\partial \theta}{\partial y}\right) = 0,$$
$$\theta(x, 0) = x(1 - x), \qquad \theta(0, y) = \theta(1, y) = \theta(x, \infty) = 0. \tag{2.21}$$

When k is constant the technique of separation of variables gives the exact solution in the form of an infinite series. MWR can be applied when k is constant as well as when k depends on temperature, and both cases are treated below. In either case the objective is to estimate the average heat flux along the boundary $y = 0$, which in dimensionless form is the Nusselt number:

$$\text{Nu} \equiv -\int_0^1 \left.\frac{\partial \theta}{\partial y}\right|_{y=0} dx. \tag{2.22}$$

The problem is symmetric about $x = \frac{1}{2}$, and the solution vanishes at $x = 0$, 1 so that we want to choose trial functions which satisfy these properties. The y dependence of the approximate solution is not specified in advance. Applying these conditions to the second-degree polynomial $a_0 + a_1 x + a_2 x^2$ gives $x(1 - x)$. The set of functions $x^i(1 - x)^i$ then provides a complete set of functions satisfying these three properties. The trial solution is then taken as

$$\theta_N = \sum_{i=1}^{N} x^i(1 - x)^i c_i(y). \tag{2.23}$$

This conveniently satisfies the boundary conditions if $c_1(0) = 1$, $c_i(0) = 0$, $i \geq 2$; $c_i(\infty) = 0$, $i \geq 1$. For the linear case in the first approximation the residual is

$$R(\theta) = \theta_{xx} + \theta_{yy} = -2c_1 + x(1 - x)c_1''. \tag{2.24}$$

Applying the Galerkin method first,

$$\int_0^1 x(1 - x)R(\theta)\,dx = 0, \qquad -2c_1(\tfrac{1}{6}) + c_1''(\tfrac{1}{30}) = 0. \tag{2.25}$$

The solution is then $c_1 = \exp(-\sqrt{10}\, y)$ and the first approximation is

$$T(x, y) = x(1 - x)\exp(-\sqrt{10}\, y). \tag{2.26}$$

Using the point $x = \frac{1}{2}$ in the collocation method gives the solution $c_1 = \exp(-\sqrt{8}\, y)$, while the point $x = \frac{1}{4}$ (and hence $\frac{3}{4}$, too, by symmetry) gives $c_1 = \exp(-\sqrt{32/3}\, y)$. The subdomain method uses

$$\int_0^1 R(T)\, dx = 0 \qquad \text{to obtain} \qquad c_1 = \exp(-\sqrt{12}\, y). \tag{2.27}$$

In the least squares method the trial solution is taken as $T = x(1 - x)e^{\lambda y}$, and the residual is minimized with respect to λ:

$$\frac{\partial}{\partial \lambda}\int_0^\infty \int_0^1 R^2\, dx\, dy = 0 \qquad \text{or} \qquad c_1 = \exp(-\sqrt{10.48}\, y). \tag{2.28}$$

All methods give similar results with slight differences in the numerical values.

Next, apply the Galerkin method to obtain a second approximation. The weighted residuals are

$$\int_0^1 x(1 - x)R(T)\, dx = 0, \qquad \int_0^1 x^2(1 - x)^2 R(T)\, dx = 0, \tag{2.29}$$

which lead to the set of equations

$$\begin{pmatrix} 0.03333 & 0.007143 \\ 0.007143 & 0.001587 \end{pmatrix}\begin{pmatrix} c_1'' \\ c_2'' \end{pmatrix} = \begin{pmatrix} 0.3333 & 0.06667 \\ 0.06667 & 0.01905 \end{pmatrix}\begin{pmatrix} c_1 \\ c_2 \end{pmatrix} \tag{2.30}$$

to solve for c_1 and c_2.† Standard techniques are used to solve this system of linear equations‡:

$$\begin{aligned} c_1 &= 0.8035\exp(-3.1416y) + 0.1965\exp(-10.1059y), \\ c_2 &= 0.9104\,[\exp(-3.1416y) - \exp(-10.1059y)]. \end{aligned} \tag{2.31}$$

† The integrals can be evaluated conveniently using the formula

$$\mathcal{F}(m, n) = \int_0^1 x^m(1 - x)^n\, dx = m!n!/(m + n + 1)!$$

This function is tabulated in Table 6.4.

‡ In this case the dependence $c_i(y) = a_i \exp(-\lambda y)$ can be tried, leading to two simultaneous, homogeneous equations for the constants a_i. A solution exists only for certain values of λ^2 which are roots of a quadratic. The general solution is then

$$c_i(y) = \sum_{j=1}^{2} a_i^{(j)} \exp(-\lambda_j y).$$

The ratio $a_2^{(j)}/a_1^{(j)}$ is determined from the equations (with λ_j), the value of $\sum_{j=1}^{2} a_i^{(j)}$ from the initial conditions. For matrix methods see Amundson (1966).

With $T = x(1 - x)e^{-\lambda y}$ for the first approximation, the heat flux (2.22) is $\mathrm{Nu} = \lambda/6$. The results of the various methods are shown in Table 2.2. The

TABLE 2.2

AVERAGE HEAT FLUX FOR LINEAR PROBLEM

Method	Nu	Φ[a]
Exact	0.542	0.105040
Galerkin, $N = 1$	0.527	0.105409
$N = 2$	0.540	0.105048
Collocation, $x = \frac{1}{2}$	0.471	0.106066
$x = \frac{1}{4}$	0.544	0.105464
Subdomain	0.577	0.105848
Least squares	0.540	0.105438

[a] Defined in Eq. (7.48).

first and second approximation for the Galerkin method differ by only 2.4%, so that we stop with the second approximation. It differs from the exact value

$$\mathrm{Nu} = \frac{16}{\pi^3} \sum_{n=0}^{\infty} \frac{1}{(2n + 1)^3} \tag{2.32}$$

derived by separation of variables (see Section 2.3) by only $\frac{1}{2}$%.

If the thermal conductivity depends on temperature, the separation of variables technique cannot be applied, but MWR is applied in the same fashion. Take the thermal conductivity function as $k = k_0 + k_1(T - T_0)$ and the residual is, in place of (2.24),

$$R = (1 + a\theta)(\theta_{xx} + \theta_{yy}) + a(\theta_x{}^2 + \theta_y{}^2). \tag{2.33}$$

The Galerkin method in the first approximation (2.25) gives

$$\begin{aligned} -2c\mathscr{F}(1, 1) + c''\mathscr{F}(2, 2) = a[c^2(2\mathscr{F}(2, 2) - 4\mathscr{F}(3, 1) - \mathscr{F}(1, 1) + 4\mathscr{F}(2, 1)) \\ - \mathscr{F}(3, 3)(cc'' + c'^2)], \end{aligned} \tag{2.34}$$

where $\mathscr{F}(m, n)$ is listed in Table 6.4. Since the residual is nonlinear, so are the ordinary differential equations (2.34). Since a is usually small we solve (2.34) using the perturbation method.

The function $c(y)$ is expanded in powers of a and substituted into (2.34):

$$c(y) = c_0(y) + ac_1(y) + a^2c_2(y) + \cdots. \tag{2.35}$$

We collect like powers of a and require that each coefficient of a^n be zero. Such a procedure leads to a set of linear equations, which can be solved sequentially, but the set becomes rapidly more complex with higher powers of a. The first two such equations are

$$a^0: \quad -2c_0\mathscr{F}_{11} + \mathscr{F}_{22}c_0'' = 0, \qquad \mathscr{F}_{mn} \equiv \mathscr{F}(m, n),$$

$$a^1: \quad -2c_1\mathscr{F}_{11} + \mathscr{F}_{22}c_1'' = [c_0^2(2\mathscr{F}_{22} - 4\mathscr{F}_{31} - \mathscr{F}_{11} + 4\mathscr{F}_{21}) - \mathscr{F}_{33}(c_0c_0'' + c_0'^2)] \tag{2.36}$$

The solution to the first equation is $c_0 = \exp(-\sqrt{10}\, y)$. This must be substituted into the next equation to solve for

$$c_1(y) = +0.11(\exp(-\sqrt{10}\, y) - \exp(-2\sqrt{10}\, y)). \tag{2.37}$$

To first order in a then the solution is

$$\theta = x(1 - x)[\exp(-\sqrt{10}\, y) + a0.11(\exp(-\sqrt{10}\, y) - \exp(-2\sqrt{10}\, y))].$$

$$\mathrm{Nu} = -\int_0^1 \left[(1 + a\theta)\frac{\partial\theta}{\partial y}\right]_{y=0} dx = \frac{\sqrt{10}}{6}[1 + 0.09a]. \tag{2.38}$$

Thus MWR can be applied to both linear and nonlinear problems when standard methods are not applicable.

2.3 Boundary Methods

A problem consists of the differential equation and boundary conditions. It may be easier to choose trial functions satisfying the differential equations than the boundary conditions, possibly due to complicated or irregular boundaries. In these cases a boundary method is used: MWR is applied on the boundary. The same two-dimensional example (with constant thermal conductivity) is used to illustrate the method, but first the exact solution is derived using separation of variables.

A separated trial solution $\theta(x, y) = X(x)Y(y)$ is substituted into the differential equation (2.21). The result is divided by XY and rearranged to give

$$Y''/Y = -X''/X = \lambda. \tag{2.39}$$

Since a function of x is equal to a function of y, for all x and y, both functions must be constant λ. Equation (2.39) can then be separated into two problems, recognizing that the three homogeneous boundary conditions also separate into $X(0) = X(1) = Y(\infty) = 0$:

$$\begin{aligned} X'' + \lambda X &= 0, \qquad & Y'' - \lambda Y &= 0, \\ X(0) = X(1) &= 0, \qquad & Y(\infty) &= 0. \end{aligned} \tag{2.40}$$

The solutions are $X = \sin n\pi x$ and $Y = \exp(-n\pi y)$. The trial solution

$$\theta_n = \exp(-n\pi y) \sin n\pi x \tag{2.41}$$

then satisfies the differential equation, and three of the boundary conditions. To fit the other boundary condition at $y = 0$ we add up all the solutions (2.41), with $\theta = \Sigma A_n \theta_n$, giving

$$x(1-x) \sim \sum_{n=1}^{\infty} A_n \sin n\pi x, \qquad \text{at} \quad y = 0. \tag{2.42}$$

The constants A_n are chosen by multiplying both sides by $\sin m\pi x$, integrating from 0 to 1 to obtain $A_n = 8/(n\pi)^3$, n odd, $A_n = 0$ for n even. This procedure is reminiscent of the Galerkin method: the residual (2.42) is made orthogonal to the trial functions. It can be shown that this scheme also minimizes the mean square error

$$I = \int_0^1 \left[x(1-x) - \sum_{n=1}^{\infty} A_n \sin n\pi x\right]^2 dx. \tag{2.43}$$

A least squares boundary method is applied next. We need trial functions satisfying the differential equations. To obtain them in this problem we would apply separation of variables, resulting in (2.41). The trial solution (symmetric about $x = \frac{1}{2}$) is thus

$$\theta_N = \sum_{i=1}^{N} A_i e^{-(2i-1)\pi y} \sin(2i-1)\pi x. \tag{2.44}$$

This trial solution satisfies the differential equation and boundary conditions, except for those along $y = 0$. The boundary residual is (for $N \to \infty$)

$$B \equiv x(1-x) - \sum_{n=1}^{\infty} A_i \sin(2i-1)\pi x. \tag{2.45}$$

In the least squares method this residual is squared and integrated along $x = 0$, 1, giving just (2.43). Thus the least squares method minimizes I, and results in the same solution as derived by separation of variables (or Galerkin's method).

Next apply the collocation method. For the first approximation the boundary residual is set to zero at the midpoint of the interval, $x = \frac{1}{2}$, giving $A_1 = \frac{1}{4}$. Then

$$\theta_1 = \tfrac{1}{4} e^{-\pi y} \sin \pi x. \tag{2.46}$$

Table 2.3 lists values of the average flux, Nu.

TABLE 2.3

HEAT TRANSFER ANALYSIS BY THE BOUNDARY METHOD

Method	Nu	Φ	$I \times 10^4$	ε
Least squares, Galerkin	0.516	0.104568	0.485	0.0107
Least squares–collocation				
$M = 1$	0.500	0.098175	0.806	0.0141
$M = 2$	0.513	0.103427	0.495	0.0112
$M = 3$	0.515	0.104220	0.486	0.0109
$M = 4$	0.516	0.104428	0.485	0.0108
Exact	0.542	0.105040	—	—

If the collocation method is applied at other collocation points we find that the solution depends on the choice of collocation points, with A_1 taking values from 0.25 to 0.318 as the collocation point ranges from 0 to 1. In order to reduce this dependence on the choice of the collocation point we can apply the least squares-collocation method. In this method the residual is evaluated at more points than there are adjustable constants. The residuals at each of these points are squared and added; the result is minimized with respect to the parameters. If the number of evaluation points equals the number of parameters, the method is called the collocation method. As the number of evaluation points increases above the number of parameters, the method approaches the least squares method. In this example evaluate (2.45) at the M evaluation points, x_i; square the results, add, and minimize with respect to A. The result is

$$A_1 = \frac{\sum_{i=1}^{M} x_i(1 - x_i) \sin \pi x_i}{\sum_{i=1}^{M} \sin^2 \pi x_i}. \tag{2.47}$$

Calculations for various M are shown in Table 2.3. Even for $M = 4$ the results approach the least squares method, giving the first term in the exact solution. Note that the numerical value of Nu is not as good as the other methods in MWR (Table 2.2), because the first term in the exact solution is not a good approximation. In the regular MWR the approximate solution attempts to account for all the terms in the exact solution, in an average way. Obviously the least squares-collocation method can be applied when the differential equation is not satisfied but the boundary conditions are, or when neither condition is satisfied.

For this problem it is easy to establish error bounds using Theorem 11.32. Consider the error $E = \theta_N - \theta$, where θ is the exact solution and θ_N is defined by (2.44). The function E satisfies the differential equation because both θ and θ_N do, and the three boundary conditions on $x = 0$, 1 and $y \to \infty$

are satisfied, too. Theorem 11.32 says that under these conditions the function E must take its maximum and minimum values on the boundary $y = 0$. The maximum of E on the boundary is thus an upper bound for the error *for all x and y*, i.e., for the whole domain. The maximum error is thus $\varepsilon = \max_{0 \le x \le 1} |x(1 - x) - A_1 \sin \pi x|$, and is tabulated in Table 2.3. The approximate solution then satisfies $|\theta(x, y) - \theta_N(x, y)| < \varepsilon$.

A boundary collocation method was first applied by Sparrow and Loeffler (1959) to determine the laminar flow between cylinders arranged in regular arrays. Many applications have been made since to heat transfer: Sparrow *et al.* (1961) used the predicted velocity field to determine the heat transfer to the laminar flow. Balcerzak and Raynor (1961) solved for the steady-state temperature distribution in several geometries: n-sided polygons, with from 3 to 10 sides; rectangles with a hole, for various aspect ratios; elliptical cross sections with confocal elliptical holes and confocal slits. In this case the authors used a conformal mapping to transform the differential equation, which was then solved by collocation on the boundary. Altmann (1958) apparently originated the idea of least squares-collocation on the boundary, and applied it to calculations in crystallography. Hulbert (1965) and Hulbert and Niedenfuhr (1965) applied it to elasticity problems in irregular geometries, such as plates with elliptical holes, star-shaped holes, etc. Ojalvo and Linzer (1965) compared the boundary collocation with the least squares-collocation on the boundary and found the latter generally gave better results in two-dimensional steady-state heat conduction and one-dimension transient heat conduction problems. Several combined flow and heat transfer problems are discussed using boundary collocation in Sections 4.1 and 4.5. Sparrow and Haji-Sheikh (1968) described applications of least squares-collocation on the boundary to both steady-state and transient heat conduction in arbitrary bodies. In that case it is necessary to choose solutions to Laplace's equation (1.9b), as trial functions, before applying least squares-collocation on the boundary. These are readily available, since both the real and imaginary parts of $z^n = (x + iy)^n$ satisfy the equation. Thus the first few functions are $1, x, y, x^2 - y^2, x^3 - 3xy^2, y^3 - 3yx^2$. Shih (1970) used the boundary collocation method to solve for the temperature distribution in a square column surrounding a heating cylinder. Details of a computer code useful in these problems is in Davis (1962).

Many solutions in two-dimensional heat transfer problems can be deduced from solutions for the velocity in ducts [Eq. (1.19), see also Section 4.1] or from solutions to the torsion problem [Eq. (1.19) with $\lambda = 2$]. A number of these solutions, as well as other references to boundary collocation in the field of elasticity (where it is called point matching) are contained in Sattinger and Conway (1965) and Leissa and Neidenfuhr (1966) and the references cited by them. For the torsion problem Higgins (1949) and Sherwood (1962) expanded the solution in terms of a Fourier series which was required to satisfy the boundary conditions using least squares (Higgins) and collocation (Sherwood). Warren (1964) compares some boundary least squares to boundary collocation for an example in the bending of rhombic plates and concludes that the boundary collocation is considerably easier to apply. Sparrow and Haji-Sheikh (1970) apply the least squares method to the integral equations governing radiative heat transfer. Leissa *et al.* (1969) compare several methods for plate bending problems including singularities: collocation (interior and boundary), least squares-collocation (interior and boundary), subdomain, Galerkin, and Rayleigh–Ritz (a variational method; see Chapter 7). The methods are compared using several criteria. The boundary collocation and boundary least squares-collocation are favored when the differential equations are easy to solve but the boundary conditions are not (irregular shape). The interior methods are preferable when the differential equation is complicated (variable coefficients or nonlinear). The collocation and least squares-collocation are straightforward to program and easy to understand and apply. The calculations done by Leissa *et al.* (1969) used a large number of terms (3 to 10 for simple problems, 20 to 40 for complicated ones). Langhaer (1969) has shown for problems with residuals which are linear functions of the parameters that the least squares, collocation, and subdomain methods are the same as $N \to \infty$, provided the collocation points and subdomains are arranged uniformly in the domain.

2.4 General Treatment of Steady-State Heat Conduction

MWR is applied to a general heat transfer problem including heat generation (depending on position and temperature) and several different boundary conditions. For example,

$$\nabla \cdot (k \nabla T) = f(\mathbf{x}, T) + \rho C_v \mathbf{u} \cdot \nabla T \qquad \text{in V}, \tag{2.48a}$$

$$T = T_1 \qquad \text{on } S_1, \tag{2.48b}$$

$$-k\mathbf{n} \cdot \nabla T = q_2 \qquad \text{on } S_2, \tag{2.48c}$$

$$-k\mathbf{n} \cdot \nabla T = h(T - T_3) \qquad \text{on } S_3. \tag{2.48d}$$

The physical properties and heat transfer coefficient h can be functions of temperature, while the velocity is a known function of position. The three boundary conditions: temperature specified, flux specified, and the radiation condition, are called boundary conditions of the first, second, and third kind.

The first step is to try the following transformation to reduce the problem to linear form:

$$\phi = \int_{T_0}^{T} k(T')\, dT', \qquad \nabla\phi = k \nabla T. \tag{2.49}$$

Under this transformation the problem reduces to

$$
\begin{aligned}
\nabla^2 \phi &= f'(\mathbf{x}, \phi) + \frac{\rho C_v}{k} \mathbf{u} \cdot \nabla \phi \qquad \text{in } V, \\
\phi &= \phi_1 \qquad \text{on } S_1, \\
-\mathbf{n} \cdot \nabla \phi &= q_2 \qquad \text{on } S_2, \\
-\mathbf{n} \cdot \nabla \phi &= h[T(\phi) - T_3] \qquad \text{on } S_3 .
\end{aligned}
\tag{2.50}
$$

This is linear provided the heat generation does not depend on temperature, either $\mathbf{u} = 0$ or $\rho C_v/k$ is constant, and there are no boundary conditions of the third kind. If this transformation fails to give a linear problem it is perhaps simpler to treat Eqs. (2.48) directly.

It is commonly thought that in MWR the trial functions must satisfy all boundary conditions. This is not true. In variational methods the boundary conditions are divided into two classes: natural and essential conditions. The trial functions need not satisfy the natural boundary conditions, since the variational principle forces them to be satisfied, whereas it is essential that the trial functions satisfy the essential boundary conditions, since the variational principle does not force them to be satisfied. The same approach can be used in MWR. This is advantageous because it is sometimes cumbersome to require each trial function to satisfy all the boundary conditions. Here (2.48b) is an essential boundary condition and the trial function is taken as

$$
T_N = T_1 + \sum_{i=1}^{N} c_i \, T_i(\mathbf{x}), \tag{2.51}
$$

where the $T_i = 0$ on S_1. Weighting functions are chosen which vanish on S_1, the weighted residual is integrated by parts and the divergence theorem is applied,† giving

$$
\begin{aligned}
\int_V w_j \nabla \cdot k \, \nabla T_N \, dV &= \int_V \nabla \cdot (w_j k \, \nabla T_N) \, dV - \int_V k \, \nabla T_N \cdot \nabla w_j \, dV \\
&= -(k \, \nabla T_N \cdot \nabla w_j) + \int_S w_j k \mathbf{n} \cdot \nabla T_N \, dS,
\end{aligned}
\tag{2.52a}
$$

$$
(R, w_j) = -(k \, \nabla T_N \cdot \nabla w_j) - (f + \rho C_v \mathbf{u} \cdot \nabla T_N, w_j) + \int_S w_j k \mathbf{n} \cdot \nabla T_N \, dS. \tag{2.52b}
$$

† The divergence theorem is

$$
\int \nabla \cdot \mathbf{A} \, dV = \int \mathbf{n} \cdot \mathbf{A} \, dS,
$$

where $\mathbf{n}$ is the outward pointing normal. Then Eq. (2.52a) follows.

Weighted boundary residuals are also formed using the same weighting functions:

$$(B_2, w_j)_{s_2} = \int_{s_2} [k\mathbf{n} \cdot \nabla T_N + q_2] w_j \, dS,$$
$$(B_3, w_j)_{s_3} = \int_{s_3} [k\mathbf{n} \cdot \nabla T_N + h(T_N - T_3)] w_j \, dS. \tag{2.53}$$

Comparison of Eqs. (2.52) and (2.53) shows that they include similar terms. The results are combined and set to zero to give the equations governing the approximate solution:

$$\int [k \nabla T_N \cdot \nabla w_j + w_j(f + \rho C_v \mathbf{u} \cdot \nabla T_N)] \, dV$$
$$= -\int_{S_2} q_2 w_j \, dS - \int_{S_3} h(T_N - T_3) w_j \, dS. \tag{2.54}$$

Convergence theorems have been proved for a variety of special cases of Eq. (2.48), including nonlinear problems (see Chapter 11). Pointwise error bounds result from use of the maximum principle and mean square residual. If the problem is linear, error bounds are easy to derive (see Chapter 11).

The two-dimensional problem has been discussed in the literature for a variety of shapes of domains. Higgins (1942, 1943) discusses hundreds of exact and approximate solutions to the problem

$$u_{xx} + u_{yy} = -2 \qquad \text{in } S, \qquad u = 0 \qquad \text{on } C \tag{2.55}$$

in domains which are rectangular, triangular, are bounded by polygons, curves and arcs, and flanges and webs. The same methods (and trial functions) can be used in the heat transfer problem because of the equivalence of the following two problems, with $u = T - T_0$:

$$T_{xx} + T_{yy} = f(x, y) \qquad \text{in } S,$$
$$u_{xx} + u_{yy} = f - \nabla^2 T_0 = g(x, y) \qquad \text{in } S, \tag{2.56}$$
$$T = T_0(x, y) \qquad \text{on } C, \qquad u = 0 \qquad \text{on } C.$$

The same problem of Eq. (2.55) arises in laminar flow in ducts, which is solved in Section 4.1.

2.5 Mass Transfer from a Sphere

The steady-state transport equation with convection, (1.7), is an interesting and difficult problem because the solution depends crucially on the size of a parameter, such as the velocity. As an example we study mass transfer from a fluid or solid sphere to a fluid in Stokes flow. The important variable is the dimensionless mass transfer coefficient, called the Sherwood number.

The flow field around the sphere must first be determined. We assume the fluid sphere retains its shape, the flow is slow, $\mathrm{Re} = (dU/\nu) \ll 1$, where d is the sphere diameter, and U is the constant velocity at infinity. The viscosities of the fluid inside and outside the sphere are μ_1 and μ. The flow field is given by the Hadamard solution (Levich, 1962, p. 395), which reduces to the Stokes solution for flow around a solid sphere when $\mu_1/\mu \to \infty$ (Levich, 1962, p. 81; Bird *et al.*, 1960, p. 132). The stream function is (in dimensionless form)

$$\psi = -\frac{1}{2} r^2 \left[1 - \frac{3}{2r}\frac{3\mu_1 + 2\mu}{3\mu_1 + 3\mu} + \frac{1}{2r^3}\frac{\mu_1}{\mu_1 + \mu}\right] \sin^2\theta \tag{2.57}$$

outside the sphere and the velocity components are given by

$$u_r = \frac{1}{r^2 \sin\theta}\frac{\partial\psi}{\partial\theta}, \qquad u_\theta = -\frac{1}{r\sin\theta}\frac{\partial\psi}{\partial r}. \tag{2.58}$$

The fluid thus flows past the sphere, inducing a circulation inside the fluid sphere due to the traction exerted on the boundary. The fluid mechanics is thus determined, and applicable for $\mathrm{Re} \gtrsim 1$.

In the mass transfer problem we assume the concentration in the sphere maintains the value c_0. Thus the mass depleted from the sphere is small compared to the total amount available. The circulation inside helps insure that the concentration is uniform. Mass is transferred to the external fluid, which has concentration c_1 far upstream of the sphere. The transport equation is then [in dimensionless form with $c = (c' - c_1)/(c_0 - c_1)$]

$$\frac{\mathrm{Pe}}{2}\left[u_r\frac{\partial c}{\partial r} + u_\theta\frac{1}{r}\frac{\partial c}{\partial\theta}\right] = \frac{1}{r^2}\frac{\partial}{\partial r}\left(r^2\frac{\partial c}{\partial r}\right) + \frac{1}{r^2\sin\theta}\frac{\partial}{\partial\theta}\left(\sin\theta\frac{\partial c}{\partial\theta}\right),$$
$$c(1,\theta) = 1, \qquad c(r,\theta) = 0 \quad \text{as} \quad r\to\infty, \quad \theta = 0. \tag{2.59}$$

The Peclet number, $Ud/\mathscr{D}$, is the product of the Reynolds number and Schmidt number $\mathrm{Pe} = \mathrm{Re}\,\mathrm{Sc}$, $\mathrm{Sc} = \nu/\mathscr{D}$. The quantity of interest is the Sherwood number

$$\mathrm{Sh} = \int_0^\pi \left(-\left.\frac{\partial c}{\partial r}\right|_{r=1}\right)\sin\theta\, d\theta. \tag{2.60}$$

The Schmidt number can take very large values, so that it is necessary to examine the solution for the full range of Peclet numbers, even though $\mathrm{Re} < 1$.

For small $\mathrm{Pe} \sim 0$, the situation corresponds to diffusion into a stagnant fluid. The solution to (2.59) for $\mathrm{Pe} = 0$ is

$$c = \frac{1}{r}, \qquad r \ge 1, \quad \mathrm{Sh} = 2.$$

This applies to both fluid and solid spheres, since in the limit of $\mathrm{Pe} = 0$ there is no motion inside the sphere either. For nonzero but small Peclet numbers, solutions have been derived for a solid sphere using the perturbation method, but the procedure is complicated, because the convection and diffusion terms are of comparable magnitudes far from the sphere. Thus an inner and outer expansion technique is used (Acrivos and Taylor, 1962). Instead we derive the solution using MWR.

For $\mathrm{Pe} \to \infty$ solutions are also known. Levich (1962) obtained the following solutions for both fluid and solid spheres:

$$\left.\begin{array}{ll} \mathrm{Sh} = 0.991\ \mathrm{Pe}^{0.33} & \text{solid} \\ \mathrm{Sh} = 0.923(U_1/U)^{0.5}\ \mathrm{Pe}^{0.5} & \text{fluid} \end{array}\right\} \mathrm{Pe} \to \infty, \tag{2.61a}$$

where

$$U_1/U = \mu/2(\mu_1 + \mu). \tag{2.61b}$$

In these cases the concentration profiles exhibit a boundary layer phenomena, with the boundary layer thickness being smallest at the stagnation point. Acrivos and Goddard (1965) have calculated the next term in the perturbation series for the solid sphere. The several asymptotic solutions provide a test of the approximate solution derived below. Griffith (1960) compares several solutions to experiment and Brian and Hales (1969) recently obtained a finite difference solution for the solid sphere but including transpiration and a changing diameter.

Bowman *et al.* (1961) and Ruckenstein (1964) have solved for intermediate Peclet numbers using the integral method. Using Levich's analysis for $\mathrm{Pe} \to \infty$ as a guide, we ignore the last term in (2.59) on the grounds that diffusion in the θ direction is negligible compared to convection in that direction. This is assumed to be true even though the steamlines do not exactly coincide with lines of constant θ. Furthermore, a boundary layer is developed whose thickness depends on θ and the Peclet number. For $\mathrm{Pe} \to 0$ the boundary layer becomes very thick. Rewrite (2.59) in vector notation for an incompressible fluid. Integrate over the boundary layer, $1 \le r \le \delta(\theta)$, along lines of constant θ,

$$\frac{\mathrm{Pe}}{2} \int_1^\delta \nabla \cdot (\mathbf{u}c) r^2\, dr = \int_1^\delta \nabla_r{}^2 c r^2\, dr = - \left.\frac{\partial c}{\partial r}\right|_{r=1}. \tag{2.62}$$

The last term results from integration and requiring $\partial c/\partial r = 0$ at $r = \delta$. The first term can be rewritten using the expression for the divergence operator in spherical coordinates, giving

$$\nabla \cdot (\mathbf{u}c) = \frac{1}{r^2} \frac{\partial}{\partial r} (r^2 u_r c) + \frac{1}{r \sin \theta} \frac{\partial}{\partial \theta} (u_\theta \sin \theta c). \tag{2.63}$$

The first term can be integrated directly and the boundary terms vanish since $u_r = 0$ at $r = 1$ and $c = 0$ at $r = \delta$. The second term can be rearranged using the stream function (2.58) and integration by parts:

$$\frac{1}{\sin\theta}\int_1^\delta \frac{\partial}{\partial\theta}(u_\theta \sin\theta\, c) r\, dr = \frac{-1}{\sin\theta}\int_1^\delta \frac{\partial}{\partial\theta}\left(c\frac{\partial\psi}{\partial r}\right) dr$$

$$= \frac{-1}{\sin\theta}\left[\frac{d}{d\theta}\int_1^\delta c\frac{\partial\psi}{\partial r}\, dr - \frac{d\delta}{d\theta}\, c\, \frac{\partial\psi}{\partial r}\bigg|_\delta\right] \tag{2.64}$$

The final term vanishes because $c(\delta, \theta) = 0$. The final equation is then

$$\frac{\text{Pe}}{2}\frac{d}{d\theta}\int_1^\delta c\frac{\partial\psi}{\partial r}\, dr = \sin\theta \frac{\partial c}{\partial r}\bigg|_{r=1}. \tag{2.65}$$

This is an ordinary differential equation for $\delta(\theta)$ once a profile has been assumed for c.

Bowman *et al.* made calculations using two different polynomial trial functions:

$$c_1 = a + (b/r), \tag{2.66a}$$

$$c_2 = a + (b/r) + cr + dr^2. \tag{2.66b}$$

The constants were chosen to satisfy

$$c(1, \theta) = 1, \qquad c(\delta, \theta) = 0, \tag{2.67}$$

whereas the two addition constants in the second trial function were determined from

$$\frac{\partial}{\partial r}\left(r^2\frac{\partial c}{\partial r}\right) = 0 \qquad \text{at} \quad r = 1, \tag{2.68a}$$

$$\frac{\partial c}{\partial r} = 0 \qquad \text{at} \quad r = \delta. \tag{2.68b}$$

Notice that the first trial function does not satisfy the restriction $\partial c/\partial r = 0$ at $r = \delta$ which was used to derive (2.65). Bowman derived (2.65) in a different way and apparently did not realize the error. Equation (2.68a) is a derived condition† obtained by evaluating the differential equation (2.59) on the boundary and noting that $u_r = \partial c/\partial\theta = 0$ there.

For intermediate Pe Eq. (2.65) is integrated numerically. For large Peclet numbers an asymptotic formula can be obtained for the solid sphere. The result is (2.61a) with the constant 0.978 when using trial function (2.66b) and

† Derived boundary conditions were first used by Duncan (1937). See Section 4.2 for an examination of their utility in boundary layer flow.

0.89 when using (2.66a). For small Peclet number Bowman *et al.* obtained $\mathrm{Sh} = 2 + 0.56\ \mathrm{Pe}$ using (2.66b). They were unable to obtain asymptotic formula for the fluid sphere case, although the numerical results confirm the form of Eqs. (2.61).

The analysis by Ruckenstein (1964) is similar except that he restricted attention to large Schmidt numbers. The concentration boundary layer is thin and the curvature effects can be ignored:

$$\frac{1}{r^2}\frac{\partial}{\partial r}\left(r^2\frac{\partial c}{\partial r}\right) \simeq \frac{\partial^2 c}{\partial y^2}, \qquad y = r - 1. \tag{2.69}$$

Ruckenstein used a trial function of the form

$$c = 1 - 2\left(\frac{y}{\delta}\right) + 2\left(\frac{y}{\delta}\right)^3 - \left(\frac{y}{\delta}\right)^4, \tag{2.70}$$

which satisfies the restrictions

$$\begin{aligned} c = 1, \qquad c_{yy} = 0 \quad &\text{at} \quad y = 0, \\ c = c_y = c_{yy} = 0 \quad &\text{at} \quad y = \delta. \end{aligned} \tag{2.71}$$

He obtained asymptotic formulas similar to Eqs. (2.61) with a constant of 1.037 for the solid and 0.895 for the fluid sphere.

Using these solutions it is possible to predict the mass transfer from a fluid sphere to another fluid for a variety of viscosity ratios. Bowman *et al.* do this for a solid in a gas or liquid, a liquid drop in a liquid, and a gas bubble in a liquid. The integral method provides a means for interpolating between regions where asymptotic results are valid, and the asymptotic results themselves provide insight into the problem and a check on the approximate solution. Higher approximations have apparently not been calculated for this problem. Wasserman and Slattery (1969) have used the Galerkin method to solve for creeping flow past a fluid globule when a trace of surfactant is present. The Galerkin method is used to solve the mass transfer problem, which influences the shape of the globule through the surface tension, which in turn influences the flow problem, making the problem exceedingly difficult to treat by anything but MWR or finite difference solutions.

2.6 Choice of Trial Functions

When applying MWR one important choice is the trial function. This choice provides the power of the method, in that known information can be incorporated into the trial solution. In low-order approximations the choice may influence the results, but higher approximations are less affected since numerical convergence is desired. The main influence then is the rate of convergence rather than the eventual solution. The first step in tackling any problem is to try several exact methods—often they suggest trial functions which satisfy some, but not all, of the conditions. Perhaps the full problem cannot be solved exactly but special cases can. The special cases provide a test of the approximate solution. The mass transfer problems in Section 2.5 provide an example, as do problems involving non-Newtonian fluids in Section 8.5. At the same time, we must be careful not to overspecify the trial function. In Section 4.3 is a case in which the solution becomes flat as one of the parameters becomes small, but the same parabolic trial function is used in all cases, leading to poor results. Such a difficulty is revealed when successive approximations are calculated.

The trial functions must be complete (see p. 10) and linearly independent. The polynomials are complete, for example, so that any continuous function can be expanded in terms of them. Several other examples are given in Chapter 11. The completeness property of a set of functions insures that we can represent the exact solution provided enough terms are used. Otherwise the successive approximations might converge to something which was not the solution.

There are two other guidelines to the choice of trial functions: examine the symmetry of the problem and apply the boundary conditions. If the boundary conditions are $y(x, z) = f(x, z)$ a convenient form of the trial functions is

$$y(x, z) = f(x, z) + \sum_{i=1}^{N} a_i y_i(x, z), \tag{2.72}$$

where the $y_i = 0$ on the boundary. The functions themselves can sometimes be found by starting with a general polynomial and applying the boundary conditions and symmetry conditions (see Sections 2.1 and 2.5). Derived boundary conditions can be used (i.e., requiring the differential equation to be satisfied on the boundary, as in Sections 2.5 and 4.2). Finally, the choice should not unduly complicate the analysis.

If the boundary conditions are of the second and third kind it is often convenient to combine the differential equation and boundary residual, as in (2.54), rather than making each term satisfy the boundary condition. Orthogonal polynomials are very useful trial functions, and can be constructed to satisfy some of the boundary conditions and made to fit others by combinations of the polynomials. The orthogonal collocation method discussed in Chapter 5 uses this approach and is useful for boundary conditions of the first, second, or third kind. The orthogonality of the polynomials gives computational advantages, although the same approximation can be expressed in terms of powers of x, if the computations could be done accurately enough.

Transcendental trial functions have been used (see Sections 4.2, 4.5, 9.5,

Schetz, 1963; and Richardson, 1968), although usually for only a first or second approximation due to the added difficulty of performing the computations. The solutions to simpler, but related, problems often provide a source of trial functions, particularly for eigenvalue problems (see Section 3.1). Specialized sets of trial functions have been developed for convective instability problems (Chapter 6) and the solutions of linear problems are used as trial functions for the nonlinear problems (Section 6.5). Trial functions for three-dimensional velocity vectors have been constructed (Section 6.2).

In time-dependent problems it is convenient to expand the solution in terms of spatial modes which satisfy the boundary conditions:

$$Y(x, t) = f(x) + \sum_{i=1}^{N} A_i(t)X_i(x). \tag{2.73}$$

The functions $A_i(t)$ are then determined by the approximate method and the initial conditions are fit approximately too (Chapter 3). Solutions may be spliced together. For example, if the solution at $z = 0$ is $T_1(x, y)$ and at $z = \infty$ it is $T_2(x, y)$, the trial function can be taken as

$$T(x, y, z) = Z_1(z)T_1(x, y) + Z_2(z)T_2(x, y) \tag{2.74}$$

and MWR is used to determine the splicing functions Z_i. The solutions T_1 and T_2 can be exact solutions, finite difference solutions, or MWR solutions. Discontinuous trial functions can be used (see Section 7.4). If the differential equation is easy to satisfy but the boundary conditions are not, then boundary methods are in order, as listed for a wide variety of geometries, with holes, etc., in Sections 2.3 and 4.1. Spline functions and finite element functions (Section 5.6) are completely general trial functions and provide a means to compare MWR to finite difference methods. Thus the examples in this book provide a number of case studies in the choice of trial functions.

EXERCISES

2.1. Verify Eq. (2.17).

2.2. Apply one of MWR to $\theta'' = x$ in $0 < x < 1$ under the boundary conditions

$$\theta(0) = 1, \qquad \theta'(1) + \text{Nu}\,\theta(1) = 0.$$

Deduce a trial function by assuming a polynomial (2.4) and applying the boundary conditions.
Answer: Collocation at $x = 0.5$, $\theta_1 = 1 - \frac{1}{2}x + A_1(x^2 - \frac{3}{2}x)$ gives $A_1 = 0.25$ when $\text{Nu} = 1$.

2.3. Deduce a general form for a trial function satisfying the boundary conditions of Exercise 2.2.

Answer: $\theta_N = 1 - \dfrac{\text{Nu}}{\text{Nu} + 1} x + \sum_{i=1}^{N} A_i \theta_i; \quad \theta_i = x^i - \dfrac{i + \text{Nu}}{1 + \text{Nu}} x.$

2.4. For $N = 2$ apply any of MWR and show that the result is the exact solution.

2.5. Apply a one-term collocation method at $x = 0.5$ to solve $u'' = \exp(u)$ in $0 < x < 1$, $u(0) = u(1) = 0$. Compare to the exact answer:

x	0.1	0.2	0.3	0.4	0.5
$u(x)$	-0.0414	-0.0733	-0.0958	-0.1092	-0.1137

Answer: With $u_1 = Ax(1 - x)$, $A = -0.4471$.

2.6. Solve $u_{xx} + u_{yy} = 0$ in $0 < x, y < 1$,

$$u(x, 0) = x(1 - x),$$
$$u(0, y) = u(1, y) = u(x, 1) = 0,$$

using a trial function of the form $u_1 = c_1(y)x(1 - x)$. Apply one of MWR to derive an ordinary differential equation for $c_1(y)$ and solve for $c_1(y)$. Answer: Using Galerkin's method, $c_1(y) = \{\exp(-\sqrt{10}\, y) - \exp[\sqrt{10}\,(y - 2)]\}/\{1 - \exp(-2\sqrt{10})\}$.

2.7. Solve Exercise 2.6 using the trial function

$$u_1 = (1 - y)x(1 - x) + A_1 y(1 - y)x(1 - x).$$

Why is this a feasible trial function? Would you expect this trial solution to give a better or worse answer than that obtained in Exercise 2.6? Answer: Using collocation method at $x = y = 0.5$, $A_1 = -1$.

2.8. Check the energy balance for the solutions in Exercises 2.6 and 2.7 to test which is the better solution. The flux entering along $y = 0$ should equal the flux leaving along $y = 1$, $x = 0$, and $x = 1$, although the equality holds only for the exact solution.

REFERENCES

Acrivos, A., and Goddard, J. D. (1965). Asymptotic Expansions for Laminar Forced-Convection Heat and Mass Transfer. Part 1. Low Speed Flows, *J. Fluid Mech.* **23**, 273–291.

Acrivos, A., and Taylor, T. (1962). Heat and Mass Transfer from Single Spheres in Stokes Flow, *Phys. Fluids* **5**, 387–394.

Altmann, S. L. (1958). The Cellular Method for Close-Packed Hexagonal Lattice II. The Computations: A Program for a Digital Computer and an Application to Zirconium Metal, *Proc. Roy. Soc. (London)* **A244**, 153–165.

Amundson, N. R. (1966). "Mathematical Methods in Chemical Engineering." Prentice-Hall, Englewood Cliffs, New Jersey.

Balcerzak, M. J., and Raynor, S. (1961). Steady State Temperature Distribution and Heat Flow in Prismatic Bars with Isothermal Boundary Conditions, *Int. J. Heat Mass Transfer* **3**, 113–125.

Bird, R. B., Stewart, W. E., and Lightfoot, E. N. (1960). "Transport Phenomena." Wiley, New York.

Bowman, C. W., Ward, D. M., Johnson, A. I., and Trass, O. (1961). Mass Transfer from Fluid and Solid Spheres at Low Reynolds Numbers, *Can. J. Chem. Eng.* **39**, 9–13.

Brian, P. L. T., and Hales, H. B. (1969). Effects of Transpiration and Changing Diameter on Heat and Mass Transfer to Spheres, *AIChE J.* **15**, 419–425.

Davis, P. J. (1962). Orthonormalizing Codes in Numerical Analysis *in* "Survey of Numerical Analysis" (J. Todd, ed.), pp. 347–379. McGraw-Hill, New York.

Duncan, W. J. (1937). Galerkin's Method in Mechanics and Differential Equations. Great Britain Aero. Res. Counc. London. Rep. and Memo, No. 1798. Reprinted in *Great Britain Air Ministry Aero. Res. Comm. Tech. Rep.* **1**, 484–516.

Griffith, R. M. (1960). Mass Transfer from Drops and Bubbles, *Chem. Eng. Sci.* **12**, 198–213.

Higgins, T. J. (1942). A Comprehensive Review of Saint-Venant's Torsion Problem. *Amer. J. Phys.* **10**, 248–259.

Higgins, T. J. (1943). The Approximate Mathematical Methods of Applied Physics as Exemplified by Application to Saint-Venant's Torsion Problem, *J. Appl. Phys.* **14**, 469–480.

Higgins, T. J. (1949). A Survey of the Approximate Solution of Two-Dimensional Physical Problems by Variational Methods and Finite Difference Procedures, *in* "Numerical Methods of Analysis in Engineering" (L. E. Grinter, ed.), Chapter 10. Macmillan, New York.

Hulbert, L. E. (1965). The Numerical Solution of Two-Dimensional Problems of the Theory of Elasticity. Bull. 198, Eng. Exp. Station, Ohio State Univ., Columbus, Ohio.

Hulbert, L. E., and Niedenfuhr, F. W. (1965). Accurate Calculations of Stress Distributions in Multiholed Plates, *J. Eng. Ind., Trans. ASME, Ser. B* **87**, 331–336.

Kantorovich, L. V. (1933). A Direct Method of Solving the Problem of the Minimum of a Double Integral (in Russian), *Izv. Akad. Nauk SSSR* **5**, 646–652.

Kantorovich, L. V. (1942). Application of Galerkin's Method to the So-Called Procedure of Reduction to Ordinary Differential Equations (in Russian with English summary), *Prikl. Math. Mech*, **6**, 31–40.

Kerr, A. D. (1968). An Extension of the Kantorovich Method, *Quart. Appl. Math.* **26**, 219–229.

Langhaar, H. L. (1969). Two Numerical Methods That Converge to the Method of Least Squares, *J. Franklin Inst.* **288**, 165–173.

Leissa, A. W., and Niedenfuhr, F. W. (1966). A Study of a Cantilevered Square Plate Subjected to a Uniform Loading, *J. Aerospace Sci.* **29**, 162–169.

Leissa, A. W., Clausen, W. E., Hulbert, L. E., and Hopper, A. T. (1969). A Comparison of Approximate Methods for the Solution of Plate Bending Problems, *AIAA J.* **7**, 920–928.

Levich, V. G. (1962). "Physicochemical Hydrodynamics." Prentice-Hall, Englewood Cliffs, New Jersey.

Ojalvo, I. U., and Linzer, F. D. (1965). Improved Point-Matching Techniques, *Quart. J. Mech. Appl. Math.* **18**, 41–56.

Poritzky, H. (1938). The Reduction of the Solution of Certain Partial Differential Equations to Ordinary Differential Equations, *Trans. Fifth Int. Congr. Appl. Mech., 5th, Cambridge, Massachusetts* pp. 700–707.

Richardson, P. D. (1968). Further Results from Use of a Transcendental Profile Function in Conduction and Convection, *Int. J. Heat Mass Transfer* **11**, 359–365.

Ruckenstein, E. (1964). On Mass Transfer in the Continuous Phase from Spherical Bubbles or Drops, *Chem. Eng. Sci.* **19**, 131–146.

Sattinger, S. S., and Conway, H. D. (1965). The Solutions of Certain Isosceles-Triangle and Rhombus Torsion and Plate Problems, *Int. J. Mech. Sci.* **7**, 221–228.

Schetz, J. A. (1963). On the Approximate Solution of Viscous Flow Problems, *J. Appl. Mech.* **30**, 263–268.

Sherwood, A. A. (1962). The Application of Fourier Series to the Solution of the Saint-Venant Torsion Problem, *Aust. J. Appl. Sci.* **13**, 285–299.

Shih, F. S. (1970). On the Temperature Field of a Square Column Embedding a Heating Cylinder, *AIChE J.* **16**, 134–138.

Sparrow, E. M., and Haji-Sheikh, A. (1968). Transient and Steady Heat Conduction in Arbitrary Bodies with Arbitrary Boundary and Initial Conditions, *J. Heat Trans., Trans. ASME, Ser. C* **90**, 103–108.

Sparrow, E. M., and Loeffler, A. L., Jr. (1959). Longitudinal Laminar Flow between Cylinders Arranged in Regular Array, *AIChE J.* **5**, 325–330.

Sparrow, E. M., Loeffler, A. L., Jr., and Hubbard, H. A. (1961). Heat Transfer to Longitudinal Laminar Flow between Cylinders, *J. Heat Transfer, Trans. ASME, Ser. C* **83**, 415–422.

Warren, W. E. (1964). Bending of Rhombic Plates, *AIAA J.* **2**, 166–168.

Wasserman, M. L., and Slattery, J. C. (1969). Creeping Flow past a Fluid Globule When a Trace of Surfactant Is Present, *AIChE J.* **15**, 533–547.

Chapter

3

Eigenvalue and Initial-Value Problems in Heat and Mass Transfer

One of the two most prolific areas of MWR application is transient heat transfer problems (the other prolific area being boundary layer flows). A great many of these applications are for one- or two-term approximations. There is no room to discuss all the results, which differ in the criterion used or the trial function, corresponding to different boundary conditions. We apply MWR to a few examples to illustrate the major ideas.

Transient heat and mass transfer problems can be solved either as eigenvalue problems or initial-value problems, and MWR is applied to both types of problems in Sections 3.1 and 3.2. The general entry-length and initial-value problem is discussed next, followed by an application: diffusion to a moving fluid. This chapter concludes with an application to heat transfer involving a phase change.

3.1 Eigenvalue Problems

Consider the linear eigenvalue problem

$$Lu + \lambda N u = 0, \tag{3.1a}$$

$$B_k u = 0, \qquad k = 1, \ldots, m, \quad \text{on boundary}, \tag{3.1b}$$

where L and N are general differential operators (oftentimes $Nu = u$). Usually the problem has a solution only for discrete (and infinitely many) values of the eigenvalue λ. The objective is to approximate the eigenvalues and eigenfunctions. We expand the trial function in a series of functions, each of which satisfies the homogeneous boundary conditions of (3.1b), giving

$$u = \sum_{i=1}^{N} c_i u_i, \qquad B_k u_i = 0 \quad \text{on boundary.} \tag{3.2}$$

The trial function is substituted into the differential equation to form the residuals, which are made orthogonal to the weighting functions, w_j:

$$\sum_{i=1}^{N} [(w_j, Lu_i) + \lambda(w_j, Nu_i)]c_i = 0. \tag{3.3}$$

This set of equations can be conveniently written in the form

$$\sum_{i=1}^{N} (A_{ji} + \lambda B_{ji})c_i = 0, \tag{3.4}$$

which is a set of N homogeneous linear equations for the constants c_i. The set has a nontrivial solution if and only if the determinant vanishes:

$$\det(A_{ji} + \lambda B_{ji}) = 0. \tag{3.5}$$

This equation is a polynomial in λ of degree N and has N roots, which are the approximations to the eigenvalues. Usually these roots are distinct and real, although this need not be the case. The Galerkin method is often preferred because of its equivalence to variational methods, when applicable (Section 7.4), and because under certain conditions the eigenvalue is stationary or insensitive to small errors in the eigenfunction approximation (Section 6.4). In addition convergence theorems are applicable (Chapter 11). Consider the example problem

$$X'' + \lambda(1 - x^2)X = 0, \qquad X(0) = X'(1) = 0 \tag{3.6}$$

and apply several criteria of MWR. This problem is complicated by the factor $(1 - x^2)$; without it the exact solution is known:

$$X_i = \sin[\tfrac{1}{2}(2i - 1)\pi x], \qquad \lambda_i = \tfrac{1}{4}(2i - 1)^2\pi^2. \tag{3.7}$$

Since these functions satisfy the boundary conditions they provide a convenient source of trial functions. Polynomials can also be used: $x^{i+1} - (i + 1)x$ satisfies the boundary conditions:

$$\begin{aligned} X &= c_1 \sin(\pi x/2) \\ R(X) &= -(\pi/2)^2 c_1 \sin(\pi x/2) + \lambda(1 - x^2)c_1 \sin(\pi x/2). \end{aligned} \tag{3.8}$$

In the Galerkin method the residual is made orthogonal to the trial function $\sin(\pi x/2)$, which gives a value of

$$\lambda = \frac{(\pi/2)^2 \int_0^1 \sin^2(\pi x/2)\,dx}{\int_0^1 (1-x^2)\sin^2(\pi x/2)\,dx} = 5.317. \tag{3.9}$$

We find a second approximation using (3.5) with

$$\begin{aligned} A_{ji} &= -(2j-1)^2(\pi/2)^2 \int_0^1 \sin[(2i-1)(\pi x/2)]\sin[(2j-1)(\pi x/2)]\,dx \\ &= -(2j-1)^2(\pi^2/8)\,\delta_{ij}, \end{aligned} \tag{3.10a}$$

$$B_{ji} = \int_0^1 (1-x^2)\sin[(2i-1)(\pi x/2)]\sin[(2j-1)(\pi x/2)]\,dx$$

$$= \begin{cases} \dfrac{1}{3} - \dfrac{1}{\pi^2(2j-1)^2}, & i = j, \\[2ex] \dfrac{(-1)^{i+j+1}[1/(i-j)^2 + 1/(i+j-1)^2]}{\pi^2}, & i \neq j. \end{cases} \tag{3.10b}$$

The roots are $\lambda_1 = 5.126$, $\lambda_2 = 45.54$.

The exact values, obtained by numerical integration (Emmert and Pigford 1954), are listed in Table 3.1. The λ_1 is close because it represents a second

TABLE 3.1

EIGENVALUES FOR EQ. (3.6)

Method		λ_1	λ_2	λ_3
Galerkin or	$N=1$	5.317		
variational,	$N=2$	5.126	45.54	
satisfying all BC	$N=3$	5.122	39.68	136.7
Exact		5.122	39.66	106.3
Collocation	$N=1$	3.29		
Subdomain	$N=1$	4.59		
Galerkin or	$N=1$	7.50		
variational,	$N=2$	5.14		
satisfying only				
essential BC				

approximation. The λ_2 is only a first approximation to the second eigenvalue and is not so close. The approximate eigenvalues are also above the exact solution, due to the equivalence of the Galerkin method and the Rayleigh–Ritz method, which gives upper bounds (Section 7.5).

Other criteria of MWR can also be applied. The collocation method gives

$$R(X)\Big|_{x=x_1} = 0 \qquad \text{or} \qquad \lambda = \frac{(\pi/2)^2}{(1 - x_1^{\,2})}. \tag{3.11}$$

The approximation ranges from $(\pi/2)^2$ to infinity depending on the choice of collocation point. The value $x_1 = \frac{1}{2}$, chosen as the midpoint of the interval, gives the value $\lambda = 3.29$. The subdomain method requires

$$\int_0^1 R(X)\, dx = 0,$$

$$\lambda = \left(\frac{\pi}{2}\right)^2 \frac{\int_0^1 \sin(\pi x/2)\, dx}{\int_0^1 (1 - x^2)\sin(\pi x/2)\, dx} = 4.59. \tag{3.12}$$

Consider the approximation when the trial functions do not satisfy the boundary condition $X'(1) = 0$. For a $2n$th-order differential equation, all boundary conditions involving at least nth-order derivatives can be treated in the manner outlined below (natural boundary conditions for variational methods, Section 7.1). The weighted residual (3.3) must be augmented by a boundary residual

$$\begin{aligned} \sum_{j=1}^{N} [(w_i, u_j'') + \lambda(w_i, (1 - x^2)u_j)]c_j &= 0, \\ \sum_{j=1}^{N} w_i(1)u_j'(1)c_j &= 0. \end{aligned} \tag{3.13}$$

The first term of the differential equation residual is integrated by parts and the boundary residual is applied:

$$\sum_{j=1}^{N} [-(w_i', u_j') - w_i(0)u_j'(0) + \lambda(w_i, (1 - x^2)u_j)]c_j = 0. \tag{3.14}$$

The trial functions must satisfy only the condition $u(0) = 0$. Apply the Galerkin method [$w_i = u_i$, hence $w_i(0) = 0$] using the trial functions $u_j = x^j$ (see Table 3.1). The results are not as good as before since one boundary condition is satisfied only approximately. The trial functions are easier to use however, and in computer calculations it is often preferable to sacrifice some accuracy (for low N) in order to facilitate the calculations; more terms are then needed.

3.2 Transient Heat and Mass Transfer

As a simple example consider a slab of thickness d which is initially at uniform concentration c_0. At time zero the side at $z = 0$ is exposed to a fluid of constant concentration c_1, while the side at $z = d$ is impervious to mass flux. The dimensionless equations are

$$c_t = c_{xx}, \qquad c(0, t) = 1, \qquad c_x(1, t) = 0, \qquad c(x, 0) = 0, \tag{3.15}$$

with dimensionless variables $c = (c' - c_0)/(c_1 - c_0)$, $x = z/d$, $t = \mathscr{D}t'/d^2$, where c' and t' are the dimensional concentration and time, respectively.

We apply MWR in a manner first suggested by Bickley (1941). The solution is expanded in a series which satisfies the boundary conditions with unknown functions of time:

$$c_N(x, t) = 1 + \sum_{i=1}^{N} A_i(t) X_i(x), \tag{3.16}$$

where $X_i(0) = X_i'(1) = 0$. Possible trial functions are the sine functions (3.7) and the polynomials of the previous section. If the sine functions and Galerkin's method are used, the result is the exact solution obtained using separation of variables. Here we use the polynomials $X_i = x^{i+1} - (i + 1)x$. Consider only the first approximation, and use the integral method. Higher approximations can be derived as a special case of the formulas derived in the next section.

The first approximation and residual are

$$c_1 = 1 + A(t)(x^2 - 2x), \qquad R = A'(x^2 - 2x) - 2A. \tag{3.17}$$

The residual is integrated from $0 \le x \le 1$ to give $A = \exp(-3t)$. To fit the initial conditions, the initial residual is satisfied in an integral sense:

$$\int_0^1 [1 + A(0)(x^2 - 2x)]\, dx = 0, \qquad A(0) = \tfrac{3}{2}. \tag{3.18}$$

The first approximation,

$$c_1 = 1 + \tfrac{3}{2} \exp(-3t)(x^2 - 2x), \tag{3.19}$$

is compared to the exact solution in Table 3.2. The approximation is acceptable only for times greater than 0.1. To improve the results higher approximations can be calculated, and convergence is assured by Theorem 11.16. Error bounds can be calculated using Theorem 11.34. Here we derive another approximate solution, valid for small times.

TABLE 3.2

SOLUTION TO UNSTEADY MASS TRANSFER PROBLEM

	$c(0.2, t)$			$c(1, t)$		
t	Exact	c_1 Eq. (3.19)	c_2 Eqs. (3.24), (3.26)	Exact	c_1	c_2
0.01	0.15	0.48	0.18	0	−0.46	0
0.04	0.48	0.52	0.51	0.0008	−0.33	0
0.10	0.66	0.60	0.66	0.05	−0.11	0.05
0.20	0.75	0.70	0.75	0.23	0.18	0.30
0.40	0.85	0.84	0.86	0.53	0.55	0.61
0.60	0.91	0.91	0.92	0.71	0.75	0.78
1.00	0.97	0.97	0.98	0.89	0.93	0.93

At small time the concentration changes occur very near the boundary, $x = 0$. Further away the concentration is nearly zero. Think of a penetration depth, $q(t)$, which defines the distance near the wall where the concentration has risen appreciably above zero. As time proceeds this penetration depth advances out into the slab until it reaches the opposite wall. At any given time the concentration field varies smoothly from the value one at the boundary $x = 0$ to the value zero at $x = q(t)$. The goal of the approximate solution is to predict $q(t)$.

Represent the concentration $c(x, t)$ by a similar profile

$$c = \phi(\eta), \qquad \eta = x/q(t). \tag{3.20}$$

The functions are required to satisfy the conditions

$$\phi(0) = 1, \qquad \phi(1) = \phi'(1) = q(0) = 0. \tag{3.21}$$

The first two conditions make the concentration one at the wall, $x = 0$, and zero at the edge of the penetration depth. The third condition is applied in order that mass not diffuse past the penetration depth, $x = q(t)$, and the fourth condition makes the penetration depth zero initially. The residual is then

$$\begin{aligned} c_t &= -\eta\phi' q'/q, \qquad & c_x &= \phi'/q, \qquad & c_{xx} &= \phi''/q^2, \\ \phi' &= d\phi/d\eta, & q' &= dq/dt & q^2 R &= -\eta\phi' q q' - \phi''. \end{aligned} \tag{3.22}$$

A trial function is chosen for $\phi(\eta)$: the simplest polynomial satisfying (3.21) is $\phi(\eta) = (1 - \eta)^2$. The integral method gives

$$\tfrac{1}{3} q q' = 2, \qquad q = \sqrt{12t}. \tag{3.23}$$

The approximate solution is then

$$c_2 = (1 - \eta)^2, \qquad \eta = x/\sqrt{12t}, \qquad t \le \tfrac{1}{12}. \tag{3.24}$$

This solution is valid until the penetration depth reaches the opposite boundary, $q = 1$, or for $t \le 0.0833$.

For larger times it is convenient to match this solution with (3.17). The constant $A(0)$ in Eqs. (3.18) is chosen not to satisfy the initial residual, but to match with Eqs. (3.24) at time $t = 0.0833$, when $q = 1$, giving

$$\begin{aligned} (1 - x)^2 &= 1 + A(\tfrac{1}{12})e^{-1/4}(x^2 - 2x), \\ A(\tfrac{1}{12}) &= e^{1/4} = 1.284. \end{aligned} \tag{3.25}$$

The result is

$$c_2 = 1 + 1.284e^{-3t}(x^2 - 2x), \qquad t \ge \tfrac{1}{12}. \tag{3.26}$$

Equations (3.24) and (3.26) provide simple approximations for the concentration over the entire time interval, as tabulated in Table 3.2. The penetration-depth solution provides an improved approximation at small times. These approximations can be improved by computing higher approximations, but these are best done after treating the general entry length and initial-value problem.

The penetration-depth type of solution is applied in the literature to simple one-dimensional problems in a first approximation. Many of the papers deal with comparisons to exact solutions and try to adjust the trial function to obtain the best agreement. Then the choice of trial function becomes little more than curve fitting. In realistic situations when no exact solution is available one must assess the error either by computing successive approximations or by examining the residual. Error bounds in terms of the residual are discussed in Section 11.6 and Theorem 11.34 applies to many of the problems treated here (including some of the nonlinear ones). Since the integral methods are widely used, however, we review their important features here. Goodman (1964) gives many applications of integral methods to nonlinear heat transfer problems.

The first step is the choice of trial function. It is written in terms of the variable $x/q(t)$, where q is the penetration depth. Polynomials are often used and suitable trial functions can be deduced as follows. The boundary conditions provide two conditions—one at each boundary, or in semi-infinite domains one at infinity. There is usually some other reasonable condition that can be imposed, such as the temperature and the flux must be zero at the edge of the penetration region. Finally, the differential equation is often applied at one or the other boundary (this is a derived boundary condition). These were the considerations leading to (3.21). There we did not use a derived condition, and Goodman suggests evaluating the differential equation at the edge of the penetration depth. This would add the condition $\phi'' = 0$ to (3.21) and would require a third-degree polynomial to satisfy all four conditions. The reader can easily verify that the result is $\phi = (1 - \eta)^3$. If this trial function is used, the penetration depth is $q = \sqrt{24t}$ and, for example, $c(0.2, 0.01) = 0.21$. This is a worse approximation than (3.24). Thus the addition of a constraint is not guaranteed to improve the results. Solutions for larger times are also possible: after the thermal wave reaches the insulating boundary, the temperature there begins to rise. Then a suitable profile is a polynomial which takes the value 1 at $x = 0$, $q_2(t)$ (unknown) at $x = 1$ and has zero slope at $x = 1$. The trial form,

$$c = A + Bx + Cx^2, \tag{3.27}$$

is easily reduced to $c = 1 + (1 - q_2)(x^2 - 2x)$ under these conditions. The unknown $q_2(t)$ is then determined by the integral method. Here too, derived conditions can be employed by invoking the differential equation at the boundaries.

Lardner and Pohle (1961) consider problems in cylindrical geometry (outside a cylinder) and show that best results are achieved for the first approximation if the trial function is taken as $c =$ (polynomial in r) $\ln r$. Richardson (1968) uses exponential trial functions, which complicate the analysis but improve the first approximation. When the initial conditions are not uniform, it is not possible to transform them to be zero. Then a penetration depth solution is not feasible, except possibly for solving the adjoint equation (Goodman, 1962). The trial function can still be taken in the form of Eq. (3.16), as was done by Thorsen and Landis (1965) and Bengston and Kreith (1970). Rozenshtok (1965) and Kumar and Narang (1967) apply similar ideas to the combined flow of heat and mass, using a penetration depth for each. Laura and Faulstich (1968) solve unsteady heat conduction in plates of polygonal shape by applying a conformal transformation (to change the domain to a circle) and then using Bessel trial functions and the collocation and Galerkin methods. Sparrow and Haji-Skeikh (1968) treat linear, transient heat conduction in arbitrary bodies using the least squares-collocation method on the boundary.

Another complication is a boundary condition of the third kind or nonlinear boundary conditions. In heat transfer problems, in place of the boundary conditions in Eq. (3.15) we might have

$$-k\, \partial\theta/\partial x = h_1[\theta^n - \theta_0{}^n] + h_2[\theta - \theta_0], \tag{3.28}$$

where h_1 refers to radiant transfer and h_2 applies to convective transport. In this case the temperature at $x = 0$ is not known. Polynomials can still be used, but with a variable $q_1(t)$ as the wall temperature. Then

$$\theta = q_1(t)\left[1 - \frac{x}{q_2(t)}\right]^2. \tag{3.29}$$

To determine the two unknown functions the trial function can be made to satisfy the boundary condition (3.28) and the integral condition used before. This leads to nonlinear differential equations which are difficult to solve. Sometimes a perturbation solution is possible for small and large times and in between the equations are integrated numerically. Examples of such computations are reported by Richardson (1964), Rafalski and Zyszkowski (1968), and Zyszkowski (1969) using polynomial trial functions and Koh (1961) using exponential trial functions. Some of these authors used Biot's Lagrangian approach, which is shown in Chapter 10 to be equivalent to the Galerkin method.

A trial solution of the type (3.16) can also be used for a boundary condition of the third kind. For example, if $-c_x = \mathrm{Nu}(c - c_0)$ is to be satisfied at $x = 0$ we can try a polynomial (3.27). Application of this boundary condition and $c_x(1, t) = 0$ gives the trial function

$$c = c_0 + B(t)[-(1/\mathrm{Nu}) + x - \tfrac{1}{2}x^2]. \tag{3.30}$$

The same technique is used when the flux is specified. Applications of this type are discussed by Tsoi (1967) for a variety of transient heat transfer problems.

These same techniques are applicable, of course, to the nonlinear problem when the physical properties depend on temperature. Indeed this is their primary application. The earliest application to such problems is by Fujita (1951) who used a two-term trial function to study diffusion with concentration-dependent diffusivity. The problem in Eq. (1.11a) was solved by Finlayson (1969) for diffusivity dependent on concentration in an exponential manner. The nonlinear problem of transient evaporation was solved by Bethel (1967) using the moments method and computing up to the fifth approximations.

3.3 Entry-Length and Initial-Value Problems

The preceding section considered first approximations to initial-value problems, in which we wished to predict the evolution in time of temperature or concentration. Another very similar problem is an entry-length problem in a duct. In this case a fluid is flowing in the z direction in rectilinear flow. At position $z = 0$ the boundary conditions on the temperature or concentration change. We then wish to predict the evolution of temperature or concentration profiles, and the related heat or mass transfer, as the fluid moves down the duct. The governing differential equations are Eq. (1.8) or Eqs. (1.11) or the corresponding equation for heat transfer. We treat the entry-length heat transfer problem and the initial-value problem follows as a special case. The two common types of boundary conditions are either temperature or heat flux fixed at the wall:

$$\left.\begin{aligned} T &= T_1 \qquad (3.31a)\\ k\mathbf{n}\cdot\nabla T &= q \qquad (3.31b)\end{aligned}\right\} \quad \text{at wall,} \quad z > 0.$$

The eigenfunction expansions follow the classical treatment of this problem as discussed by Sellars *et al.* (1956), Sparrow *et al.* (1957), and Siegel *et al.* (1958). The initial-value problem with a boundary condition of the third kind can be treated using the equations in Section 2.4, augmented by the time-dependent terms as illustrated in Section 3.2.

Constant Temperature Case

Consider the problem in a general cylindrical region with axis (and generator) z and cross section $A(x, y)$, which is bounded by a closed curve C. The equations then represent, for example, flow in a circular cylinder, a rectangular or polygonal duct. As derived in Chapter 1, Eq. (1.8) has no second derivative in the z direction; thus axial conduction is neglected. This is known to be a good approximation provided the Peclet number is above 100 (Hsu, 1967; LeCroy and Eraslan, 1969). Turbulent flow is included provided we have empirical laws for the velocity and thermal conductivity variations with position. The dimensionless equations are

$$u(r)\frac{\partial y}{\partial z} = 2\nabla\cdot(f(r)\,\nabla y) \qquad \text{in } A, \qquad z > 0, \tag{3.32a}$$

$$y = g(r) \qquad \text{at} \quad z = 0, \tag{3.32b}$$

$$y = 0 \qquad \text{on } C, \quad z > 0, \tag{3.32c}$$

where the dimensionless variables are $y = (T - T_1)/T_0 - T_1)$, T_0 is the inlet temperature which takes the form $T' = T_0 g(r)$, $z = 2z'/\text{Pe}\,D$, $r = r'/R$, $u(r) = u'(r)/u'_{av}$, $\text{Pe} = \rho C_p u'_{av} D/k_0$, the thermal conductivity takes the form $k' = k_0 f(r)$, r represents (x, y), and R is a characteristic dimension in the cross section A, $D = 2R$ and $u'_{av} = \int u'\,dA/\int dA$. The gradient operator is made dimensionless using the standard R. Clearly if $u(r) = 1$ this is an initial-value problem. We apply separation of variables by writing the solution in the form $y(z, r) = Z(z)R(r)$. Following the treatment of (2.39) the problem is reduced to an eigenvalue problem:

$$\nabla\cdot(f(r)\,\nabla R_n) + \tfrac{1}{2}\lambda^2 u(r)R_n = 0 \qquad \text{in } A, \qquad R_n = 0 \qquad \text{on } C. \tag{3.33}$$

Once we have solved for the eigenfunctions and eigenvalues the complete solution is

$$y(z, r) = \sum_{n=0}^{\infty} A_n R_n(r) \exp\{-\lambda_n^2 z\},$$

$$A_n = \int g(r)u(r)R_n \, dA \Big/ \int u(r)R_n^2 \, dA. \tag{3.34}$$

The $u(r)$ in the definition of A_n is required because the eigenfunctions R_n are orthogonal with that weighting function. We are interested in several additional properties of the solution. The mixing-cup temperature is defined as

$$y_m(z) = \sum_{n=1}^{\infty} A_n \exp\{-\lambda_n^2 z\} \frac{\int u(r)R_n(r) \, dA}{\int u(r) \, dA} \tag{3.35}$$

and represents the average temperature if the fluid were discharged into a cup and mixed completely. The Nusselt number is a dimensionless heat transfer coefficient $\mathrm{Nu} = hD/k$, with the heat transfer coefficient given by the empirical definition

$$-k'\mathbf{n} \cdot \nabla T\Big|_c = h[T_m(z) - T_i], \qquad -[f\mathbf{n} \cdot \nabla y]_c = (\mathrm{Nu}/2)y_m. \tag{3.36}$$

The Nusselt number depends on z (as well as possibly on x and y) and a suitable average is

$$\langle \mathrm{Nu} \rangle = \frac{1}{L} \int_0^L \mathrm{Nu}(z) \, dz. \tag{3.37}$$

For simple geometries, such as circular cylinders or planar geometries the problem simplifies to

$$\frac{1}{r^{a-1}} \frac{d}{dr} \left(f(r) r^{a-1} \frac{dR_n}{dr} \right) + \tfrac{1}{2}\lambda_n^2 u(r) R_n = 0, \qquad 0 < r < 1, \tag{3.38a}$$

$$R(1) = 0, \qquad R'(0) = 0, \tag{3.38b}$$

with $a = 1$ or 2 for planar or cylindrical geometry. The solution is then simpler:

$$y_m = \sum_{n=1}^{\infty} 2B_n \exp(-\lambda_n^2 z) / \left(\lambda_n^2 \int_0^1 u(r) r^{a-1} \, dr\right), \tag{3.39a}$$

$$B_n = -A_n f(1) R_n'(1), \tag{3.39b}$$

$$\mathrm{Nu} = 2 \sum_{n=1}^{\infty} B_n \exp(-\lambda_n^2 z) / y_m, \tag{3.39c}$$

$$\mathrm{Nu}_\infty = \lambda_1^2 \int_0^1 u r^{a-1} \, dr \tag{3.39d}$$

$$= \lambda_1^2/2 \qquad \text{(cylindrical geometry).} \tag{3.39e}$$

The last quantity is the asymptotic Nusselt number when the temperature is fully developed. This completes the formal solution of the entry-length problems. The difficult part of the solution is of course the solution of the eigenvalue problem, which is often solved approximately using MWR or finite difference methods.

In the weighted-residual methods we expand the solution

$$R(r) = \sum_{i=1}^{N} c_i S_i(r), \tag{3.40}$$

where the S_i are known functions which satisfy the boundary conditions and the c_i are undetermined constants. The approximate solution is determined by the weighted residuals

$$\sum_i (A_{ji} + \tfrac{1}{2}\lambda^2 B_{ji})c_i = 0, \tag{3.41a}$$

$$A_{ji} = \int_A w_j \nabla \cdot (f(r) \nabla S_i)\, dA = -\int_A f(r) \nabla w_j \cdot \nabla S_i \, dA, \tag{3.41b}$$

$$B_{ji} = \int_A w_j u(r) S_i \, dA. \tag{3.41c}$$

The last term in (3.41b) holds for weighting functions which vanish on C. We have integrated by parts and used the divergence theorem (see p. 29). The approximate kth eigenfunction is

$$R_k = \sum_{i=1}^{N} c_i^{(k)} S_i(r), \qquad k = 1, \ldots, N, \tag{3.42}$$

where the $c_i^{(k)}$ satisfy

$$\sum_{i=1}^{N} (A_{ji} + \tfrac{1}{2}\lambda_k^2 B_{ji})c_i^{(k)} = 0, \qquad k = 1, \ldots, N. \tag{3.43}$$

The solution then takes the form

$$y = \sum_{k=1}^{N} A_k R_k(r) \exp(-\lambda_k^2 z)$$

$$= \sum_{k,i=1}^{N} A_k c_i^{(k)} S_i(r) \exp(-\lambda_k^2 z). \tag{3.44}$$

The A_k are chosen to satisfy the weighted initial residual

$$\int_A w_j u(r) \left[g(r) - \sum_{k=1}^{N} A_k R_k \right] dA = 0. \tag{3.45}$$

We next snow that in the Galerkin method ($w_j = S_j$) the approximate eigenfunctions are orthogonal. When $w_j = S_j$ the matrices A and B in Eqs. (3.41b) and (3.41c) are symmetric in i and j. Consider (3.43) for the kth eigenfunction and a similar equation for the lth eigenfunction:

$$\sum_{i=1}^{N} (A_{ji} + \tfrac{1}{2}\lambda_l^2 B_{ji}) c_i^{(l)} = 0. \tag{3.46}$$

Multiply (3.43) by $c_j^{(l)}$ and sum over j; multiply (3.46) by $c_j^{(k)}$ and sum over j. Subtract the two equations, using $A_{ij} = A_{ji}$, $B_{ij} = B_{ji}$ to obtain

$$\tfrac{1}{2}(\lambda_k^2 - \lambda_l^2) \sum_{i,j} B_{ji} c_i^{(k)} c_j^{(l)} = 0. \tag{3.47}$$

When $k \neq l$, and assuming the eigenvalues are distinct ($\lambda_k \neq \lambda_l$ when $k \neq l$), the only solution is

$$\sum_{i,j} B_{ji} c_i^{(k)} c_j^{(l)} = \int u(r) R_k R_l \, dA = 0, \qquad k \neq l \tag{3.48}$$

and the eigenfunctions are orthogonal with weighting function $u(r)$. The orthogonality holds only when the matrices A and B are symmetric and thus is not valid for all MWR. The solution for A_k can be simplified by multiplying (3.45) by $c_j^{(l)}$, summing over j, and using the orthogonality condition:

$$A_l \int_A u(r) R_l^2 \, dA = \int_A g(r) u(r) R_l \, dA. \tag{3.49}$$

The mixing-cup temperature and Nusselt numbers are

$$y_m = \sum_{k=1}^{N} 2A_k \exp(-\lambda_k^2 z) \int u(r) R_k \, dA \Big/ \int u(r) \, dA, \tag{3.50a}$$

$$B_k = -A_k f(1) R_n'(1), \tag{3.50b}$$

$$\mathrm{Nu} = 2 \sum_{k=1}^{N} B_k \exp(-\lambda_k^2 z)/y_m, \tag{3.50c}$$

$$\mathrm{Nu}_\infty = \lambda_1^2 \int u(r) \, dA. \tag{3.50d}$$

Equations (3.50a)–(3.50d) apply to the simpler problem (3.38a). Equation (3.50d) does not follow from Eqs. (3.50a)–(3.50c) since the approximate eigenfunctions do not necessarily satisfy (3.38a) integrated over the region. Instead it follows from the definition (3.39d). Equations (3.47)–(3.49) apply only to the Galerkin method, or an equivalent method (such as the variational method introduced in Chapter 7). This completes the formal solution to the problem, and computations in specific cases are treated in the next section as well as in Section 7.5. Convergence of the Galerkin method is assured by Theorems 11.15 through 11.18 and error bounds can be calculated using Theorem 11.34.

Constant Flux Case

Consider the similar eigenvalue problem when the boundary condition is constant heat flux. The dimensionless equation is the same (3.32a) but with different boundary conditions (3.31b). The nondimensional temperature is defined as $y = k_0(T - T_0)/qR$, and the dimensionless boundary conditions are

$$y(r, 0) = 0, \qquad \mathbf{n} \cdot \nabla y = 1, \quad \text{on } C. \tag{3.51}$$

In this case, it is assumed that the inlet temperature is a constant. In the fully developed region (large z) the temperature increases linearly with distance. Separate this asymptotic solution by writing $y = y_1 + y_2$. The function y_1 satisfies the equations

$$\begin{aligned} u(r)\frac{\partial y_1}{\partial z} &= 2\,\nabla \cdot (f(r)\,\nabla y_1) \qquad \text{in } A, \quad z > 0, \\ \mathbf{n} \cdot \nabla y &= 1, \qquad \text{on } C. \end{aligned} \tag{3.52}$$

A solution is

$$y_1 = A_\infty z + \Psi(r), \tag{3.53}$$

where A_∞ is a constant. The function $\Psi(r)$ satisfies

$$\begin{aligned} 2\,\nabla \cdot (f(r)\,\nabla \Psi) &= A_\infty u(r), \qquad \text{in } A, \\ \partial \Psi/\partial n &= 1, \qquad \text{on } C. \end{aligned} \tag{3.54}$$

The constant A_∞ is determined such that the average energy of the stream equals the total energy input:

$$\begin{aligned} 2z \int \mathbf{n} \cdot \nabla y \, dC = 2z \int dC &= \int y_1 u(r)\, dA, \\ 2 \int dC &= A_\infty \int u \, dA, \\ \int \Psi u \, dA &= 0. \end{aligned} \tag{3.55}$$

The function y_2 satisfies the same differential equation (3.32a), but different boundary and initial conditions:

$$\partial y_2/dn = 0, \qquad \text{on } C, \tag{3.56a}$$

$$y_2(r, 0) = -\Psi(r). \tag{3.56b}$$

To solve this eigenvalue problem we expand the solution in the form (3.40) but now the trial functions satisfy the zero normal derivative condition on the wall. The equations governing the approximate eigenfunctions are the same [Eqs. (3.41)–(3.44)] and the constants $\bar{A}_k$ are chosen to satisfy the initial conditions:

$$\bar{A}_l = -\int_A \Psi(r)u(r)R_l\, dA \bigg/ \int_A u(r)R_l^2\, dA. \tag{3.57}$$

Equation (3.57) assumes the weighting function $w_i = S_i$, corresponding to the Galerkin method, and the fact that the approximate eigenfunctions are orthogonal. The complete solution is then

$$y = A_\infty z + \Psi(r) + \sum_{l=1}^{N} \bar{A}_l R_l \exp(-\bar{\lambda}_l^2 z), \tag{3.58a}$$

$$y_m(z) = A_\infty z + \left[\int_A \Psi(r)u(r)\, dA + \sum_{l=1}^{N} \bar{A}_l \exp(-\bar{\lambda}_l^2 z) \int u(r)R_l\, dA\right] \bigg/ \int u(r)\, dA, \tag{3.58b}$$

$$\mathrm{Nu} = \frac{1}{y_m - y_{\mathrm{wall}}} = \frac{2}{(2/\mathrm{Nu}_\infty) + \sum_{n=1}^{\infty} \bar{B}_n \exp(-\bar{\lambda}_n^2 z)}, \tag{3.58c}$$

$$\mathrm{Nu}_\infty = \frac{1}{16 \sum_{n=1}^{\infty} B_n/\lambda_n^4} = -\frac{1}{\Psi(1)}, \tag{3.58d}$$

where $\bar{B}_n$ and $\bar{\lambda}_n$ refer to the solution of the eigenvalue problem with boundary conditions (3.56) and the B_n and λ_n refer to the solution to Eqs. (3.38). Equation (3.58c) is derived by Sellars *et al.* (1956) and Sparrow *et al.* (1957). Spatially varying wall temperature and wall flux can be expressed in terms of the solutions for the constant wall temperature or flux, but it is probably simpler either to integrate the equations numerically or to derive the approximate solution for the case of interest. Applications of these results are discussed in Section 3.4 and Section 7.5. In particular for small distances an asymptotic solution may be more convenient (see Sections 3.4, 5.2).

Relation between Eigenfunction Expansion and Initial-Value Treatment

The above treatment uses separation of variables to reduce the problem to an eigenvalue problem, and the solution is expressed in terms of an expansion of the eigenfunctions. The initial-value problem in Section 3.2 is solved using an expansion which leads to simple first-order ordinary differential equations. Such a treatment can also be applied to nonlinear problems, whereas the eigenfunction expansion is limited to linear problems. It is of interest, therefore, to show that the two methods are identical for linear problems, provided the same expansion trial functions are used.

Consider the problem of Eqs. (3.32), which can be thought of as an entry-length problem or as a transient heat or mass transfer problem. The case of a constant flux or other boundary condition can be treated similarly. Expand the solution

$$y(z, r) = \sum_{i=1}^{N} Z_i(z) S_i(r). \tag{3.59}$$

We show that the solution derived using (3.59) is just (3.44). Application of MWR to Eqs. (3.32) gives

$$\tfrac{1}{2} \sum_{i=1}^{N} B_{ji} Z_i' = \sum_{i=1}^{N} A_{ji} Z_i, \tag{3.60}$$

where the matrices A and B are defined in (3.41b) and (3.41c). MWR applied to the initial residual gives†

$$\sum_{i=1}^{N} B_{ji} Z_i(0) = \int w_j u(r) g(r)\, dA \equiv b_j. \tag{3.61}$$

We wish to establish the following equivalence:

$$d_i(z) \equiv \sum_{k=1}^{N} A_k c_i^{(k)} \exp(-\lambda_k^2 z) = Z_i(z). \tag{3.62}$$

Equation (3.45) for the eigenfunction expansion can be rewritten as

$$\sum_{k=1}^{N} A_k \sum_{i=1}^{N} c_i^{(k)} B_{ji} = b_j \equiv \int_A w_j u(r) g(r)\, dA. \tag{3.63}$$

Interchanging the order of summation gives

$$\sum_{i=1}^{N} B_{ji} \sum_{k=1}^{N} A_k c_i^{(k)} = \sum_{i=1}^{N} B_{ji} d_i(0) = b_j, \tag{3.64}$$

which verifies that $d_i(0) = Z_i(0)$. We can evaluate $\tfrac{1}{2}\sum B_{ji} d_i' - \sum A_{ji} d_i$ to show that it is zero:

$$\begin{aligned} &\tfrac{1}{2} \sum_{i=1}^{N} B_{ji} d_i' - \sum_{i=1}^{N} A_{ji} d_i \\ &\quad = \sum_{i,k=1}^{N} \{-\tfrac{1}{2} B_{ji} \lambda_k^2 A_k c_i^{(k)} - A_{ji} A_k c_i^{(k)}\} \exp(-\lambda_k^2 z) \\ &\quad = -\sum_{k=1}^{N} A_k \exp(-\lambda_k^2 z) \sum_{i=1}^{N} \{[A_{ji} + \tfrac{1}{2}\lambda_k^2 B_{ji}] c_i^{(k)}\} = 0 \end{aligned} \tag{3.65}$$

by Eq. (3.46). Thus $d_i(z)$ and $Z_i(z)$ satisfy identical equations and are therefore the same, so that the two methods give identical results.

† Here the use of the weighting function $u(r)$ is not so obvious but is motivated by the eigenfunction expansion. If it is omitted the two methods give different results.

Laplace Transform Method

Another way to solve Eqs. (3.32) is to take the Laplace transform, solve the resulting equation approximately, and take the inverse Laplace transform. We show that this gives the same result as derived above. Take the Laplace transform of Eqs. (3.32),

$$\tfrac{1}{2}u(s\Psi - g) = \nabla \cdot (f\nabla\Psi), \qquad \text{in } A, \tag{3.66a}$$

$$\Psi = 0, \qquad \text{on } C, \tag{3.66b}$$

where the transform is defined as

$$\begin{aligned} \mathscr{L}\{y\} &= \Psi(s, r) = \int_0^\infty e^{sz} y(z, r)\, dz, \\ \mathscr{L}\left\{\frac{\partial y}{\partial z}\right\} &= s\mathscr{L}\{y\} - y(0). \end{aligned} \tag{3.67}$$

Equation (3.66a) is solved by the trial function

$$\Psi = \sum_{i=1}^{N} D_i S_i(r). \tag{3.68}$$

The weighted residual is [(3.41), (3.63)]

$$\tfrac{1}{2}s \sum_{i=1}^{N} B_{ji} D_i - \tfrac{1}{2}b_j = \sum_{i=1}^{N} A_{ji} D_i. \tag{3.69}$$

The Laplace transform of Eq. (3.60) is

$$\tfrac{1}{2}s \sum_{i=1}^{N} B_{ji} \mathscr{L}\{Z_i\} - \tfrac{1}{2} \sum_{i=1}^{N} B_{ji} Z_i(0) = \sum_{i=1}^{N} A_{ji} \mathscr{L}\{Z_i\} \tag{3.70}$$

and through use of (3.64) this is just (3.69). Thus the solutions are the same.

The use of the Laplace transform and the Galerkin method was apparently initiated by Weiner (1955). Later Dicker and Friedman (1963a, b) applied the same ideas to transient heat transfer problems in nonseparable domains, such as trapezoids, and ellipses with circular holes in them, and they recognized that the solution was the same as that obtained with an eigenfunction expansion. Erdogan (1963)† applied the Laplace transform and Galerkin methods to transient heat transfer in a cylinder. In the discussion following, Goodman solved the same problem using the Galerkin method as in (3.59).

† The author has a sentimental attachment to this article since it was through it that he was introduced to Crandall's book and the Method of Weighted Residuals. The article was discovered during a casual perusal of the literature!

The results were different, and the authors argued over whose method was best, without realizing that the difference was due to the fact that Erdogan used the integral $\int (\ )\, dr$ whereas Goodman used the integral $\int (\ ) r\, dr$.

The entry-length problem is more complicated when axial conduction is included. LeCroy and Eraslan (1969) apply the Galerkin method to the entrance region of a magnetohydrodynamic channel, including viscous dissipation, Joule heating, and a nonzero net current. Eraslan and Eraslan (1969) add the Hall effect. They consider both constant temperature and constant heat flux. The equations for the former are

$$\text{Pe}\, u(r) \frac{\partial y}{\partial z} = \frac{\partial^2 y}{\partial r^2} + \frac{\partial^2 y}{\partial z^2} + \Psi(r),$$

$$y(r, 0) = y_0(r), \qquad y(\pm 1, z) = y_w, \tag{3.71}$$

$$\lim_{z \to \infty} y(r, z) = y_\infty,$$

where y_∞ is the known solution to

$$\frac{d^2 y_\infty}{dr^2} + \Psi(r) = 0, \qquad y_\infty(\pm 1) = y_w. \tag{3.72}$$

The viscous dissipation and Joule heating are represented by $\Psi(y)$, which is a known function since the velocity profile is given. The solution is constructed in the form

$$y(r, z) = y_\infty(r) + \sum_{i=1}^{N} A_n R_n \exp(-\beta_n z/\text{Pe}), \tag{3.73}$$

which leads to the eigenvalue problem

$$\frac{d^2 R_n}{dr^2} + \left[\beta_n u(r) + \frac{\beta_n^{\,2}}{\text{Pe}^2}\right] R_n = 0, \qquad R_n(\pm 1) = 0. \tag{3.74}$$

It is clear that for large Peclet numbers this problem reduces to the one solved above. The same trial functions can be used to solve Eqs. (3.33) and (3.74) since the boundary conditions are the same. In this case the authors use $S_i = \cos(2i - 1)\pi y/2$. The eigenfunctions are constructed in a manner similar to that used above. In this case, however, they are not orthogonal due to the presence of the term $\beta_n^{\,2}/\text{Pe}^2$ in the equation. The results compared well to the numerical calculations reported by Hsu (1967) for the special case with no magnetic effects.

3.4 Mass Transfer to a Moving Fluid

To illustrate an entry-length problem we consider absorption of a gas by a liquid flowing down a flat inclined plate (Fig. 3.1). At $z = 0$ the liquid is

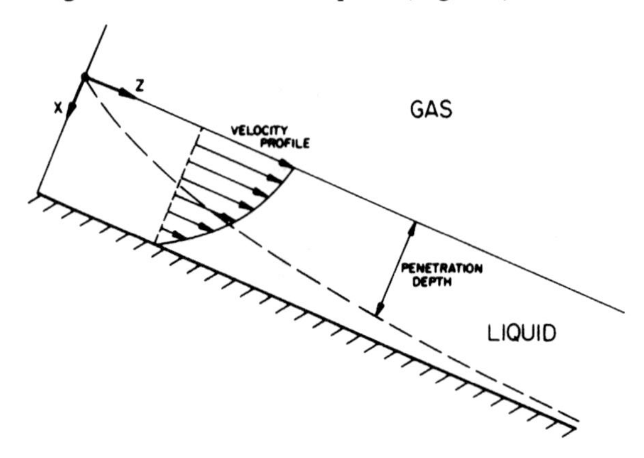

Fig. 3.1. Diffusion into a falling liquid film (Finlayson, 1969; reprinted from *British Chemical Engineering* with permission of the copyright owner).

exposed to a gas which contains a chemical species which is absorbed into the liquid. The equilibrium concentration of the liquid is c_1. The governing equations are forms of Eqs. (3.32) with $y = (c' - c_1)/(c_0 - c_1)$, $c_0 = 0$, $x = x'/d$, $u(r) = u'(r)/'_{\text{av}}$, $\text{Pe} = u_{\text{av}}\, d/\mathscr{D}$, $z = z'/\text{Pe}\, d$, and slightly different boundary conditions:

$$\frac{3}{2}(1 - x^2)\frac{\partial y}{\partial z} = \frac{\partial^2 y}{\partial x^2} \tag{3.75a}$$

$$y = 0 \quad \text{at} \quad x = 0, \qquad \frac{\partial y}{\partial x} = 0 \quad \text{at} \quad x = 1, \tag{3.75b}$$

$$y = 1 \quad \text{at} \quad z = 0. \tag{3.75c}$$

The Sherwood number is

$$\text{Sh} = \frac{+(\partial y/\partial x)\big|_{x=0}}{y_m}. \tag{3.76}$$

The positive sign is used here since $-\mathbf{n} \cdot \nabla y = +(\partial y/\partial x)$ because $\mathbf{n} = -\mathbf{e}_x$. The trial functions are chosen to satisfy the boundary conditions, and one choice is:

$$S_i = \sin \tfrac{1}{2}(2i - 1)\pi x. \tag{3.77}$$

The eigenfunctions are given by (3.42) and (3.43), with A and B evaluated in Eqs. (3.10). Furthermore, we normalize the eigenfunctions so that

$$\int_0^1 u R_l^2 \, dx = \sum_{i,j=1}^{N} B_{ji} c_i^{(l)} c_j^{(l)} = 1. \tag{3.78}$$

The formula (3.49) for A_k is then simpler:

$$A_l = \int_0^1 u R_l \, dx = \sum_{i=1}^{N} c_i^{(l)} b_i,$$

$$b_i = \tfrac{3}{2} \int_0^1 (1 - x^2) S_i \, dx = \frac{3}{(2i-1)\pi} + \frac{24}{(2i-1)^3 \pi^3} \left\{ 1 - \frac{(2i-1)\pi}{2} (-1)^{i+1} \right\}. \tag{3.79}$$

The mean concentration and flux are then

$$y_m = \sum_{k=1}^{N} A_k^2 \exp(-\lambda_k z),$$

$$\left. \frac{\partial y}{\partial x} \right|_{x=0} = \sum_{k=1}^{N} A_k \exp(-\lambda_k z) \left[\sum_{i=1}^{N} c_i^{(k)} \frac{(2i-1)\pi}{2} \right]. \tag{3.80}$$

Calculations are done for $N = 3$, 6, and 10 and are shown on Fig. 3.2. The

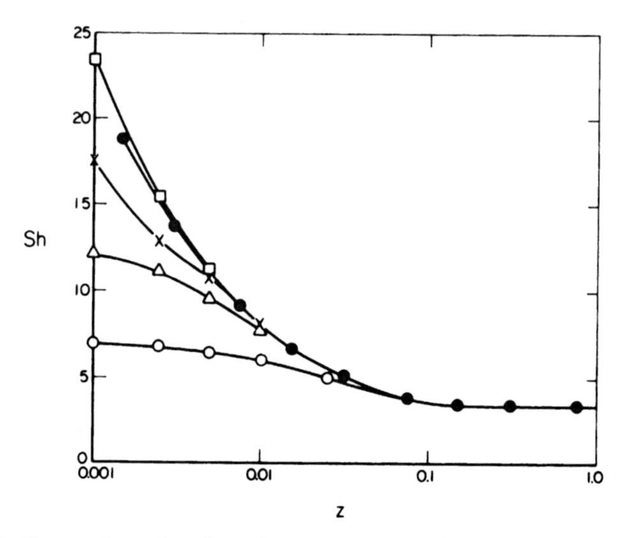

Fig. 3.2. Sherwood number dependence on position (● exact, $N = 20$; approximate ○$N = 3$, △ $N = 6$, × $N = 10$; □ penetration depth).

comparison solution is a numerical one (Rotem and Neilson, 1969).† Using only a few terms gives a poor approximation near the entrance, as was the case with initial-value problems (Section 3.2). The number of terms needed for a good approximation then depends on z.

The penetration depth concept can be applied to this problem to obtain results valid for small z. It is simpler to solve the problem in terms of $c = c'/c_1$ rather than y:

$$\frac{3}{2}(1 - x^2)\frac{\partial c}{\partial z} = \frac{\partial^2 c}{\partial x^2},$$

$$c = 1 \quad \text{at} \quad x = 0, \qquad \frac{\partial c}{\partial x} = 0 \quad \text{at} \quad x = 1, \tag{3.81}$$

$$c = 0 \quad \text{at} \quad z = 0.$$

The trial function is taken as $c = \phi(\eta)$, with ϕ a second degree polynomial in $\eta = x/q$. The boundary conditions (3.21) are applied to give $\phi = (1 - \eta)^2$. The residual is then

$$\tfrac{2}{3}q^2 R = -\eta\phi' q q' + \eta^3 \phi' q^3 q' - \tfrac{4}{3}. \tag{3.82}$$

The integral method is applied to give $q(z)$:

$$q^2 = \tfrac{10}{3}[1 - (1 - 4.8z)^{1/2}], \qquad z \leq 17/160. \tag{3.83}$$

The mean concentration c_m and Sherwood number are then

$$c_m = q\int_0^1 (1 - \eta^2 q^2)(1 - \eta)^2\, d\eta \Big/ \int_0^1 (1 - x^2)\, dx = q(1 - q^2/10)/2,$$

$$\text{Sh} = \frac{2/q}{1 - q(1 - q^2/10)/2}. \tag{3.84}$$

This result is plotted on Fig. 3.2 and is a good approximation for small z. It is clear that an eigenfunction expansion for large z and a penetration-depth for solution for small z gives results valid over the full range of z. Other penetration-depth type solutions have been derived for heat transfer problems (Sparrow and Siegel, 1958; Yang, 1962; Novotny and Eckert, 1965; Lyman, 1964; Schechter, 1967, p. 205) as well as for developing flow problems discussed in Section 4.1. See also Section 5.2 for the use of asymptotic solutions.

† Professor Z. Rotem kindly provided tabular values of the data plotted in the paper, which also includes axial diffusion. The results for no axial diffusion were reported earlier by Emmert and Pigford (1954).

3.5 Heat Transfer Involving a Phase Change

Another transport problem of interest is heat transfer when there is a phase change. We illustrate the integral method with a simple example from Goodman (1958, 1964). It involves melting a semi-infinite solid when the surface is changed from the melting temperature to some higher temperature at time zero. The liquid–solid interface propagates into the semi-infinite region as time proceeds, and is one of the features of interest. Assume the interface is sharp and the physical properties are constant. We study the simple case when the temperature in the solid remains constant (until it melts).

The differential equations are

$$T_t = \alpha T_{xx}, \qquad 0 \le x \le s(t), \tag{3.85a}$$

$$\left.\begin{aligned} T_x &= -(\rho\lambda/k)\, ds/dt \\ T &= 0 \end{aligned}\right\} x = s(t), \tag{3.85b}$$

$$T = T_s \qquad \text{at} \quad x = 0, \tag{3.85c}$$

where λ is the latent heat, $s(t)$ is the position of the interface, and the melting temperature is taken as zero whereas the surface temperature is T_s. We derive an integral heat balance by integrating (3.85a) with respect to x over the region from 0 to $s(t)$:

$$\frac{d\theta}{dt} = -\frac{\alpha\rho L}{k}\frac{ds}{dt} - \alpha\frac{\partial T}{\partial x}(0, t), \qquad \theta = \int_0^{s(t)} T(x, t)\, dx. \tag{3.86}$$

A temperature profile must be assumed for $T(x, t)$. It is convenient to make the trial function satisfy the boundary conditions, but in a slightly different form. Differentiate (3.85c) with respect to time:

$$\frac{\partial T}{\partial x}\frac{ds}{dt} + \frac{\partial T}{\partial t} = 0, \qquad \text{at} \quad x = s. \tag{3.87}$$

Combining this with (3.85a) and (3.85b), we get

$$(T_x)^2 = \frac{\rho\lambda\alpha}{k} T_{xx}, \qquad \text{at} \quad x = s(t). \tag{3.88}$$

This provides one condition on the temperature profile. Two other conditions are that the temperature take the prescribed values at $x = 0$ and $s(t)$. A suitable polynomial is

$$T = a(x - s) + b(x - s)^2. \tag{3.89}$$

The quantities a and b are then

$$a = \frac{\rho L \alpha}{ks}\,[1 - (1 + \mu)^{1/2}], \qquad bs^2 = as + T_s, \qquad \mu = \frac{2T_s k}{\rho L \alpha}. \tag{3.90}$$

The integral equation and its solution are

$$2s\frac{ds}{dt} = K^2,$$

$$K = \left[12\alpha \frac{1 - (1 + \mu)^{1/2} + \mu}{5 + (1 + \mu)^{1/2} + \mu}\right]^{1/2}, \tag{3.91}$$

$$s = Kt^{1/2}.$$

This solution compares favorably with the exact solution (Carslaw and Jaeger, 1959, Goodman, 1958): the largest error is 7% for $\mu = 2.8$, and the error is less for smaller μ. Other variations of the problem can be handled using integral methods, and several of these are summarized in Goodman (1964). Altman (1961) considers pulselike energy inputs. Poots (1962) has treated a two-dimensional problem and Hrycak (1967) treated a stratified medium and found results which compared favorably with the experiments. Libby and Chen (1965) study the growth of a deposited layer on cold surfaces, and Muehlbauer and Sunderland (1965) review several approximate methods applied to freezing or melting problems. Horvay (1965) studies the freezing of a spherical nucleus in an undercooled melt, Tien and Yen (1966) study melting with known free convection. These authors used either the integral method or the method of moments. Goldfarb and Ereskovskii (1966) and Lardner (1967) used Biot's method, which is equivalent to Galerkin's method (Chapter 10). Theofanous, Isbin and Fauske (1970) used the integral method to study convective diffusion and bubble dissolution.

EXERCISES

3.1. Solve Eq. (3.6) using $X_i = x^{i+1} - (i+1)x$ and the Galerkin method for $N = 1$ and 2.
Answer: $N = 1: \lambda = 5.185$; $N = 2: \lambda = 5.161, 42.81$.

3.2. Consider the eigenvalue problem

$$y'' + \lambda(1 - x^2)y = 0, \qquad y(0) = y(1) = 0.$$

Obtain an approximation for the first eigenvalue using the Galerkin method for $N = 1$ and 2. Use as trial functions the exact eigenfunctions for $y'' + \lambda y = 0$ under the same boundary conditions.
Answer: $N = 1: \lambda = 13.76$; $N = 2: \lambda = 13.49, 63.5$.

3.3. Apply the Galerkin method to

$$y'' + \lambda y = 0, \qquad y(0) = y(1) = 0$$

using the trial function

$$y = c_1 x(1 - x) + c_2 x^2(1 - x)^2.$$

Table 6.4, p. 156, may be useful in calculating the integrals. The approximate eigenvalue is an approximation to π^2 and the approximate eigenfunction is an approximation to $\sin \pi x$.
Answer: $N = 2$: $\lambda_1 = 9.86975$.

3.4. Solve Exercise 3.3 using a finite difference method, i.e.,

$$y_i'' \sim (y_{i+1} - 2y_i + y_{i-1})/(\Delta x)^2.$$

For $N = 1$, $\Delta x = 0.5$ and the equation is

$$\frac{y(1) - 2y(0.5) + y(0)}{(0.5)^2} + \lambda y(0.5) = 0.$$

Thus $\lambda_1 = 8$. For $N = 2$, $\Delta x = 0.333$ and an equation is written at $x = 0.333$ and at $x = 0.667$. Calculate λ for $N = 2$ and 3 and compare the accuracy to that found in Exercise 3.3.
Answer: $N = 2$: $\lambda_1 = 9$; $N = 3$: $\lambda_1 = 9.37$.

3.5. Solve the problem

$$\frac{\partial c}{\partial t} = \frac{\partial^2 c}{\partial x^2} - kc^2, \qquad c(0, t) = 1, \quad c(x, 0) = c(\infty, t) = 0,$$

for small times using a trial function of the form $c = \phi(\eta)$, $\eta = x/q(t)$. What boundary conditions must ϕ satisfy?
Answer: Integral method, with $\phi = (1 - \eta)^2$, gives

$$q(t) = \{10[1 - \exp(-6kt/5)]/k\}.$$

What does a collocation method give?

3.6. Consider unsteady diffusion of methemoglobin in a layer of water. Initially the concentration is 10 gm/100 ml. For $t > 0$, one side ($x = 0$) is maintained at a concentration of 30 gm/100 ml while the other side ($x = d = 1$ mm) is kept at 100 gm/100 ml. The diffusity of methemoglobin in water is approximately linear over this range, taking the value of 5×10^{-7} cm^2/sec at 10 gm/100 ml and 1.5×10^{-7} cm^2/sec at 30 gm/100 ml (see Keller, 1968). Deduce the importance of taking into account the concentration dependence of the diffusivity.

To do this, solve the problem approximately using a one-term MWR for two cases: actual $\mathscr{D}(c)$, and an average $\mathscr{D}$. Even though the exact solution is known when using an average $\mathscr{D}$, the best strategy is to compare the two approximate solutions, one for $\mathscr{D}(c)$ and one for $\mathscr{D}_{av}$. If the approximate solution for $\mathscr{D}(c)$ is compared to the exact solution for $\mathscr{D}_{av}$, any disagreement can be ascribed to two causes: errors in the approximate solution and the effect of the nonlinearity. If two approximate solutions are compared, the difference is more likely due to the nonlinearity. A solution valid for small time can be easily derived using a trial solution as in Exercise 3.5. For large time the steady-state results can be used (see Section 2.1).

Answer: With $c = (1 - \eta)^2$, $\eta = x/q(t)$, applying the collocation method at $\eta = \frac{1}{2}$ gives $q = 2.1\sqrt{t}$ for $\mathscr{D}(c)$ and $q = 2.3\sqrt{t}$ for $\mathscr{D}_{av}$, where $c = (c' - 0.1)/0.2$, $x = x'/d$, $t = t'(5 \times 10^{-7})/d^2$.

3.7. Consider the problem

$$\rho C_v \left(\frac{\partial T}{\partial t} + \mathbf{u} \cdot \nabla T\right) = \nabla \cdot (k \nabla T) - f(x, T)$$

subject to the interpretation of variables as in Section 2.4 and boundary conditions Eqs. (2.48b)–(2.48d). Derive the general formula for the weighted residual when the trial functions do not satisfy the radiation boundary condition (2.48d). To do this combine the boundary and differential equation residuals as done in Section 2.4.

Answer: Add the term $\int w_j \rho C_v \, \partial T_N/\partial t \, dV$ to the left-hand side of Eq. (2.54).

3.8. Consider the entry-length problem in a cylinder with a fluid in laminar flow [Eqs. (3.38) with $a = 2$]. The velocity profile is $u(r) = 2(1 - r^2)$. Oftentimes plug flow is assumed: $u(r) = 1$. Estimate the effect of this assumption on the asymptotic Nusselt number of Eq. (3.39e) by solving the eigenvalue problem using a one-term MWR.

Answer: Applying the Galerkin method, $R = 1 - r^2$, gives $\mathrm{Nu}_\infty = 4$ when $u = 2(1 - r^2)$ and $\mathrm{Nu}_\infty = 6$ when $u = 1$.

3.9. Rework the example in Section 3.4 when $\frac{3}{2}(1 - x^2)$ is replaced by 1. Calculate a first approximation to the asymptotic Sherwood number and a solution valid for small z. How significant is the approximation of a uniform flow rate?

Answer: Integral method gives $q^2 = 12z$. The approximate eigenvalue for Eqs. (3.75) comes from Section 3.1, $N = 1$, $\mathrm{Sh}_\infty = \lambda_1{}^2 \simeq \frac{2}{3} \times 5.317 = 3.54$; when $u = 1$. $\mathrm{Sh}_\infty = 2.47$.

3.10. Consider the eigenvalue problem of Eqs. (3.38) for plane geometry ($a = 1$) and a velocity profile corresponding to a power law fluid.

$$u(r) = a(1 - r^{(n+1)/n}), \qquad a = 1\Big/\int_0^1 (1 - r^{(n+1)/n})\, dr.$$

Calculate the asymptotic Nusselt number when $n = 0.5$. How does this compare to the asymptotic Nusselt number when $n = 1$ (Newtonian fluid)?
Answer: $\mathrm{Nu}_\infty = 4.1$ for the power law fluid and $\mathrm{Nu}_\infty = 3.9$ for the Newtonian fluid, using the Galerkin method, $N = 1$, and $R = 1 - r^2$. Note that these Nusselt numbers are in terms of the channel width, $2h$, whereas the equivalent diameter is $4h$.

REFERENCES

Altman, M. (1961). Some Aspects of the Melting Solution for a Semi-Infinite Slab, *Chem. Eng. Prog. Symp. Ser.* **57**, No. 32, 16–23.

Bengston, H. H., and Kreith, F. (1970). Approximate Solution of Heat-Conduction Problems in Systems with Nonuniform Initial Temperature Distribution, *J. Heat Trans., Trans. ASME Ser. C* **92**, 182–184.

Bethel, H. E. (1967). A Generalized Galerkin–Kantorovich Treatment of Transient Evaporation through a Finite Region, *Int. J. Heat Mass Transfer* **10**, 1509–1520.

Bickley, W. G. (1941). Experiments in Approximating to the Solution of a Partial Differential Equation, *Phil. Mag.* [7] **32**, 50–66.

Carslaw, H. S., and Jaeger, J. C. (1959). "Heat Conduction in Solids," 2nd ed. Oxford Univ. Press, London and New York.

Dicker, D., and Friedman, M. B. (1963a). Solutions of Heat-Conduction Problems with Nonseparable Domains, *J. Appl. Mech., Trans. ASME, Ser. E* **85**, 493–499.

Dicker, D., and Friedman, M. B. (1963b). Heat Conduction in Elliptical Cylinders and Cylindrical Shells, *AIAA J.* **1**, 1139–1145.

Emmert, R. E., and Pigford, R. L. (1954). A Study of Gas Absorption in Falling Liquid Films, *Chem. Eng. Prog.* **50**, 87–93.

Eraslan, A. H., and Eraslan, N. F. (1969). Heat Transfer in Magnetohydrodynamic Channel Flow, *Phys. Fluids* **12**, 120–128.

Erdogan, F. (1963). On the Approximate Solutions of Heat Conduction Problems, *J. Heat Transfer, Trans. ASME Ser. C* **85**, 203–208.

Finlayson, B. A. (1969). Applications of the Method of Weighted Residuals and Variational Methods, I, II, *Brit. Chem. Eng.* **14**, 53–57, 179–182.

Fujita, H. (1951). *Mem. Coll. Agric., Kyoto Univ.* **59**, 31 (see Crank, J. (1956). "The Mathematics of Diffusion." Oxford Univ. Press (Clarendon), London and New York.

Goldfarb, E. M., and Ereskovskii, O. S. (1966). Biot's Variational Method in Thermal Conductivity Problems Involving a Change in Phase State with a Plane Phase Separation Boundary, *High Temp.* **4**, 628–632.

Goodman, T. R. (1958). The Heat-Balance Integral and Its Application to Problems Involving a Change in Phase, *Trans. ASME* **80**, 335–342.

Goodman, T. R. (1962). The Adjoint Heat-Conduction Problem for Solids, *Proc. U.S. Nat. Cong. Appl. Mech. 4th, Berkeley, California*, pp. 1257–1262.

Goodman, T. R. (1964). Application of Integral Methods to Transient Nonlinear Heat Transfer, *Advan. Heat Transfer* **1**, 51–122.

Horvay, G. (1965). The Tension Field Created by a Spherical Nucleus Freezing into Its Less Dense Undercooled Melt, *Int. J. Heat Mass Transfer* **8**, 195–243.

Hrycak, P. (1967). Heat Conduction with Solidification in a Stratified Medium, *AIChE J.* **13**, 160–164.
Hsu, C. J. (1967). An Exact Mathematical Solution for Entrance Region Laminar Heat Transfer with Axial Conduction, *Appl. Sci. Res.* (A) **17**, 359–376.
Keller, K. H. (1968). Mass Transport Phenomena in the Human Circulatory System, *Chem. Eng. Educ.* **2** (1), 20–26.
Koh, J. C. Y. (1961). One-Dimensional Heat Conduction with Arbitrary Heating Rate and Variable Properties, *J. Aerosp. Sci.* **28**, 989–991.
Kumar, I. J., and Narang, H. N. (1967). A Boundary-Layer Method in Porous Body Heat and Mass Transfer, *Int. J. Heat Mass Transfer* **10**, 1095–1107.
Lardner, T. J. (1967). Approximate Solutions to Phase-Change Problems, *AIAA J.* **5**, 2079–2080.
Lardner, T. J., and Pohle, F. V. (1961). Application of the Heat Balance Integral to Problems of Cylindrical Geometry, *J. Appl. Mech.* **28**, 310–312.
Laura, P. A., and Faulstich, A. J., Jr. (1968). Unsteady Heat Conduction in Plates of Polygonal Shape, *Int. J. Heat Mass Transfer* **11**, 297–303.
LeCroy, R. C., and Eraslan, A. H. (1969). The Solution of Temperature Development in the Entrance Region of an MHD Channel by the B. G. Galerkin Method, *J. Heat Trans., Trans ASME, Ser. C* **91**, 212–220.
Libby, P. A., and Chen, S. (1965). The Growth of a Deposited Layer on a Cold Surface, *Int. J. Heat Mass Transfer* **8**, 395–402.
Lyman, F. A. (1964). Heat Transfer at a Stagnation Point When the Free-Stream Temperature Is Suddenly Increased, *Appl. Sci. Res.* (A) **13**, 65–80.
Muehlbauer, J. C., and Sunderland, J. E. (1965). Heat Conduction with Freezing or Melting, *Appl. Mech. Rev.* **18**, 951–959.
Novotny, J. L., and Eckert, E. R. G. (1965). Integral Analysis of the Flow of a Heat-Generating Fluid in the Entrance Region of a Parallel-Plate Channel, *J. Heat Trans., Trans. ASME, Ser. C* **87**, 313–314.
Poots, G. (1962). An Approximate Treatment of a Heat Conduction Problem Involving a Two-Dimensional Solidification Front, *Int. J. Heat Mass Transfer* **5**, 339–348.
Rafalski, P., and Zyszkowski, W. (1968). Lagrangian Approach to the Nonlinear Heat-Transfer Problem, *AIAA J.* **6**, 1606–1608.
Richardson, P. D. (1964). Unsteady One-Dimensional Heat Conduction with a Nonlinear Boundary Condition, *J. Heat Transfer, Trans. ASME, Ser. C* **86**, 298–299.
Richardson, P. D. (1968). Further Results from Use of a Transcendental Profile Function in Conduction and Convection, *Int. J. Heat Mass Transfer* **11**, 359–365.
Rotem, Z., and Neilson, J. E. (1969). Exact Solution for Diffusion to Flow down an Incline, *Can. J. Chem. Eng.* **47**, 341–346.
Rozenshtok, Yu., L. (1965). Application of Boundary-Layer Theory to the Solution of Problems with Coupled Heat and Mass Transfer, *J. Eng. Phys.* **8**, 483–486.
Schechter, R. S. (1967). "The Variational Method in Engineering." McGraw-Hill, New York.
Sellars, J. R., Tribus, M., and J. S. Klein (1956). Heat Transfer to Laminar Flow in a Round Tube or Flat Conduit—the Graetz Problem Extended, *Trans. ASME* **78**, 441–448.
Siegel, R., Sparrow, E. M., and Hallman, T. M. (1958). Steady Laminar Heat Transfer in a Circular Tube with Prescribed Wall Heat Flux, *Appl. Sci. Res.* (A) **7**, 386–392.
Sparrow, E. M., and Haji-Sheikh, A. (1968). Transient and Steady Heat Conduction in Arbitrary Bodies with Arbitrary Boundary and Initial Conditions, *J. Heat Trans., Trans. ASME, Ser. C* **90**, 103–108.

Sparrow, E. M., and Haji-Sheikh, A. (1970). The Solution of Radiative Exchange Problems by Least Square Techniques, *Int. J. Heat Mass Transfer* **13**, 647–650.

Sparrow, E. M., and Siegel, R. (1958). Thermal Entrance Region of a Circular Tube under Transient Heat Conditions, *Proc. U.S. Nat. Cong. Appl. Mech.* 3*rd*, *Brown Univ. Providence*, *R. I.*, pp. 817–826.

Sparrow, E. M., Hallman, T. M., and Seigel, R. (1957). Turbulent Heat Transfer in the Thermal Entrance Region of a Pipe with Uniform Heat Flux, *Appl. Sci. Res.* (A) **7**, 37–52.

Theofanous, T. G., Isbin, H. S., and Fauske, H. K. (1970). An Integral Method for Convective Diffusion-Bubble Dissolution, *AIChE J.* **16**, 688–690.

Thorsen, R., and Landis, F. (1965). Integral Methods in Transient Heat Conduction Problems with Non-Uniform Initial Conditions, *Int. J. Heat Mass Transfer* **8**, 189–192.

Tien, C., and Yen, Y. C. (1966). Approximate Solution of a Melting Problem with Natural Convection, *Chem. Eng. Prog. Symp. Ser.* **62**, No. 64, 166–172.

Tsoi, P. V. (1967). Approximate Method for Solving Problems in Non-Stationary Heat Conduction, *High Temp.* **5**, 937–947.

Weiner, J. H. (1955). A Method for the Approximate Solution of the Heat Equation, WADC Tech. Rep. 54-427 (AD 97343).

Yang, K. T. (1962). Laminar Forced Convection of Liquids in Tubes with Variable Viscosity, *J. Heat Transfer, Trans. ASME, Ser. C*, **84**, 353–362.

Zyszkowski, W. (1969). The Transient Temperature Distribution in One-Dimensional Heat-Conduction Problems with Nonlinear Boundary Conditions, *J. Heat Transfer, Trans. ASME, Ser. C* **91**, 77–82.

Chapter

4

Applications to Fluid Mechanics

MWR can be used to predict flow phenomena as well as heat and mass transfer. This chapter begins with a discussion of laminar flow through ducts, possibly the simplest fluid mechanical problem, and one for which the method of collocation (point matching) and least squares-collocation have been developed. The integral method was developed for boundary layer flow past flat plates, and is discussed in Section 4.2. Successive approximations are considered in Section 4.3; while the last two sections consider entry-length problems in which the velocity and the temperature field are developing. Finally natural convection is studied: the flow and energy equations are strongly coupled.

4.1 Laminar Flow in Ducts

Consider a fluid in rectilinear flow in the z direction in a duct with cross section A, bounded by curve C. The governing equation is (1.19), giving

$$\begin{aligned} \mu\,\nabla^2 u' &= -\Delta p/L \qquad \text{in } A, \\ u' &= 0 \qquad \text{on } C. \end{aligned} \tag{4.1}$$

We first consider the flow through a rectangular duct, and then discuss some applications of point matching to polygonal ducts. The nondimensional form of Eq. (4.1) is

$$u_{xx} + \varepsilon^2 u_{yy} = -1, \qquad \varepsilon = a/b,$$
$$u = 0 \quad \text{on} \quad x = \pm 1 \quad \text{and on} \quad y = \pm 1, \tag{4.2}$$

where $x = x'/a$, $y = y'/b$, $u = u'/u_0$; $u_0 = \Delta p a^2/\mu L$, where a and b are the half-width in the x and y directions. The trial functions must vanish on $x = \pm 1$, $y = \pm 1$ and be symmetric about $x = 0$ and about $y = 0$. Begin with the most general polynomial of degree 2 in both x and y (including cross products) and apply these conditions:

$$u(x, y) = c(1 - x^2)(1 - y^2). \tag{4.3}$$

We apply the collocation method here and anticipate the results of the next chapter somewhat to guide the choice of collocation points. The reader may have become aware by now that the calculation of the integrals needed in higher approximations is a tedious task. The orthogonal collocation method simplifies that task considerably without losing the accuracy usually associated with the Galerkin and integral methods. The formalism of the orthogonal collocation method is not needed in the first approximation, except for the collocation points. Thus we judiciously choose the collocation points to correspond to the first approximation using the orthogonal collocation method, and leave higher approximations to later.

The residual is evaluated at the point $x_1 = 0.447$, $y_1 = 0.447$:

$$-2c[(1 - y_1{}^2) + \varepsilon^2(1 - x_1{}^2)] = -1. \tag{4.4}$$

The velocity and flow rate are then

$$u = (1 - x^2)(1 - y^2)/(1.6(1 + \varepsilon^2)), \tag{4.5a}$$

$$\phi = \int_{-a}^{a}\int_{-b}^{b} u'(x', y')\, dx'\, dy' = 4abu_0 \int_0^1 \int_0^1 u\, dx\, dy$$
$$= \frac{10}{9}\frac{a^3 b^3\, \Delta p}{(a^2 + b^2)\mu L}. \tag{4.5b}$$

Compared to the exact solution (derived using separation of variables) for a square duct the flow rate is 1.4% low and the velocity at the center is 6% high. It is shown in Section 5.6 that this solution is the same one derived using the Galerkin method with the same trial functions. This results from the choice of collocation points. Furthermore, in Section 7.4 it is shown that for this problem the Galerkin method is equivalent to the Rayleigh–Ritz method (a variational method), and the variational method is known to give lower bounds on the flow rate (Section 7.3). Thus the solution for flow rate

(4.5b) is a lower bound on the exact answer. After reading Section 5.1 the reader can easily compute higher approximations using the orthogonal collocation method. This has been done by Villadsen and Stewart (1967), who report that the third approximation agrees with the exact result to five significant figures.

Boundary Collocation

The same problem can be solved using the boundary methods introduced in Section 2.3. Here we apply the collocation method to flow through a square duct. As trial functions we must find solutions to Eq. (4.2) with $\varepsilon = 1$. The trial functions must be symmetric: $(x, y) \leftrightarrow (-x, y)$, $(x, y) \leftrightarrow (x, -y)$, $(x, y) \leftrightarrow (y, x)$. The inhomogeneous solution to (4.2) can be found by taking a polynomial in x and y, substituting it into (4.2) and determining what relation the constants must have. Trying a second-order polynomial in x and y (with the symmetry properties) gives $u_0 = -(x^2 + y^2)/4$. We next need a series of functions satisfying the homogeneous equation [-1 in Eq. (4.2) replaced by 0]. These solutions to Laplace's equation are provided by the real and imaginary parts of $(x + iy)^n$. The terms with the appropriate symmetry properties are (see p. 27)

$$1 \quad \text{and} \quad x^4 - 6x^2y^2 + y^4. \tag{4.6}$$

The trial function is then

$$u = c_1 - \tfrac{1}{4}(x^2 + y^2) + c_2(x^4 - 6x^2y^2 + y^4). \tag{4.7}$$

We apply the collocation method using the roots to the polynomials in Section 5.1 with $w = 1$. For the first approximation the boundary collocation points are then (1, 0.577) and (0.577, 1), but only one pair is needed because of the symmetry. Evaluating the boundary residual gives

$$0 = c_1 - \tfrac{1}{4}(1 + \tfrac{1}{3}), \qquad c_1 = \tfrac{1}{3}. \tag{4.8}$$

For the second approximation the boundary residual is evaluated at the points (1, 0.861) and (1, 0.340), giving the solution $c_1 = 0.2944$, $c_2 = -0.0486$. The corresponding flow rates and center velocity are in Table 4.1.

Error bounds can be calculated for these solutions using Theorem 11.32. The error $\varepsilon = u_N - u$ satisfies the homogeneous differential equation so that Theorem 11.32 applies. The maximum and minimum values of ε throughout the region occur on the boundary. Since $u = 0$ there, the maximum and minimum can be calculated and these bounds apply to the interior region.

TABLE 4.1

FLOW RATE AND CENTERLINE VELOCITY FOR SQUARE DUCT

Method	$QL/\Delta pa^4$	$u(0, 0)$	ε
Orthogonal collocation			
$N=1$	0.5556	0.3125	—
$N=2$	0.5622	0.2949	—
Boundary collocation			
$N=1$	0.6667	0.3333	0.083
$N=2$	0.5630	0.2944	0.012
Exact	0.5623	0.2947	—

Thus

$$|u_N - u| \leq \varepsilon \qquad \text{in } A,$$
$$\varepsilon \equiv \max_{c} |u_N| = \max_{0 \leq y \leq 1} |u_N(1, y)|. \tag{4.9}$$

The error bounds, listed in Table 4.1, are quite conservative compared to the actual error at the center, but they can be improved with higher approximations. Error bounds can also be calculated using the mean square residual, Eq. (11.72).

A quantity of frequent interest for engineering systems is the friction factor. It is dimensionless and is defined as

$$f = \frac{F}{LC0.5\rho\langle u'\rangle^2}, \tag{4.10}$$

where F is the force exerted on the conduit of length L, circumference C. In terms of the Reynolds number, $\text{Re}' = \rho\, d'\langle u'\rangle/\mu$, it can be written $(F = A\,\Delta p,\, u = u'/u_0,\, u_0 = \Delta p\, d'^2/(\mu L))$

$$f\,\text{Re}' = K' = \frac{2A}{Cd'}\frac{1}{\langle u\rangle}, \tag{4.11}$$

where A is the area, d' is the dimension used to make the equations dimensionless and $\langle u\rangle$ is the average velocity for the solution to (4.2). The Reynolds number can be defined in terms of an equivalent diameter, $\text{Re} = \rho d\langle u'\rangle/\mu$, $d = 4A/C = 4\times$ hydraulic radius:

$$f\,\text{Re} = K = K'd/d'. \tag{4.12}$$

The constant K depends less on geometry than does K'. Sparrow (1962) has applied boundary collocation to flow in an isosceles triangular duct. Shih (1967) used the same method for star-shaped conduits and regular polygons.

Ratkowsky and Epstein (1968) and Hagen and Ratkowsky (1968) use least squares-collocation to compute values for flow through a duct with a regular polygon boundary and a circular core as well as in the inverse case of a polygon inside a circle. From 5 to 16 expansion functions were necessary and from 2 to 3 times as many collocation points were used as there were expansion functions. Values of K are listed in Table 4.2 for various geometries.

TABLE 4.2

FRICTION FACTOR–REYNOLDS NUMBER PRODUCT FOR LAMINAR FLOW THROUGH DUCTS

Shape parameter	$f\,\mathrm{Re} = K$	Shape parameter	$f\,\mathrm{Re} = K$
Regular polygon with n sides[a]		Ellipse with κ = ratio of semiaxes[c]	
$n = 3$	13.3	$\kappa = 0.2$	18.6
$n = 4$	14.2	$\kappa = 0.4$	17.3
$n = 6$	15.0	$\kappa = 0.6$	16.5
$n = 8$	15.4	$\kappa = 0.8$	16.1
$n = 18$	15.9	$\kappa = 1.0$ (circle)	16.0
$n = \infty$ (circle)	16.0		
Rectangle with aspect ratio κ[b]		Star conduit formed with n circles[d]	
$\kappa = 0.0$ (flat plates)	24.0	$n = 3$	6.50
$\kappa = 0.2$	19.0	$n = 4$	6.61
$\kappa = 0.4$	16.3		
$\kappa = 0.6$	15.0		
$\kappa = 0.8$	14.4		
$\kappa = 1.0$	14.2		

[a] Ratkowsky and Epstein (1968). Used by permission of the copyright owner, the Chemical Institute of Canada.

[b] Shih (1967). Used by permission of the copyright owner, the Chemical Institute of Canada.

[c] From "Fluid Dynamics and Heat Transfer" by Knudsen and Katz. Copyright 1958, McGraw-Hill Book Co. Used with permission of McGraw-Hill Book Co.

[d] Shih (1967). Conduit formed by placing n circles tangent to adjacent circles in a symmetrical array about the origin.

Ramacharyulu (1967) has employed the Galerkin method to solve for the laminar flow of an Oldroyd, non-Newtonian fluid through a triangular duct.

Nonlinear Problems

Consider heat transfer to a fluid in laminar flow in regular polygonal ducts. Cheng (1967) predicts the velocity profile and temperature distribution, including heat generation and viscous dissipation, which make the problem nonlinear. The governing equations are

$$\nabla^2 u \equiv \frac{1}{r}\frac{\partial}{\partial r}\left(r\frac{\partial u}{\partial r}\right) + \frac{1}{r^2}\frac{\partial^2 u}{\partial \phi^2} = -\frac{1}{\mu}\frac{\Delta p}{L}, \tag{4.13a}$$

$$\nabla^2 \theta = \frac{1}{\alpha}\frac{\Delta t}{L}u - \frac{Q}{k} - \frac{\mu}{k}\left[\left(\frac{\partial u}{\partial r}\right)^2 + \left(\frac{1}{r}\frac{\partial u}{\partial \phi}\right)^2\right], \tag{4.13b}$$

$$u = \theta = 0 \qquad \text{on boundary}, \tag{4.13c}$$

where the pressure gradient and temperature gradient are constants in fully developed flow.

To solve (4.13a) using boundary collocation, the velocity is expanded in terms of functions automatically satisfying the differential equation (found by separation of variables),

$$u = \frac{a^2}{\mu}\frac{\Delta p}{L}\left[-\frac{1}{4}\left(\frac{r}{a}\right)^2 + \sum_{i=0}^{N} A_i\left(\frac{r}{a}\right)^{in} \cos in\phi\right]. \tag{4.14}$$

To satisfy the boundary conditions, the velocity is set equal to zero at N collocation points, which are located on the boundary of the regular polygon at equal angular intervals for $0 \leq \phi \leq \pi/n$. The velocity vanishes at other points on the boundary by symmetry. This velocity is then substituted into (4.13b) which is solved for θ. The solution takes the form

$$\theta = -\frac{Q}{4k}\left(\frac{r}{a}\right)^2 + \sum_{i=0}^{N} D_i\left(\frac{r}{a}\right)^{in} \cos in\phi + f(A_i, r, \phi), \tag{4.15}$$

where the function f is a nonlinear function of the A_j, powers of r, and trigonometric functions of ϕ. Thus the f is a known function of position. For arbitrary heat generation, rather than the constant generation assumed here, the heat generation function is expanded in the same trigonometric series appearing in the velocity expression. The thermal boundary condition is satisfied by applying boundary collocation at the same N points used for the velocity. Cheng reports solutions for $N = 3$, 5, and 10 which differed by less than 0.5%. The error in temperature on the boundary was less than 10^{-4} times the centerline temperature. Viscous dissipation increases the Nusselt number. The effect is more pronounced as the number of sides of the ducts increases. Cheng (1969) has solved the same problem with uniform peripheral heat flux, since the boundary collocation method is easily extended to all types of boundary conditions. Casarella *et al.* (1967) treat a similar problem without the heat generation or viscous dissipation. They use conformal mapping techniques, followed by collocation in radius and subdomain in arcs, and study various shapes: cardioid, corrugated, square, and hexagons.

4.2 Boundary Layer Flow past a Flat Plate

We study the flow of a fluid past a flat plate. Due to the viscous drag exerted by the plate, a boundary layer develops in which the velocity varies from zero at the plate to the main stream velocity far from the plate. As the fluid advances down the plate the boundary layer thickness increases. The objective is to predict the drag on the flat plate. The integral method was developed for this problem in 1921 by the German aeronautical engineer Pohlhausen and has been widely used ever since. We shall see however, that the approach usually used in the field of fluid mechanics really is just the first approximation using MWR. We also see that the use of additional derived boundary conditions (4.24) does not necessarily improve the results. Convergence to the exact solution is possible only by calculating successive approximations, as in Section 4.3.

We consider the simplest problem: flow past a flat plate with uniform velocity, which is governed by Eqs. (1.20) and (1.21) with $U = \text{constant}$:

$$\begin{gathered} u_x + v_y = 0, \\ u u_x + v u_y = \nu u_{yy}, \\ u = 0 \quad \text{at} \quad y = 0, \\ u = U \quad \text{at} \quad y \to \infty, \quad \text{and} \quad x = 0, \quad \text{all } y. \end{gathered} \tag{4.16}$$

The problem has a similarity transformation and is a function of the variable $\eta = y/\delta(x)$, where $\delta(x)$ is the boundary layer thickness, $u(x, y) = U\phi(\eta)$. Transform the differential equation and boundary conditions using

$$\begin{aligned} &\frac{\partial \eta}{\partial x} = -\frac{\eta}{\delta}\delta', && \frac{\partial \eta}{\partial y} = \frac{1}{\delta}, \\ &\frac{\partial u}{\partial x} = -U\phi'\frac{\eta}{\delta}\delta', && \frac{\partial u}{\partial y} = \frac{U}{\delta}\phi', \\ &\frac{\partial^2 u}{\partial y^2} = \frac{U}{\delta^2}\phi'', && v = -\int_0^y \frac{\partial u}{\partial x}\,dy = U\delta'\int_0^\eta \eta\phi'\,d\eta. \end{aligned} \tag{4.17}$$

The problem is then

$$\delta\delta'\left\{-\phi\phi'\eta + \phi' \int_0^\eta \eta\phi' \, d\eta\right\} - \frac{\nu}{U}\phi'' = 0, \tag{4.18}$$

$$\phi(0) = \delta(0) = 0, \qquad \phi(\infty) = 1.$$

Equation (4.18) can be written $F(\eta) = G(x)$, so that both functions must be constant:

$$\delta\delta' = \frac{\nu}{U}\alpha, \tag{4.19a}$$

$$\phi'' = \alpha\left\{-\phi\phi'\eta + \phi' \int_0^\eta \eta\phi' \, d\eta\right\} = -\alpha\phi' \int_0^\eta \phi \, d\eta. \tag{4.19b}$$

The solution for the boundary layer thickness is

$$\delta = (2\nu\alpha x/U)^{1/2}. \tag{4.20}$$

The second equation can be simplified by defining $f'(\eta) = \phi(\eta), f(0) = 0$ and the problem reduces to

$$f''' + \alpha f f'' = 0. \tag{4.21}$$

Blasius has solved this equation for $\alpha = \frac{1}{2}$ using an infinite series expansion (see Schlichting, 1960). We apply the integral method by integrating (4.19b) from $\eta = 0$ to $\eta = 1$:

$$\int_0^1 R(\phi; \eta) \, d\eta = \int_0^1 \left[\alpha\phi' \int_0^\eta \phi \, d\eta' + \phi''\right] d\eta = 0. \tag{4.22}$$

The approximate solution is (4.20) with α determined by the trial function assumed for $\phi(\eta)$. A quantity of interest is the shear stress at the wall

$$\tau = \mu \left.\frac{\partial u}{\partial y}\right|_{y=0} = \frac{\mu U}{(\nu x/U)^{1/2}} \frac{\phi'(0)}{(2\alpha)^{1/2}}. \tag{4.23}$$

The trial function for ϕ is determined by applying the boundary conditions and various derived conditions to the polynomial expansion $\phi = \sum a_i \eta^i$. We consider several trial functions to illustrate the effect of additional constraints:

$$\begin{aligned}
&N = 1: \quad \phi(0) = 0, \quad \phi(1) = 1; \quad \text{boundary conditions.}\\
&N = 2: \quad \phi'(1) = 0; \quad \text{continuity of velocity gradient at } \eta = 1.\\
&N = 3: \quad \phi''(0) = 0; \quad \text{collocation at } \eta = 0.\\
&N = 4: \quad \phi''(1) = 0; \quad \text{collocation at } \eta = 1.\\
&N = 5: \quad \phi'''(0) = 0; \quad \text{derivative of residual at } \eta = 0.
\end{aligned} \tag{4.24}$$

These results are shown in Table 4.3 as well as those obtained with exponential and error function trial functions. Notice that the approximations converge towards the exact solution as various conditions in Eqs. (4.24) are added

TABLE 4.3

APPROXIMATE SOLUTION TO BOUNDARY LAYER PROBLEMS

	$\phi'(0)/(2\alpha)^{1/2}$	$\|R\|$	E
$N=1, \phi=\eta$	0.289	0.35	0.040
$N=2, \phi=2\eta-\eta^2$	0.365	0.16	0.029
$N=3, \phi=3/2\eta-1/2\eta^3$	0.323	0.14	0.016
$N=4, \phi=2\eta-2\eta^3+\eta^4$	0.343	0.067	0.011
$N=5, \phi=\frac{5}{3}(\eta-\eta^4)+\eta^5$	0.322	0.067	0.010
$\phi=1-e^{-\eta}$	0.500	0.37	0.056
$\phi=\text{erf}\,\eta$	0.363	0.090	0.023
Exact	0.332	—	—

but then bypass it. This illustrates the same feature discussed in Section 3.2: as additional constraints are added the answer may not become better. In contrast the successive approximations calculated using the subdomain or moments methods show convergence as conditions are added (see Section 4.3).

Shown in Table 4.3 are also the mean square residuals:

$$\int_0^\infty R^2(y)\,dy = \frac{U^2\nu^2}{(2\nu x/U)^{3/2}}\frac{1}{\alpha^{3/2}}\int_0^\infty R^2(\eta)\,d\eta, \qquad \|R\| = \left\{\alpha^{-3/2}\int_0^\infty R^2(\eta)\,d\eta\right\}^{1/2}. \tag{4.25}$$

For many problems it can be shown that the mean square error is bounded in terms of the mean square residual (Section 11.6). While this has not been shown for the boundary layer problem, the results of Table 4.3 indicate that the mean square residual is a fair indicator of the accuracy. Also shown is a measure of the pointwise error,

$$E = \tfrac{1}{4}\sum_{i=1}^{4}|\phi^*(\xi_i)-\phi(\xi_i)|, \tag{4.26}$$

where ϕ^* and ϕ are the approximate and exact solution, respectively, and $\xi_i = y(U/\nu x)^{1/2}$ is equal to 1.0, 2.0, 3.0, and 4.0.

4.3 Laminar Boundary Layers

The integral method presented above gives a first approximation which is adequate for many purposes. We consider next successive approximations which give precise answers for more complicated problems. The form of

MWR applied here is due to Bethel (1967) and Abbott and Bethel (1968). The equations are (1.20) and (1.21) with a bar denoting the dimensional quantity:

$$\begin{aligned} &\bar{u}_x + \bar{u}_y = 0, \\ &\bar{u}\bar{u}_x + \bar{v}\bar{u}_y = U_e U_e' + \nu \bar{u}_{yy}, \\ &\bar{u}(x, 0) = \bar{v}(x, 0) = 0, \\ &\bar{u}(x, y) \to U_e(x) \qquad \text{as} \quad y \to \infty. \end{aligned} \tag{4.27}$$

The Göertler transformation changes the independent variables:

$$\begin{aligned} \xi &= \int_0^{x/L} u \, d(x/L), \\ \eta &= \mathrm{Re}^{1/2} \int_0^{y/L} u \, d(y/L) = \mathrm{Re}^{1/2} Uy/L, \\ u &= \bar{u}/U_e, \qquad v = \bar{v}\mathrm{Re}^{1/2}/U_e, \end{aligned} \tag{4.28}$$

where $\mathrm{Re} = U_\infty L/\nu$ is the Reynolds number and $U = U_e/U_\infty$. (Note that η is a function of both x and y.) The system of equations (4.27) becomes

$$u_\xi + w_\eta = 0, \tag{4.29a}$$

$$uu_\xi + wu_\eta = \frac{1}{U}\frac{dU}{d\xi}(1 - u^2) + u_{\eta\eta}, \tag{4.29b}$$

subject to

$$\begin{aligned} &u(\xi, 0) = w(\xi, 0) = 0, \\ &u(\xi, \eta) = 1 \qquad \text{as} \quad \eta \to \infty, \\ &w = v + \frac{1}{U}\frac{dU}{d\xi}\eta u. \end{aligned} \tag{4.30}$$

The two equations are reduced to a single one by multiplying the continuity equation (4.29a) by u and adding it to (4.29b):

$$F(u, w) = (u^2)_\xi + (uw)_\eta - \frac{1}{U}\frac{dU}{d\xi}(1 - u^2) - u_{\eta\eta} = 0. \tag{4.31}$$

The weighted integral of (4.31) is set equal to zero:

$$\int_0^\infty H_i(u) F(u, w) \, d\eta = 0, \qquad i = 1, 2, \ldots, N. \tag{4.32}$$

This equation can be integrated by parts to obtain

$$\frac{d}{d\xi}\int_0^\infty h_i u\,d\eta + [h_i w]_0^\infty - \frac{1}{U}\frac{dU}{d\xi}\int_0^\infty h_i'(1-u^2)\,d\eta - \int_0^\infty h_i'\frac{\partial^2 u}{\partial \eta^2}\,d\eta = 0, \quad (4.33)$$

where $h_i' = dh_i/du = H_i(u)$, $\partial h_i/\partial\xi = H_i\,\partial u/\partial\xi$, and $\partial h_i/\partial\eta = H_i\,\partial u/\partial\eta$. If we choose $h_i(u)$ such that

$$\lim_{\substack{\eta\to\infty\\ u\to 1}} h_i(u) = 0, \quad (4.34)$$

the term involving w drops out of (4.33). If we choose $h_i(u) = 1 - u$, we obtain von Kármán's integral momentum equation

$$\frac{d}{d\xi}\int_0^\infty (1-u)u\,d\eta + \frac{1}{U}\frac{dU}{d\xi}\int_0^\infty (1-u^2)\,d\eta + \int_0^\infty \frac{\partial^2 u}{\partial\eta^2}\,d\eta = 0, \quad (4.35)$$

or in the usual formulation

$$\frac{d}{dx}(U_e^2\,\delta_2) + \delta_1 U_e\frac{dU_e}{dx} = \frac{\tau_w}{\rho},$$

$$\tau_w = \mu\left.\frac{\partial \bar{u}}{\partial y}\right|_{y=0}, \qquad \delta_1 = \int_0^\infty (1-u)\,dy, \qquad \delta_2 = \int_0^\infty u(1-u)\,dy. \quad (4.36)$$

First consider the solution of these equations by the method of subdomains. In the Russian literature this method is called the method of integral relations. Let us suppose we wish to derive the solution to

$$\frac{\partial}{\partial\xi}P(u, w, \xi, \eta) + \frac{\partial}{\partial\eta}Q(u, w, \xi, \eta) = F(u, w, \xi, \eta) \quad (4.37)$$

in the region $0 \le \eta \le \delta(\xi)$. Then in place of Eq. (4.33) we have

$$\frac{d}{d\xi}\int_0^{\delta(\xi)} f_n P\,d\eta - \left[\frac{d\delta}{d\xi}f_n P\right]_0^\delta + [Qf_n]_0^\delta - \int_0^\delta Q\frac{\partial f_n}{\partial\eta}\,d\eta = \int_0^\delta Ff_n\,d\eta. \quad (4.38)$$

Next divide the region of y into N regions:

$$\eta_n(x) = \frac{n}{N}\,\delta(\xi), \qquad n = 1, 2, \ldots, N-1. \quad (4.39)$$

The weighting function is then chosen for the subdomain method:

$$f_n(\eta) = \begin{cases} 1, & \eta_{n-1} \le \eta \le \eta_n, \\ 0, & \text{otherwise.} \end{cases} \quad (4.40)$$

Equation (4.33) is then

$$\frac{d}{d\xi}\int_{\eta_{n-1}}^{\eta_n} P\,d\eta - \frac{d\eta_n}{d\xi}P_n + \frac{d\eta_{n-1}}{d\xi}P_{n-1} + Q_n - Q_{n-1} = \int_{\eta_{n-1}}^{\eta_n} F\,d\eta, \quad (4.41)$$

which is integrated numerically in ξ. Applications of the method of integral relations to subsonic flow about bodies of revolution, flow through nozzles and flow of gases with shock waves are discussed in the reviews by Belotserkovskii and Chuschkin (1964, 1965). In addition they discuss possible extensions to three-dimensional cases. South and Newman (1965) also present applications to supersonic gas flow over pointed bodies.

A Galerkin method has also been used for this problem, first applied by Dorodnitsyn (1962, 1964), and later studied in more detail by Bethel (1967) and Abbott and Bethel (1968). Equations (4.33) are solved by expanding the unknown function u in a series of known functions $u_i(\xi, \eta)$. The boundary conditions at infinity must be approached asymptotically, however, so that the choice of trial functions may be difficult. To avoid this problem Dorodnitsyn (1962) introduced a new dependent variable

$$\theta(\xi, u) = [\partial u/\partial \eta]^{-1}, \tag{4.42}$$

which is the inverse of the shear stress, such that the new independent variables are (ξ, u) in place of (ξ, η). This replaces a variable which is bounded and defined over a semi-infinite region with one which is defined over a finite region but includes a singularity at $u = 1$. The asymptotic form of the singularity is known to be such that $\theta(\xi, u) = O(1/(1-u))$ as $u \to 1$. The equations then become

$$\frac{d}{d\xi}\int_0^1 h_i u\theta\, du - \frac{1}{U}\frac{dU}{d\xi}\int_0^1 h_i'(1-u^2)\theta\, du + \frac{h_i'(0)}{\theta(\xi,0)} + \int_0^1 h_i''\frac{du}{\theta} = 0. \tag{4.43}$$

The trial function Bethel uses is

$$\theta(\xi, u) = \frac{1}{(1-u)}\sum_{j=1}^{N} c_j(\xi)u^{j-1} \tag{4.44}$$

for accelerating flows and

$$\theta(\xi, u) = \frac{1}{(\alpha(\xi)+u)^{1/2}(1-u)}\sum_{j=1}^{N-1} c_j(\xi)u^{j-1} \tag{4.45}$$

for flows giving rise to separation. Combining Eqs. (4.43) and (4.44) gives, with $h_i = (1-u)^i$ and $\mathscr{F}(m, n)$ defined in Table 6.4,

$$\sum_{j=1}^{N} c_j'\mathscr{F}(j, i-1) + \frac{1}{U}\frac{dU}{d\xi}\sum_{j=1}^{N} c_j\, i[\mathscr{F}(j-1, i-1) + \mathscr{F}(j, i-1)]$$

$$+\frac{i}{c_1} + I(i) = 0, \tag{4.46}$$

and the last integral is evaluated numerically. Bethel calls this method the Galerkin–Kantorovich–Dorodnitsyn method, after the people who introduced the weighting function, the reduction to ordinary differential equations, and the application to boundary layer flows.

Bethel (1967) considers three problems which have similar solutions: stagnation point flow, flow past a flat plate, and a mildly retarded flow. Typical values of the solution are given in Table 4.4 for the flat plate case. Bethel then

TABLE 4.4

WALL SHEAR STRESS PARAMETER FOR FLOW PAST A FLAT PLATE[a]

N	1	2	4	6	10	Exact
$\tau_w(\mathrm{Re}\xi)^{1/2}/\rho U_e^2$	0.50000	0.33017	0.33251	0.33218	0.33209	0.33206

[a] Bethel (1967). Used with permission of the copyright owner, the Royal Aeronautical Society.

considers three nonsimilar flow problems. We present here the results for predicting the separation point of Tani flows, in which case $U = 1 - (x/L)^a$. Table 4.5 shows predictions of the separation point found by Bethel, using a

TABLE 4.5

AXIAL LOCATION OF SEPARATION POINTS FOR TANI FLOWS[a]

a	Exact	Galerkin ($N = 4$)	Finite difference ($N = 100$–300)
1	0.120	0.12035	0.12033
2	0.271	0.27214	0.2724
4	0.462	0.46271	0.4624
8	0.640	0.64122	0.641

[a] Bethel (1967). Used with permission of the copyright owner, the Royal Aeronautical Society.

four-term MWR, as well as finite difference calculations using from 100 to 300 terms (Schoenauer, 1964). It is clear that a relatively low-order approximation gives very accurate results. For reversed flow problems Bethel (1969) develops a similar method using a different trial function where the flow is reversed.

Koob and Abbott (1968) have employed a modification of the above technique to solve the time-dependent laminar boundary layer equations. The problem concerns a semi-infinite plate suddenly injected into a uniform flow, with the plane of the plate being parallel to the undisturbed flow direction. In this case the equations are

$$u_\xi + u_\eta = 0, \qquad u_\tau + uu_\xi + vu_\eta = u_{\eta\eta}. \tag{4.47}$$

For time $\tau = 0+$ an asymptotic solution (Stokes flow) is known:

$$u = \text{erf}(\eta/2\tau^{1/2}). \tag{4.48}$$

For large times near the leading edge the solution reduces to the Blasius solution, a function of η and ξ, which is the length down the plate. For intermediate η, ξ, and τ the flow is fully time-dependent and two-dimensional. Koob and Abbott (1968) choose trial functions which reduce to the known asymptotic solutions when applicable. The MWR is then used to predict a deviation from those asymptotic results, in much the same way as is done for transient mass transfer in Chapter 5 and natural convection in Section 4.4. The authors use MWR as described previously to eliminate the η variable, they use the method of lines to eliminate the ξ variable, and numerically integrate the final set of ordinary differential equations in time. The method of lines is essentially a finite difference method which in this case uses a variable step size. Koob and Abbott find that three terms in η are sufficient to obtain accurate results (from the MWR part) and 10–20 terms are needed in the ξ direction in the method of lines. The combined method reduced a partial differential equation in three variables to a set of ordinary differential equations in one variable. The numerical solutions provide a convenient "interpolation" between the known asymptotic results.

MWR has also been applied to turbulent boundary layers, but few of the methods have been developed to give successive approximations, and the methods themselves are enmeshed in the assumptions which must be made concerning the character of the turbulence. At the AFOSR–IFP–Stanford conference on computations of turbulent boundary layers, the same problems were worked using several different methods and the results were compared. Detailed comparisons are available in Kline *et al.* (1968) and are summarized in Kline *et al.* (1969). These authors conclude that the integral methods are just as accurate and faster than finite difference methods. Three-dimensional problems are becoming more important, so that one area of future work should include development of MWR for three-dimensional laminar and turbulent boundary layers. Spaulding (1967) prefers finite difference methods with a variable grid size for these cases.

Other Applications

While it is not feasible to list all applications of MWR, we do give a few representative applications which illustrate the type of problems which have been tackled. Magnetohydrodynamic boundary layer theory is treated by Hugelman and Haworth (1965) and Heywood and Moffatt (1965).

Computations have been done for non-Newtonian flows, such as the power law fluid, or Ostwald–de Waele fluid. The constitutive relation (for simple geometries) is

$$\tau_{yz} = -K \left| \frac{\partial u}{\partial y} \right|^{n-1} \frac{\partial u}{\partial y}, \tag{4.49}$$

where $n = 1$ gives a Newtonian fluid. The integral method gives an equation like (4.36) except that the expression for the shear stress at the wall is now

$$\tau_w = -K \left[\left| \frac{\partial u}{\partial y} \right|^{n-1} \frac{\partial u}{\partial y} \right]_{y=0}. \tag{4.50}$$

Bizzell and Slattery (1962) assume a trial function of the form

$$\frac{u}{U} = a + b\eta + c\eta^2 + d\eta^3 + e\eta^4 \tag{4.51}$$

and choose the constants in order to satisfy the conditions

$$\begin{aligned} u &= 0, \qquad \frac{\partial \tau}{\partial y} = -U \frac{dU}{dx}, \qquad \text{at} \quad y = 0, \\ u &= U, \qquad \frac{\partial u}{\partial y} = \frac{\partial^2 u}{\partial y^2} = 0, \qquad \text{at} \quad y = \delta. \end{aligned} \tag{4.52}$$

The second condition arises from collocation on the boundary. The other conditions are the usual ones requiring continuity of the velocity profile. The trial function then simplifies to

$$u/U = (2\eta - 2\eta^3 + \eta^4) + (E/6)(\eta - 3\eta^2 + 3\eta^3 - \eta^4). \tag{4.53}$$

Notice that the non-Newtonian character of the fluid does not have a great influence on the trial function: the same general form is used as in the Newtonian case. The parameter E is related to the boundary layer thickness which is determined by integration of (4.36). Earlier, Acrivos *et al.* (1960, see also 1965) compared results of the integral method to exact solutions and found that the approximate solution was reasonably accurate for n near 1 (i.e., nearly a Newtonian fluid), the wall shear stress was predicted fairly well, but that the velocity gradient at the wall was not predicted well, especially for small n (<0.5). More recently Fox *et al.* (1969) studied the power law fluid in a boundary layer flow set up by a moving continuous flat sheet. They found similar results: the shear stress was predicted fairly well (15%) for values of n studied ($0.1 \le n \le 2$) but the velocity gradient at the wall was not predicted well for low n. The reason is very apparent when looking at the velocity profiles: for small n the velocity profiles are much flatter than is assumed in the above applications of the integral method. In fact Fox *et al.* used the same profile for all n. For flow between flat plates it can be shown that the steady-state velocity profile is

$$u \sim 1 - y^{(n+1)/n}. \tag{4.54}$$

For $n = 1$ this is a parabola. For small n, however, it becomes very flat for small y and rapidly drops to zero at $y = 1$. Clearly, a profile which gives adequate results for a Newtonian fluid will not give adequate results when the velocity profile is very flat, as is the case for small n. Thus the simple, first approximation should be carefully examined when extended to more general situations, such as for a non-Newtonian fluid. The trial functions may have to be revised to obtain good results for the first approximation, and the mean square residual might be calculated to check the approximation.

Combined Heat and Momentum Transfer

It is clear that the same procedure used by Bethel can be applied to the boundary layer equations with heat transfer. Several applications in the literature use the method of integral relations (subdomain method), including for compressible flow. A few of these are Poots (1960), who uses the integral method, Pallone (1961), Kuby *et al.* (1967), and Crawford and Holt (1968), who use the method of integral relations. See also the book by Walz (1969). Sinha (1963) uses the integral method to study forced convection past a flat plate with temperature-dependent viscosity and thermal conductivity.

4.4 Natural Convection

Natural convection phenomena is difficult to analyze because the energy and momentum equations are coupled. Energy can be transported by conduction, which induces a velocity through the buoyancy mechanism, which in turn affects the convection of energy. The extent of the coupling is affected by the Prandtl number: for liquid metals with small Prandtl number, $\mathrm{Pr} = \nu/\kappa$, the thermal conduction proceeds rapidly and the velocity develops quickly. To illustrate the application of MWR to this case we consider the laminar flow over a vertical flat plate when the temperature of the plate is suddenly raised to a uniform temperature as treated by Heinisch *et al.* (1969).

The dimensionless boundary layer equations are

$$\begin{aligned} u_x + v_y &= 0, \\ u_t + uu_x + vu_y &= \mathrm{Pr}^2\,\theta + \mathrm{Pr}\,u_{yy}, \\ \theta_t + u\theta_x + v\theta_y &= \theta_{yy}, \end{aligned} \tag{4.55}$$

where $u = u'd/G^{1/2}\kappa$, $v = v'd/\kappa G^{1/4}$, $t = t'G^{1/2}\kappa/d^2$, $\theta = (T - T_\infty)/(T_w - T_\infty)$, $x = x'/d$, $y = y'G^{1/4}/d$, and G is the Grashof number, $\alpha g d^3(T_w - T_\infty)/\nu^2$, where α is the thermal expansion coefficient, g the acceleration of gravity, and $\kappa = k/\rho C_p$. The dependence on the y coordinate, perpendicular to the plate, is assumed in the form of a boundary layer, and the thermal and momentum boundary layers are assumed equal. This is a reasonable assumption since the problem is for free convection and the velocity responds to changes in temperature, which occur only within the boundary layer. For small times the heat transfer occurs only by conduction and the velocity is small. For this case a similarity solution is known, and the form of this solution provides the motivation for assuming the following trial function:

$$u = u_1(x, \tau)\eta \left[\frac{\exp(-\eta^2)}{\sqrt{\pi}} - \eta \operatorname{erfc} \eta\right], \tag{4.56}$$

$$\theta = \operatorname{erfc} \eta, \qquad \eta = y/\delta(x, \tau).$$

Equations (4.55) are then integrated from $0 \le y \le \delta$ to obtain

$$\begin{gathered} \frac{1}{6}\frac{\partial u_1\delta}{\partial \tau} + c\,\frac{\partial u_1{}^2\delta}{\partial x} = \Pr^2 \delta - \Pr \frac{u_1}{\delta}, \\ \frac{\partial \delta}{\partial \tau} + d\,\frac{\partial u_1\delta}{\partial x} = \frac{2}{\delta}, \end{gathered} \tag{4.57}$$

$$c = (3\sqrt{2}/20) - (1/5), \qquad d = (\sqrt{2} - 1)/6.$$

The small time solution reduces to the exact similarity solution. For steady state we obtain

$$\begin{gathered} \delta = \Pr^{-1/2}\left(\frac{8}{3d}\right)^{1/4}\left[\frac{10c}{3d} + \Pr\right]^{1/4} x^{1/4}, \\ u_1 = \Pr\left(\frac{8}{3d}\right)^{1/2}\left[\frac{10c}{3d} + \Pr\right]^{-1/2} x^{1/2}. \end{gathered} \tag{4.58}$$

Heinisch *et al.* compare this solution with exact results for two Prandtl numbers (Pr = 1 and 0.003) and find good agreement. The next step is to eliminate the x dependence. Multiply Eqs. (4.57) by the weighting functions f_i and g_i and integrate from $x = 0$ to 1. The result is

$$\begin{gathered} \frac{1}{6}\frac{d}{d\tau}\int_0^1 f_i u_1 \delta\,dx + c f_i u_1{}^2\delta\Big|_{x=1} - c\int_0^1 f_i' u_1{}^2\delta\,dx \\ = \Pr^2 \int_0^1 f_i \delta\,dx - \Pr \int_0^1 \frac{f_i u_1}{\delta}\,dx, \\ \frac{d}{d\tau}\int_0^1 g_i\delta\,dx + d g_i u_1\delta\Big|_{x=1} - d\int_0^1 g_i' u_1\delta\,dx = 2\int_0^1 \frac{g_i}{\delta}\,dx. \end{gathered} \tag{4.59}$$

The trial functions for u_1 and δ and the weighting functions are taken as

$$u_1 = \sum_{i=1}^{n} a_i x^{i/2}, \qquad \delta = \sum_{i=1}^{n} b_i x^{i/4},$$
$$f_i = x^i, \qquad g_i = x^{1/2}. \tag{4.60}$$

This choice provides a good first approximation, and is preferable to other choices in that the resulting ordinary differential equations are stable and converge readily as the number of terms is increased. For the first approximation the equations are

$$\frac{dab}{d\tau} = \frac{22}{3}\,\mathrm{Pr}\left(b\,\mathrm{Pr} - \frac{a}{b}\right) - \frac{55}{6}\,ca^2 b,$$
$$\frac{db}{d\tau} = \frac{14}{5b} - \frac{21}{20}\,dab, \tag{4.61}$$

which are integrated numerically. The effect of decreasing Prandtl number is to increase the dimensionless time to reach steady state. The solutions exhibit an overshoot, in that the boundary layer thickness goes above its steady-state value about 8% before approaching steady state. The authors comment that the existence of the overshoot is not firmly substantiated, although a finite difference method (method of lines) gives the same result. The MWR was used in this case to piece together two regimes: one in which the solution is a function of y and t alone, and one in which the solution is a function of y and x alone.

Other applications to natural convection have used the integral method to calculate a first approximation (Ostrach and Thornton, 1958; Lemlich, 1963; Yang and Jerger, 1964; Riley, 1964; Lemlich and Steinkamp, 1964; and Dickson and Traxler, 1966). The first approximation does not always give accurate results, as demonstrated by the application to nonlinear convective diffusion (Nakano *et al.*, 1967) and forced and free convection (Acrivos, 1958, 1966). In the latter case the momentum boundary layer thickness was taken equal to the thermal boundary layer thickness. According to Acrivos (1966) this is a poor assumption for combined free and forced convection in fluids with $\mathrm{Pr} \to \infty$.

4.5 Coupled Entry-Length Problems

In the entry-length problems discussed in Section 3.3 the velocity was fully developed and known. It is possible of course for the experimental situation to be one in which the velocity and temperature profiles develop together. Then the momentum and energy equations are coupled. If the physical properties are constant, then the coupling is weak, since the momentum equation can be solved independently of the energy equation. To illustrate the weakly-coupled problem and particularly the choice of trial function, we study the flow in a flat duct as solved using the integral method by Sparrow (1955).

The governing equations are similar to the boundary layer equations, in that the velocity boundary layer is confined to a region near the boundary, and this region grows as the fluid proceeds down the duct:

$$\begin{aligned} u_x + v_y &= 0, \\ uu_x + vu_y &= U_1 U_1' + \nu u_{yy}, \\ ut_x + vt_y &= \kappa t_{yy}. \end{aligned} \tag{4.62}$$

These are integrated over the boundary layer thicknesses, δ for momentum and Δ for energy, and are rearranged as in Eq. (4.35):

$$\frac{d}{dx}\int_0^\delta (U_1 - u)u\,dy + \frac{dU_1}{dx}\int_0^\delta (U_1 - u)\,dy = \nu \left.\frac{\partial u}{\partial y}\right|_{y=0}, \tag{4.63a}$$

$$\frac{d}{dx}\int_0^\Delta (t_1 - t)u\,dy = \kappa \left.\frac{\partial t}{\partial y}\right|_{y=0}, \tag{4.63b}$$

$$\frac{d}{dx}\int_0^1 u(x, y)\,dy = -\int_0^1 \frac{\partial v}{\partial y}\,dy = 0. \tag{4.63c}$$

Velocity and temperature profiles are chosen to satisfy

$$\begin{aligned} &u = 0, \quad t = t_w, \quad t_{yy} = 0, \quad &&\text{at} \quad y = 0, \\ &u = U_1, \quad u_y = 0, \quad &&\text{at} \quad y = \delta, \\ &t = t_1, \quad t_y = 0, \quad &&\text{at} \quad y = \Delta. \end{aligned} \tag{4.64}$$

The condition on the second derivative of temperature arises from collocation on the boundary. Functions satisfying these conditions are

$$u = U_1\left[2\left(\frac{y}{\delta}\right) - \left(\frac{y}{\delta}\right)^2\right], \qquad 0 \le y \le \delta, \tag{4.65a}$$

$$\frac{t - t_w}{t_1 - t_w} = \frac{3}{2}\left(\frac{y}{\Delta}\right) - \frac{1}{2}\left(\frac{y}{\Delta}\right)^3, \qquad 0 \le y \le \Delta. \tag{4.65b}$$

The integral equations provide three equations for the variables δ, Δ, and U_1. One complication is that, as with many boundary layer problems, the equations have a singularity near the leading edge. This can be handled by deriving a series solution for small x, and using this solution to move past the entrance after which the integration can proceed. In this problem $U_1 - 1 = ax^{1/2}$. The velocity expression used in the energy equation depends on whether the velocity boundary layer is inside or outside the thermal boundary layer. Once the approximate solution is derived the Nusselt number can be calculated as a function of Reynolds and Prandtl numbers.

The trial function used above is appropriate for boundary conditions of the first kind If the flux is given on the boundary a different trial function is assumed. Siegel and Sparrow (1959) treated the same problem with wall flux specified. The right-hand side of Eq. (4.63b) is then replaced by the given flux, q. The trial function for temperature is deduced from the fully developed profile, which is

$$T - T(0) = \frac{qa}{k}\frac{5}{8}\left[1 - \frac{8}{5}\left(\frac{y}{a}\right) + \frac{4}{5}\left(\frac{y}{a}\right)^3 - \frac{1}{5}\left(\frac{y}{a}\right)^4\right]. \qquad (4.66a)$$

Siegel and Sparrow (1959) take the temperature profile as

$$T - T_e = \frac{q\Delta}{k}\frac{5}{8}\left[1 - \frac{8}{5}\left(\frac{y}{\Delta}\right) + \frac{4}{5}\left(\frac{y}{\Delta}\right)^3 - \frac{1}{5}\left(\frac{y}{\Delta}\right)^4\right], \quad 0 \le y \le \Delta, \qquad (4.66b)$$

$$T - T_e = 0 \qquad \text{for} \quad \Delta \le y \le a.$$

This profile merges smoothly with the fully developed profile as $\Delta \to a$. Kumskov and Sidorov (1969) treat a similar problem but for combined convection and radiation.

The development of the velocity profile was first done using the integral method by Schiller (1922). Later, Campbell and Slattery (1963) showed that more accurate results are achieved if the pressure drop is calculated from the kinetic energy balance including viscous dissipation. The solutions have been extended to non-Newtonian fluids in pipes (Bogue, 1959), but comparisons to more exact results show that the approximations for non-Newtonian fluids are not accurate for low n (Collins and Schowalter, 1963). The analysis of power law fluids in straight channels by Kapur and Gupta (1963, 1964) probably suffers the same deficiency, since the profile assumed for velocity is the same type as assumed for a Newtonian fluid. The results are probably poor because a parabolic profile is assumed, despite the fact that the profile develops into a flat one for small n, Eq. (4.54). Takhar (1968) used the integral method to study the entry-length flow in a vertical cooled pipe when the natural convection effects are important. Chen *et al.* (1970) applied the integral method and Campbell and Slattery's method to the entrance region for a Bingham plastic, non-Newtonian fluid. In this case the velocity profile is more appropriate since it is flat in the central core and parabolic elsewhere.

There is one further note of caution. A common engineering objective is to predict the entry length. Since the velocity and temperature profiles approach their fully developed profile in an asymptotic manner, extremely small errors in velocity or temperature will result in large errors in the entry length. In the integral method the entry length is uniquely defined: when the boundary layer thickness reaches the center of the duct. In the analytic solutions the entry length is defined, somewhat arbitrarily, as the distance it takes to reach 99% of the fully developed maximum value. Sometimes 98% or 95% values are used. Clearly the definition is arbitrary, so that comparisons to approximate solutions can be made as good or as bad as desired simply by changing the criterion. For example, Schiller (1922) obtained $x_e/a \text{ Re} = 0.0575$, whereas the more exact analysis by Collins and Schowalter (1963) gives 0.122 (99%). If a 98% criterion is used, much better agreement is obtained.

Coupled entry-length problems cannot be handled using separation of variables, because the velocity profile depends on both axial and transverse coordinates and the energy equation is not separable. The momentum equation is amenable to separation of variables leading to eigenvalue problems of the type discussed in Section 3.3. Fleming and Sparrow (1969) solve for flow in ducts of arbitrary cross section using least squares-collocation on the boundary. Another approach to entry-length problems is to apply orthogonal collocation. Such an approach is readily suited to the problem with temperature-dependent viscosity and thermal conductivity and the equations are developed in Section 5.1. They can easily be integrated on a computer.

Coupled Heat and Momentum Transport

Iqbal *et al.* (1970) study the effect of buoyancy on forced convection in vertical regular polygonal ducts. The equations are

$$\nabla^2 V + \text{R}\,\phi = -L, \qquad \nabla^2\phi - V = F, \tag{4.67}$$

where F is a known function and L is determined such that the average velocity is one. The Rayleigh number R is the product of the Grashof number and Prandtl numbers. They expand the solution in Bessel functions which satisfy the differential equation and apply collocation (point matching) on the boundary. The boundary conditions are either temperature fixed or flux fixed, and polygons with three, four, five, and six sides are considered.

4.6 Steady-State Flow Problems

The full Navier–Stokes equations have also been solved using MWR, although the methods are not as highly developed as for laminar boundary layer flows. Here we illustrate calculations for the stream function in two cases: one for small Reynolds number, in which case the equations are linear, and one for larger Reynolds number, including the nonlinear terms.

Consider two-dimensional flow in a rectangular cavity bounded on three sides by solid boundaries. The fourth side is a free surface on which a constant shear stress is imposed. The fluid is governed by the Navier–Stokes equation, see Eq. (1.18), under the additional assumption of no time dependence, no body forces, and small Reynolds number, or slow flow, so that the inertial terms can be neglected. The curl of the remaining equation is

$$0 = \nabla^2(\nabla \times \mathbf{u}). \tag{4.68}$$

In two-dimensional flow we can represent the velocity by a stream function, and the incompressibility condition, $\nabla \cdot \mathbf{u} = 0$, is satisfied automatically:

$$u = -\partial\psi/\partial y, \qquad v = \partial\psi/\partial x. \tag{4.69}$$

Employing this representation in Eq. (4.68) gives the biharmonic equation

$$\nabla^4\psi \equiv \nabla^2(\nabla^2\psi) = 0. \tag{4.70}$$

General solutions to this equation are constructed from solutions to Laplace's and Poisson's equation. Any solution to

$$\nabla^2 f = 0 \tag{4.71}$$

also satisfies (4.70). Furthermore (4.70) can be rewritten

$$\nabla^2 f = 0, \qquad f = \nabla^2\psi. \tag{4.72}$$

Thus if f is the general solution to Laplace's equation (4.71), then ψ is the solution to (4.70). In the case of polar coordinates these general solutions can be found using separation of variables:

$$\psi = a_0 + r^2 b_0 + \sum_{n=1}^{\infty} r^n[(a_n + r^2 b_n)\cos n\theta + (c_n + r^2 d_n)\sin n\theta]. \tag{4.73}$$

For Cartesian coordinates the solution to Laplace's equation are the real and imaginary parts of $(x + iy)^n$ and the solutions to the biharmonic equation are the real and imaginary parts of $(x - iy)(x + iy)^n$.

Equation (4.70) is solved subject to the boundary conditions on the solid walls:

$$\begin{aligned} \psi = \psi_y = 0, &\quad \text{at} \quad y = \pm\tfrac{1}{2}, \\ \psi = \psi_x = 0, &\quad \text{at} \quad x = \varepsilon. \end{aligned} \tag{4.74}$$

At the open end of the cavity the shear stress is constant:

$$\psi = 0, \qquad \psi_x = 1, \qquad \text{at} \quad x = 0. \tag{4.75}$$

Only functions with even powers of y need be included because of the symmetry about $y = 0$. Calculations done by Ratkowsky and Rotem (1968) used 30 functions and applied least squares collocation at 60 to 90 boundary collocation points. The solution for various aspect ratios can then be displayed as streamlines in the rectangular cavity.

Flow problems are more difficult when the nonlinear terms must be included. Consider flow past a sphere. Stokes obtained the solution for zero Reynolds number, when the inertial terms are absent. Kawaguti (1955) extended the solution to higher Reynolds number (up to 70) by including the inertial terms. Later Hamielec and Johnson (1962) and Hamielec *et al.* (1963) used the same method for $\mathrm{Re} \leq 500$. The stream function is written as

$$\begin{aligned}\psi = &\left(\frac{1}{2}r^2 + A_1 r + A_2 + \frac{A_3}{r} + \frac{A_4}{r^2}\right)\sin^2\theta \\ &+ \left(B_1 r + B_2 + \frac{B_3}{r} + \frac{B_4}{r^2}\right)\sin^2\theta\cos\theta.\end{aligned} \tag{4.76}$$

The eight constants are determined by applying the four boundary conditions

$$\mathbf{e}_r \cdot \mathbf{u} = \mathbf{e}_\theta \cdot \mathbf{u} = 0, \qquad \text{at} \quad r = 1, \text{ all } \theta. \tag{4.77}$$

Set the differential equation to zero for $r = 1$, all θ (giving two more conditions), and make the differential equation orthogonal to two Legendre polynomials. The solutions are thus mixed MWR, using Legendre weighting functions as well as derived boundary conditions. This is a related version of MWR. It is inappropriate to call the procedure a Galerkin method. These solutions were later compared to finite difference calculations (Hamielec *et al.*, 1967). The gross features were well predicted, but details were not, chiefly due to using only two angular terms in the trial function. In regions behind the sphere the angular variations were large enough to require more angular terms. Flumerfelt and Slattery (1965) use a variational principle for the equations and their "adjoints" (see Section 9.2) to include the inertial terms. Nakano and Tien (1967) apply the same method to flow of a power-law fluid past a fluid sphere, with internal circulation, and later Nakano and Tien (1970) extend the results to include the inertial terms for $5 < \mathrm{Re} < 25$. Snyder and Stewart (1966) used the Galerkin method to solve for the velocity and pressure profiles for Newtonian creeping flow in a packed bed of spheres.

Other Applications

The molecular theory of shock waves has been treated by MWR, and several standard methods, such as that by Mott-Smith (1951) are revealed as different applications of MWR by Lin (1968). These methods are usually either a method of moments or Galerkin method. The least squares method has also been used (Narasimha and Deshpande, 1969) and time-varying weighting functions have been suggested to improve the convergence (Holway, 1967). Radin and Mintzer (1966) expand the distribution function in orthogonal polynomials and apply a Galerkin method (unnamed). They find a solution useful for high Mach numbers, where the Grad solution breaks down.

Application of MWR to stagnation flows is widespread. Jain (1962) uses the collocation method, and Hugelman (1965) uses the integral method. Non-Newtonian fluids are treated by Sharma (1959), Rajeshwari and Rathna (1962), Srivastava (1966) as well as references listed there and Srivastava and Maiti (1966).

In turbulent flow Saffman (1969) interprets the Wiener–Hermite expansion as an approximate method in which certain integrals of the Navier–Stokes equations are satisfied. Orszag (1971) treats laminar flow in the Fourier-transform domain. He finds that a Galerkin solution involving N^p terms, where p is the number of dimensions, is equivalent to a finite difference solution with $(2N)^p$ grid points.

EXERCISES

4.1. Calculate the constant K (Table 4.2) for laminar flow through rectangular ducts using the approximate solution for $N = 1$ [Eq. (4.5a)] and various aspect ratios. The result for $\varepsilon = \kappa = 0$ is 20% high. How appropriate is the trial function, Eq. (4.3), for $\varepsilon = 0$?
Answer: $\varepsilon = 1$, $K = 14.4$; $\varepsilon = 0$, $K = 28.8$.

4.2. For laminar flow through a triangular duct, show why the trial function,

$$u(x, y) = \prod_{i=1}^{3} (a_i + b_i x + c_i y),$$

is appropriate, where the lines $a_i + b_i x + c_i y = 0$, $i = 1, 2, 3$, define the boundaries of the triangle.

4.3. Calculate K for laminar flow through a triangular duct using the first term of the trial function, Eq. (4.14) and boundary collocation. The center of the duct is at $r=0$, and a is the length of the diagonal from the center to a corner.
Answer: Applying collocation at $r/a = \sqrt{7}/4$ gives $K = 10.7$.

4.4. For laminar flow through a duct with an elliptical cross section, deduce a trial function.
Answer: $u(x, y) = ((x^2/a^2) + (y^2/b^2) - 1)P(x, y)$, where P is a polynomial having the symmetry $P(x, y) = P(-x, y) = P(-x, -y)$.

4.5. Consider boundary layer flow past a flat plate. Apply the integral method to Eq. (4.19b) using the trial function for $N = 3$ in Table 4.3. Then apply the collocation method using the same trial function.
Answer: $\alpha = 7.42$.

4.6. Consider entry-length problems when the velocity profile is fully developed. For flow in a flat duct [see Eq. (4.65a)],

$$u = U_1[2(y/a) - (y/a)^2],$$

where $y = 0$ at the wall and $y = a$ at the center. Determine $\Delta(x)$ using the trial function (4.65b) and the integral method. How does this approach differ from that used in Section 3.4 to derive a solution for small z? Answer: For small x, $\Delta^3 = 45\kappa xa/4U_1$.

4.7. Rework Exercise 4.6 for the boundary condition: heat flux specified [see Eqs. (4.67)].

REFERENCES

Abbott, D. E., and Bethel, H. E. (1968). Application of the Galerkin–Kantorovich–Dorodnitsyn Method of Integral Relations to the Solution of Steady Laminar Boundary Layers, *Ing.-Arch.* **37**, 110–124.

Acrivos, A. (1958). Combined Laminar Free- and Forced-Convection Heat Transfer in External Flows, *AIChE J.* **4**, 285–289.

Acrivos, A. (1966). On the Combined Effect of Forced and Free Convection Heat Transfer in Laminar Boundary Layer Flows, *Chem. Eng. Sci.* **21**, 343–352.

Acrivos, A., Shah, M. J., and Petersen, E. E. (1960). Momentum and Heat Transfer in Laminar Boundary-Layer Flows of Non-Newtonian Fluids past External Surfaces, *AIChE J.* **6**, 312–317.

Acrivos, A., Shah, M. J., and Petersen, E. E. (1965). On the Solution of the Two-Dimensional Boundary-Layer Flow Equations for a Non-Newtonian Power-Law Fluid, *Chem. Eng. Sci.* **20**, 101–105.

Belotserkovskii, O. M., and Chushkin, P. I. (1964). The Numerical Method of Integral Relations, NASA TT F-8356, Clearinghouse for Fed. Sci. Tech. Info., Springfield, Virginia.

Belotserkovskii, O. M., and Chushkin, P. I. (1965). The Numerical Solution of Problems in Gas Dynamics, *in* "Basic Developments in Fluid Dynamics" (M. Holt, ed.), pp. 1–126. Academic Press, New York.

Bethel, H. E. (1967). On a Convergent Multi-Moment Method for the Laminar Boundary Layer Equations, *Aeronaut. Q.* **18**, 332–353; (1968) **19**, 402.

Bethel, H. E. (1969). An Improved Reversed-Flow Formulation of the Galerkin–Kantorovich–Dorodnitsyn Multi-Moment Integral Method, *Aeronaut Q.* **20**, 191–202.

Bizzell, G. D., and Slattery, J. C. (1962). Non-Newtonian Boundary Layer Flow, *Chem. Eng. Sci.* **17**, 777–782.

Bogue, D. C. (1959). Entrance Effects and Prediction of Turbulence in Non-Newtonian Flow, *Ind. Eng. Chem.* **51**, 874–878.

Campbell, W. D., and Slattery, J. C. (1963). Flow in the Entrance of a Tube, *J. Basic Eng., Trans. ASME, Ser. D* **85**, 41–46.

Casarella, M. J., Laura, P. A., and Chi, M. (1967). On the Approximate Solution of Flow and Heat Transfer through Non-Circular Conduits with Uniform Wall Temperature, *Brit. J. Appl. Phys.* **18**, 1327–1335.

Chen, S. S., Fan, L. T., and Hwang, C. L. (1970). Entrance Region Flow of the Bingham Fluid in a Circular Pipe, *AIChE J.* **16**, 293–299.

Cheng, K. C. (1967). Dirichlet Problems for Laminar Forced Convection with Heat Sources and Viscous Dissipation in Regular Polygonal Ducts, *AIChE J.* **13**, 1175–1180.

Cheng, K. C. (1969). Laminar Forced Convection in Regular Polygonal Ducts with Uniform Peripheral Heat Flux, *J. Heat Transfer, Trans. ASME, Ser. C* **91**, 156–157.

Collins, M., and Schowalter, W. R. (1963). Behavior of Non-Newtonian Fluids in the Entry Region of a Pipe, *AIChE J.* **9**, 804–809.

Crawford, D. R., and Holt, M. (1968). Method of Integral Relations as Applied to the Problem of Laminar Free Mixing, *AIAA J.* **6**, 372–324.

Dickson, P. R., and Traxler, J. J. (1966). Free Convection on a Vertical Plate with Concentration Gradients, *AIAA J.* **3**, 1511–1512 (Tech. Notes).

Dorodnitsyn, A. A. (1962). General Method of Integral Relations and its Application to Boundary-Layer Theory, *Advan. Aeronaut. Sci.* **3**, 207–219.

Dorodnitsyn, A. A. (1964). Exact Numerical Methods in the Boundary-Layer Theory, *in* "Fluid Dynamic Transactions" (W. Fiszdon, ed.), Vol. 1. Pergamon Press, Oxford.

Fleming, D. P., and Sparrow, E. M. (1969). Flow in the Hydrodynamic Entrance Region of Ducts of Arbitrary Cross Section, *J. Heat Transfer, Trans. ASME, Ser. C* **91**, 345–354.

Flumerfelt, R. W., and Slattery, J. C. (1965). A Widely Applicable Type of Variational Integral-II. Newtonian Flow past a Sphere, *Chem. Eng. Sci.* **20**, 157–163.

Fox, V. G., Erickson, L. E., and Fan, L. T. (1969). The Laminar Boundary Layer on a Moving Continuous Flat Sheet Immersed in a Non-Newtonian Fluid, *AIChE J.* **15**, 327–333.

Hagen, S. L., and Ratkowsky, D. A. (1968). Laminar Flow in Cylindrical Ducts Having Regular Polygonal Shaped Cores, *Can. J. Chem. Eng.* **46**, 387–388.

Hamielec, A. E., and Johnson, A. I. (1962). Viscous Flow around Fluid Sphere at Intermediate Reynolds Numbers, *Can. J. Chem. Eng.* **40**, 41–45.

Hamielec, A. E., Storey, S. H., and Whitehead, S. M. (1963). Viscous Flow around Sphere at Intermediate Reynolds Numbers (II), *Can. J. Chem. Eng.* **41**, 246–251.

Hamielec, A. E., Hoffman, T. J., and Ross, L. L. (1967). Numerical Solution of the Navier–Stokes Equations for Flow past Spheres with Fnite Radial Mass Efflux at the Surface, *AIChE J.* **13**, 212–219.

Heinisch, R. P., Viskanta, R., and Singer, R. M. (1969). Approximate Solution of the Transient Free Convection Laminar Boundary Layer Equations, *Z. Angew. Math. Phys.* **20**, 19–33.

Heywood, J. B., and Moffatt, W. C. (1965). Validity of Integral Methods in MHD Boundary-Layer Analyses, *AIAA J.* **3**, 1565–1567.

Holway, L. H., Jr. (1967). Time-Varying Weight Functions and the Convergence of Polynomial Expansions, *Phys. Fluids* **10**, 35–48.

Hugelman, R. D. (1965). Application of Polhausen's Method to Stagnation-Point Flow, *AIAA J.* **3**, 2158–2159.

Hugelman, R. D., and Haworth, D. R. (1965). An MHD Boundary-Layer Compatibility Condition, *AIAA J.* **3**, 1367–1369.

Iqbal, M., Ansari, S. A., and Aggarawala, B. D. (1970). Effect of Buoyancy on Forced Convection in Vertical Regular Polygonal Ducts, *J. Heat Trans., Trans. ASME, Ser. C* **92**, 237–244.

Jain, M. K. (1962). On Extremal Point Collocation Method for Fluid Flow Problems, *Appl. Sci. Res.* (A) **11**, 177–188.

Kapur, J. N., and Gupta, R. C. (1963). Two-Dimensional Flow of Power Law Fluids in the Inlet Length of a Straight Channel, *Z. Angew. Math. Mech.* **43**, 135–141.

Kapur, J. N., Gupta, R. C. (1964). Two-Dimensional Flow of Power Law Fluids in the Inlet Length of a Straight Channel, *Z. Angew. Math. Mech.* **44**, 277–284.

Kawaguti, M. (1955). The Critical Reynolds Number for the Flow past a Sphere, *J. Phys. Soc. Japan* **10**, 694–699.

Kline, S. J., Sovran, G., Morkovin, M. V., and Cockrell, D. J. (1968). Proceedings. Computation of Turbulent Boundary Layers. AFOSR–IFP–Stanford Conf. Aug. 18–25, Vol. I. Methods, Predictions, Evaluation and Flow Structure.

Kline, S. J., Moffatt, H. K., and Morkovin, M. V. (1969). Report on the AFOSR–IFP–Stanford Conference on Computation of Turbulent Boundary Layers, *J. Fluid Mech.* **36**, 481–484.

Knudsen, J. G., and Katz, D. L. (1958). "Fluid Dynamics and Heat Transfer." McGraw-Hill, New York.

Koob, S. J., and Abbott, D. E. (1968). Investigation of a Method for the General Analysis of Time-Dependent Two-Dimensional Laminar Boundary Layers, *J. Basic Eng., Trans. ASME, Ser. D* **90**, 563–571.

Kuby, W., Foster, R. M., Byron, S. R., and Holt, M. (1967). Symmetrical, Equilibrium Flow past a Blunt Body at Superorbital Reentry Speeds, *AIAA J.* **5**, 610–617.

Kumskov, V. T., and Sidorov, V. S. (1969). Concerning Calculation of Combined Heat Transfer at the Entrance Section of Tubes, *Heat Transfer Sov. Res.* **1**, 5, 136–140.

Lemlich, R. (1963). Natural Convection to Isothermal Flat Plate with a Spatially Non-Uniform Acceleration, *IEC Fund.* **2**, 157–159.

Lemlich, R., and Steinkamp, J. S. (1964). Laminar Natural Convection to an Isothermal Flat Plate with a Spatially Varying Acceleration, *AIChE J.* **10**, 445–447.

Lin, C. J., (1968). Statistical Mechanical Theory of Shock Waves and Suspensions, Ph.D. Thesis, Univ. of Washington, Seattle, Washington.

Mott-Smith, H. M. (1951). The Solution of the Boltzmann Equation for a Shock Wave, *Phys. Rev.* **82**, 885–892.

Nakano, Y., and Tien, C. (1967). Approximate Solutions of Viscous Incompressible Flow around Fluid Spheres at Intermediate Reynolds Numbers, *Can. J. Chem. Eng.* **45**, 135–140.

Nakano, Y., and Tien, C. (1970). Viscous Incompressible Non-Newtonian Flow around Fluid Sphere at Intermediate Reynolds Numbers *AIChE J.* **16**, 569–574.

Nakano, Y., Tien, C., and Gill, W. (1967). Nonlinear Convective Diffusion: A Hyperfiltration Approach, *AIChE J.* **13**, 1092–1098.

Narasimha, R., and Deshpande, S. M. (1969). Minimum-Error Solutions of the Boltzmann Equation for Shock Structure, *J. Fluid Mech.* **36**, 555–570.

Orszag, S. A. (1971). Numerical Simulation of Incompressible Flows within Simple Boundaries: Accuracy, *J. Fluid Mech.* **49**, 75–112.

Ostrach, S., and Thornton, P. R. (1958). On the Stagnation of Natural-Convection Flows in Closed-End Tubes, *Trans. ASME* **80**, 363–366.

Pallone, A. (1961). Nonsimilar Solutions of the Compressible-Laminar Boundary-Layer Equations with Applications to the Upstream-Transpiration Cooling Problem, *J. Aero. Sci.* **28**, 449–456.

Poots, G. (1960). A Solution of the Compressible Laminar Boundary Layer Equations with Heat Transfer and Adverse Pressure Gradient, *Quart. J. Mech. Appl. Math.* **13**, 57–84.

Radin, S. H., and Mintzer, D. (1966). Orthogonal Polynomial Solution of the Boltzmann Equation for a Strong Shock Wave, *Phys. Fluids* **9**, 1621–1633.

Rajeshwari, G. K., and Rathna, S. L. (1962). Flow of a Particular Class of Non-Newtonian Visco-Elastic and Visco-Inelastic Fluids near a Stagnation Point, *Z. Angew. Math. Phys.* **13**, 43–57.

Ramacharyulu, N. C. P. (1967). Slow Steady-State Flow of an Idealized Elasticoviscous Incompressible Fluid through a Straight Tube Whose Cross-Section is an Equilateral Triangle, *Ind. J. Pure Appl. Phys.* **5**, 610–613.

Ratkowsky, D. A., and Epstein, N. (1968). Laminar Flow in Regular Polygonal Shaped Ducts with Circular Centered Cores, *Can. J. Chem. Eng.* **46**, 22–26.

Ratkowsky, D. A., and Rotem, Z. (1968). Viscous Flow in a Rectangular Cut Out. *Phys. Fluids* **11**, 2761–2763.

Riley, N. (1964). Magnetohydrodynamic Free Convection, *J. Fluid Mech.* **18**, 577–586.

Saffman, P. G. (1969). Application of the Wiener–Hermite Expansion to the Diffusion of a Passive Scalar in a Homogeneous Turbulent Flow, *Phys. Fluids* **12**, 1786–1798.

Schechter, R. S. (1967). "The Variational Method in Engineering." McGraw-Hill, New York.

Schiller, L. (1922). Die Entwichlung der laminaren Geschwindigkeitsverteilung und ihre Bedeutung für Zähigkeitmessungen, *Z. Angew. Math. Mech.* **2**, 96–106.

Schlichting, H. (1960). "Boundary Layer Theory." McGraw-Hill, New York.

Schoenauer, W. (1964). A Differencing Method for Solution of the Boundary Layer Equations for Stationary, Laminar, Incompressible Flow (In German), *Ing. Arch.* **33**, 173–189.

Sharma, S. K. (1959). Flow of a Visco-Elastic Liquid near a Stagnation Point, *J. Phys. Soc. Japan* **14**, 1421–1425.

Shih, F. S. (1967). Laminar Flow in Axisymmetric Conduits by a Rational Approach, *Can. J. Chem. Eng.* **45**, 285–294.

Siegel, R., and Sparrow, E. M. (1959). Simultaneous Development of Velocity and Temperature Distributions in a Flat Duct with Uniform Wall Heating, *AIChE J.* **5**, 73–75.

Sinha, P. C. (1963). Forced Convection Flow past a Flat Plate with Variable Viscosity and Thermal Conductivity, *J. Phys. Soc. Japan* **27**, 478–483.

Snyder, L. J., and Stewart, W. E. (1966). Velocity and Pressure Profiles for Newtonian Creeping Flow in a Regular Packed Bed of Spheres, *AIChE J.* **12**, 167–173.

South, J. C. Jr., and Newman, P. A. (1965). Application of the Method of Integral Relations to Real-Gas Flows past Pointed Bodies, *AIAA J.* **3**, 1645–1652.

Sparrow, E. M. (1955). Analysis of Laminar Forced Convection Heat Transfer in Entrance Region of Flat Rectangular Ducts, NACA TN 3331.

Sparrow, E. M. (1962). Laminar Flow in Isosceles Triangular Ducts, *AIChE J.* **8**, 599–604.

Spalding, D. B. (1967). Theories of the Turbulent Boundary Layer, *Appl. Mech. Rev.* **20**, 735–740.

Srivastava, A. C. (1966). Unsteady Flow of a Second-Order Fluid near a Stagnation Point, *J. Fluid Mech.* **24**, 33–39.

Srivastava, A. C., and Maiti, M. K. (1966). Flow of a Second-Order Fluid past a Symmetrical Cylinder, *Phys. Fluids* **9**, 462–465.

Takhar, H. S. (1968). Entry-Length Flow in a Vertical Cooled Pipe, *J. Fluid Mech.* **34**, 641–650.

Villadsen, J. V., and Stewart, W. E. (1967). Solution of Boundary-Value Problems by Orthogonal Collocation, *Chem. Eng. Sci.* **22**, 1483–1501.

Walz, A. (1969). "Boundary Layers of Flow and Temperature," MIT Press, Cambridge, Massachusetts.

Yang, K. T., and Jerger, E. W. (1964). First-Order Perturbations of Laminar Free-Convection Boundary Layers on a Vertical Plate, *J. Heat Transfer, Trans. ASME, Ser. C.* **86**, 107–115.

Chapter

5

Chemical Reaction Systems

Many of the problems in Chapters 1–4 are solved only in a first approximation. A first approximation can be useful in that many qualitative features of the solution are revealed, but more precise answers are often desired. These can be calculated using MWR. In higher approximations the choice of method—or weighting function in MWR—is not crucial, since all methods give similar results. Another criterion can then be introduced, namely, that of convenience in doing the computations. Even the choice of trial functions is not as crucial in higher approximations, provided they satisfy certain criteria, such as a completeness and linear independence. The criterion of convenience can be introduced in the choice of trial functions, too, so that the ease of computation becomes the primary goal. Improved results can be obtained merely at the expense of additional computation without reformulation or additional intervention by the analyst. This goal is achieved below by the orthogonal collocation method, which thereby competes directly with finite difference calculations. Comparisons of computation time in representative applications below indicate that the collocation method is faster than finite difference methods, and this makes MWR of real importance as a computational tool.

The orthogonal collocation method is first described, and applied to a variety of chemical engineering problems. The chapter is concluded with a comparison of collocation methods to methods using spline functions as approximating functions as well as comparison to finite element calculations.

5.1 Orthogonal Collocation

Many of the problems involving chemical reaction have nonlinear terms of the form

$$\exp\{\gamma(T-1)/T\}, \tag{5.1}$$

which arise from the temperature dependence of the chemical reaction rate. Clearly, if the temperature is expanded in a series of functions and MWR is applied, it would be difficult to derive the values of integrals involving terms like (5.1). A quadrature method can be employed to ease this problem, but we use a more direct approach. In the collocation method it is only necessary to evaluate the residual at the collocation points. This is easily done, so that the collocation method appears to be the simplest form of MWR for such problems. In higher approximations the choice of collocation points is not crucial, but a choice is possible which makes the calculations both convenient and accurate. In the orthogonal collocation method the collocation points are taken as the roots to orthogonal polynomials. Such a procedure was first advanced by Lanczos (1938, 1956) and was developed further for the solution of ordinary differential equations using Chebyshev series by Clenshaw and Norton (1963), Norton (1964), and Wright (1964). These applications were primarily for initial-value problems. Horvay and Spiess (1954) used polynomials which were orthogonal on the boundary. Villadsen and Stewart (1967) made a major advance when they developed orthogonal collocation for boundary-value problems. They chose the trial functions to be sets of orthogonal polynomials which satisfied the boundary conditions and the roots to the polynomials gave the collocation points. Thus the choice of collocation points is no longer arbitrary and the low-order collocation results are more dependable. A major simplification is that the solution can be derived not in terms of the coefficients in the trial function but in terms of the value of the solution at the collocation points. The whole problem is then reduced to a set of matrix equations which are easily generated and solved on the computer. Accurate quadrature formulas are provided, which is especially important if the primary information desired from the solution is some integrated property, and the polynomials are easily generalized to planar, cylindrical, or spherical geometries as well as to a wide variety of boundary conditions. The book by Villadsen (1970) gives a comprehensive discussion of the method and illustrations of applications to chemical engineering problems.

Consider the solution expressed in the form

$$y(x) = y_0(x) + \sum_{i=1}^{N} a_i y_i(x), \tag{5.2}$$

where the y_i are known functions of position. Usually the solution is expressed by listing values of the coefficients a_i. Let us, however, pick a set of N points $\{x_j\}$ and evaluate (5.2):

$$y(x_j) = y_0(x_j) + \sum_{i=1}^{N} a_i y_i(x_j), \qquad j = 1, \ldots, N. \tag{5.3}$$

Knowing the coefficients a_i we can calculate $y(x_j)$. Knowing $y(x_j)$, $j = 1, \ldots, N$ we can also find the coefficients a_i by inverting the matrix equation (5.3):

$$a_i = [y_i(x_j)]^{-1}[y(x_j) - y_0(x_j)]. \tag{5.4}$$

Such an inverse exists provided the trial functions are linearly independent and the x_j are distinct: i.e., that det $y_i(x_j) \neq 0$. We assume that this is true. Equation (5.4) says that if we know the solution at N points we can solve for the coefficients a_i and hence find the solution at any other point x. Furthermore, methods of solution to find the coefficients can be reformulated to find the solution in terms of $y(x_j)$, i.e., the solution at the set of chosen points (Vichnevetsky, 1969). This approach is particularly convenient for the collocation method.

In previous chapters one common form of trial function is

$$y_i = x^{i-1}. \tag{5.5}$$

These polynomials can be arranged in systems of polynomials which have convenient orthogonality properties. Define the polynomial $P_m(x)$ as a linear combination of powers of x with the highest power being m:

$$P_m(x) = \sum_{j=0}^{m} c_j x^j. \tag{5.6}$$

Define the coefficients by requiring the successive polynomials to be orthogonal to all polynomials of order less than m, with some weighting function $w(x) \geq 0$:

$$\int_a^b w(x)P_n(x)P_m(x)\,dx = 0, \qquad n = 0, 1, \ldots, m-1. \tag{5.7}$$

This specifies the polynomials to within a multiplicative constant, which we determine by requiring the first coefficient to be 1. We illustrate the ideas by choosing $w(x) = 1$, $a = 1$, $b = -1$. The first polynomial is $P_0(x) = 1$. The second one, $P_1(x) = a + bx$, is determined from (5.7):

$$\int_{-1}^{1} (a + bx)1 \, dx = 0. \tag{5.8}$$

After performing the integration we find $a = 0$, and b is arbitrary; we choose $b = 1$. Then $P_1 = x$. Take

$$P_2(x) = a + bx + cx^2 \tag{5.9}$$

and determine the constants from Eq. (5.7), making P_2 orthogonal to both P_0 and P_1. The polynomials of the resulting set are called Legendre polynomials and are summarized in Table 5.1. The polynomial $P_m(x)$ has m roots

TABLE 5.1

LEGENDRE POLYNOMIALS AND ROOTS

Legendre polynomials	Roots
$P_0 = 1$	
$P_1 = x$	$x_j = 0$
$P_2 = 1 - 3x^2$	$x_j = \pm 1/\sqrt{3} = \pm 0.577$

in the interval $a \le x \le b$ which can serve as collocation points. Note that the system of polynomials is completely specified by the orthogonality condition Eq. (5.7), once we have made choices of the weighting function $w(x)$ and the interval of integration.

Next, construct polynomials which have additional convenient properties. Consider problems in which the solution is sought on the domain $0 \le x \le 1$ and is required to be symmetric about $x = 0$. Then it can be expanded in terms of powers of x^2. Also consider a boundary condition which is of the first kind—that is, the value of the solution on the boundary is specified. More general boundary conditions are treated below. A possible choice of trial function is

$$y(x) = y(1) + (1 - x^2) \sum_{i=1}^{N} a_i P_{i-1}(x^2). \tag{5.10}$$

Throughout this chapter N is the number of interior collocation points. The orthogonal polynomials in (5.10) are constructed using an orthogonality condition like (5.7). Thus define the polynomials by the condition

$$\int_0^1 w(x^2) P_j(x^2) P_i(x^2) x^{a-1} \, dx = C_{(i)} \delta_{ij}, \qquad j = 1, 2, \ldots, i - 1, \tag{5.11}$$

where $a = 1, 2, 3$, for planar, cylindrical, or spherical geometry. The first coefficient of each polynomial is taken as one. This completely defines the trial function in Eq. (5.10). Substitute (5.10) into the differential equation to form the residual, which is set to zero at the N collocation points x_j, which are the roots to the Nth polynomial, $P_N(x^2) = 0$ at x_j. This provides N equations to solve for the N coefficients a_i.

The computer programs are simpler, however, if they are written in terms of the solution at the collocation points, $y(x_j)$, rather than a_i. Since $P_{N-1}(x^2)$ is a polynomial of degree $N-1$ in x^2, the trial function (5.10) is a polynomial of degree N in x^2. Rewrite (5.10) as

$$y(x) = \sum_{i=1}^{N+1} d_i x^{2i-2}. \tag{5.12}$$

Take the first derivative and the Laplacian of this expression and evaluate them at the collocation points:

$$\begin{aligned} y(x_j) &= \sum_{i=1}^{N+1} x_j^{2i-2} d_i, \\ \left.\frac{dy}{dx}\right|_{x_j} &= \sum_{i=1}^{N+1} \left.\frac{dx^{2i-2}}{dx}\right|_{x_j} d_i, \\ \left.\nabla^2 y\right|_{x_j} &= \sum_{i=1}^{N+1} \left.\nabla^2(x^{2i-2})\right|_{x_j} d_i. \end{aligned} \tag{5.13}$$

These can be rewritten in matrix notation as follows. Note that the $N+1$ collocation point is $x = 1$ and square matrices have $(N+1) \times (N+1)$ elements:

$$\begin{gathered} \bar{y} = \bar{Q}\bar{d}, \qquad \overline{\frac{dy}{dx}} = \bar{C}\bar{d}, \qquad \overline{\nabla^2 y} = \bar{D}\bar{d}, \\ Q_{ji} = x_j^{2i-2}, \qquad C_{ji} = \left.\frac{dx^{2i-2}}{dx}\right|_{x_j}, \qquad D_{ji} = \left.\nabla^2(x^{2i-2})\right|_{x_j}. \end{gathered} \tag{5.14}$$

Solving for $\bar{d}$ we can rewrite the first derivative and Laplacian as

$$\overline{\frac{dy}{dx}} = \bar{C}\bar{Q}^{-1}\bar{y} \equiv \bar{A}\bar{y}, \qquad \nabla^2 y = \bar{D}\bar{Q}^{-1}\bar{y} \equiv \bar{B}\bar{y}. \tag{5.15}$$

Thus the derivatives are expressed in terms of the value of the function at the collocation points.

To evaluate integrals accurately we use the quadrature formula:

$$\int_0^1 f(x^2)x^{a-1}\,dx = \sum_{j=1}^{N+1} W_j f(x_j). \tag{5.16}$$

To determine W_i, evaluate (5.16) for $f_i = x^{2i-2}$:

$$\int_0^1 x^{2i-2} x^{a-1}\, dx = \sum_{j=1}^{N+1} W_j x_j^{2i-2} = \frac{1}{2i-2+a} \equiv f_i,$$
$$\bar{W}\bar{Q} = \bar{f}, \qquad \bar{W} = \bar{f}\bar{Q}^{-1}. \tag{5.17}$$

The information presented by Kopal (1955, p. 390) can be used to show that this integration is exact for functions f which are polynomials of degree $2N$ in x^2, provided the interior collocation points are the roots to $P_N(x^2)$ defined by $w = 1 - x^2$ in (5.11).

Let us apply this procedure to a specific problem:

$$\frac{1}{x^{a-1}} \frac{d}{dx}\left(x^{a-1} \frac{dy}{dx}\right) + g(x^2)y = f(x^2),$$
$$\frac{dy}{dx} = 0 \quad \text{at} \quad x = 0, \qquad y(1) = y_1. \tag{5.18}$$

The residuals at the collocation points are simply

$$\sum_{i=1}^{N+1} B_{ji} y_i + g(x_j^2) y_j = f(x_j^2), \qquad y_{N+1} = y_1. \tag{5.19}$$

If the boundary condition is

$$-\frac{dy}{dx} = ay + b \quad \text{at} \quad x = 1, \tag{5.20}$$

we have the additional equation to determine y_{N+1}:

$$-\sum_{i=1}^{N+1} A_{N+1,\,i} y_i = a y_{N+1} + b. \tag{5.21}$$

If we need the solution for some point other than a collocation point it is given by (5.12) where

$$\bar{d} = \bar{Q}^{-1}\bar{y}. \tag{5.22}$$

In particular the value $x = 0$ is important in chemical reactor calculations:

$$y(0) = d_1 = \sum_{i=1}^{N+1} [Q^{-1}]_{1i}\, y(x_i). \tag{5.23}$$

The solution of these equations is a simple matter once the matrix B is calculated. Examination of Eqs. (5.13)–(5.15) reveals that to calculate B we need know only the collocation points, x_i. These are roots of polynomials, which can be deduced from the tables given by Stroud and Secrest (1966), who give the roots to thirty significant figures. In addition they are listed in Villadsen (1970), with a general program for their calculation. Important values are listed in Table 5.2, while the matrices for $N = 1, 2$ are in Table 5.3.

TABLE 5.2

ROOTS OF POLYNOMIALS DEFINED BY EQ. (5.11)[a]

	$a = 1$, planar geometry		$a = 2$, cylindrical geometry		$a = 3$, spherical geometry	
	$w = 1$	$w = 1 - x^2$	$w = 1$	$w = 1 - w^2$	$w = 1$	$w = 1 - w^2$
$N = 1$	0.57735 02692	0.44721 35955	0.70710 67812	0.57735 02692	0.77459 66692	0.65465 36707
$N = 2$	0.33998 10436	0.28523 15165	0.45970 08434	0.39376 51911	0.53846 93101	0.46884 87935
	0.86113 63116	0.76505 53239	0.88807 38340	0.80308 71524	0.90617 98459	0.83022 38963
$N = 3$	0.23861 91861	0.20929 92179	0.33571 06870	0.29763 72952	0.40584 51514	0.36311 74638
	0.66120 93865	0.59170 01814	0.70710 67812	0.63989 59794	0.74153 11856	0.67718 62795
	0.93246 95142	0.87174 01485	0.94196 51451	0.88750 18095	0.94910 79123	0.89975 79954
$N = 4$	0.18343 46425	0.16527 89577	0.26349 92300	0.23896 48430	0.32425 34234	0.29575 81356
	0.52553 24099	0.47792 49498	0.57446 45143	0.52615 87342	0.61337 14327	0.56523 53270
	0.79666 64774	0.73877 38651	0.81852 94874	0.76393 09081	0.83603 11073	0.78448 34737
	0.96028 98565	0.91953 39082	0.96465 96062	0.92749 13130	0.96816 02395	0.93400 14304
$N = 5$	0.14887 43390	0.13655 29329	0.21658 73427	0.19952 40765	0.26954 31560	0.24928 69301
	0.43339 53941	0.39953 09410	0.48038 04169	0.44498 69862	0.51909 61292	0.48290 98211
	0.67940 95683	0.63287 61530	0.70710 67812	0.66179 66532	0.73015 20056	0.68618 84691
	0.86506 33667	0.81927 93216	0.87706 02346	0.83394 50062	0.88706 25998	0.84634 75647
	0.97390 65285	0.94489 92722	0.97626 32447	0.94945 50617	0.97822 86581	0.95330 98466
$N = 6$	0.12523 34085	0.11633 18689	0.18375 32119	0.17122 04053	0.23045 83160	0.21535 39554
	0.36783 14990	0.34272 40133	0.41157 66111	0.38480 98228	0.44849 27510	0.42063 80547
	0.58731 79543	0.55063 94029	0.61700 11402	0.58050 38245	0.64234 93394	0.60625 32055
	0.76990 26742	0.72886 85991	0.78696 22564	0.74744 33215	0.80157 80907	0.76351 96900
	0.90411 72564	0.86780 10538	0.91137 51660	0.87705 97825	0.91759 83992	0.88508 20442
	0.98156 06342	0.95993 50453	0.98297 24091	0.96278 01781	0.98418 30547	0.96524 59265

[a] For a given N the collocation points $x_1, \ldots, x_N$ are listed above; $x_{N+1} = 1.0$.

TABLE 5.3
MATRICES FOR ORTHOGONAL COLLOCATION FOR POLYNOMIALS IN TABLE 5.2 WITH $w = 1 - x^2$

Planar geometry ($a = 1$)

$N = 1$ $\quad W = \begin{pmatrix} 0.8333 \\ 0.1667 \end{pmatrix}$ $\quad A = \begin{pmatrix} -1.118 & 1.118 \\ -2.500 & 2.500 \end{pmatrix}$ $\quad B = \begin{pmatrix} -2.5 & 2.5 \\ -2.5 & 2.5 \end{pmatrix}$

$N = 2$ $\quad W = \begin{pmatrix} 0.5549 \\ 0.3785 \\ 0.0667 \end{pmatrix}$ $\quad A = \begin{pmatrix} -1.753 & 2.508 & -0.7547 \\ -1.371 & -0.6535 & 2.024 \\ 1.792 & -8.791 & 7 \end{pmatrix}$

$B = \begin{pmatrix} -4.740 & 5.677 & -0.9373 \\ 8.323 & -23.26 & 14.94 \\ 19.07 & -47.07 & 28 \end{pmatrix}$

Cylindrical geometry ($a = 2$)

$N = 1$ $\quad W = \begin{pmatrix} 0.375 \\ 0.125 \end{pmatrix}$ $\quad A = \begin{pmatrix} -1.732 & 1.732 \\ -3 & 3 \end{pmatrix}$ $\quad B = \begin{pmatrix} -6 & 6 \\ -6 & 6 \end{pmatrix}$

$N = 2$ $\quad W = \begin{pmatrix} 0.1882 \\ 0.2562 \\ 0.0556 \end{pmatrix}$ $\quad A = \begin{pmatrix} -2.540 & 3.826 & -1.286 \\ -1.378 & -1.245 & 2.623 \\ 1.715 & -9.715 & 8 \end{pmatrix}$

$B = \begin{pmatrix} -9.902 & 12.30 & -2.397 \\ 9.034 & -32.76 & 23.73 \\ 22.76 & -65.42 & 42.67 \end{pmatrix}$

Spherical geometry ($a = 3$)

$N = 1$ $\quad W = \begin{pmatrix} 0.2333 \\ 0.1 \end{pmatrix}$ $\quad A = \begin{pmatrix} -2.291 & 2.291 \\ -3.5 & 3.5 \end{pmatrix}$ $\quad B = \begin{pmatrix} -10.5 & 10.5 \\ -10.5 & 10.5 \end{pmatrix}$

$N = 2$ $\quad W = \begin{pmatrix} 0.0949 \\ 0.1908 \\ 0.0476 \end{pmatrix}$ $\quad A = \begin{pmatrix} -3.199 & 5.015 & -1.816 \\ -1.409 & -1.807 & 3.215 \\ 1.697 & -10.70 & 9 \end{pmatrix}$

$B = \begin{pmatrix} -15.67 & 20.03 & -4.365 \\ 9.965 & -44.33 & 34.36 \\ 26.93 & -86.93 & 60 \end{pmatrix}$

Table 5.3 can be used to solve for low-order approximations, and the roots listed in Table 5.2 can be used to calculate the matrices for higher-order approximations.

The procedure for using these roots is very simple.

> Read N, and $X(J)$ roots.
> Calculate Q, C, D, F matrices using (5.14) and (5.17).
> Invert Q. (5.24)
> Calculate A, B, W, matrices using (5.15) and (5.17).
> Write or punch the results.

The reason for including polynomials with both $w = 1$ and $w = 1 - x^2$ is that low-order approximations are best using $w = 1 - x^2$, whereas for many chemical engineering problems faster convergence is obtained with $w = 1$.

To illustrate the simplicity of the method we apply it to a problem treated earlier. Consider the problem (4.2). The problem is symmetric about $x = 0$ and $y = 0$ so that it is permissible to expand the trial function in polynomials in x^2 and y^2, defined by Eq. (5.11). We choose the polynomials with $w = 1 - x^2$, since this choice corresponds to a variational method (giving lower bounds on the flow rate) (see Sections 4.1, 5.6, and 7.3). The roots and matrices come from Tables 5.2 and 5.3, with $a = 1$, and $w = 1 - x^2$. The trial function is an extension of Eq. (5.10) to two dimensions, with $u(x, 1) = u(1, y) = 0$:

$$u(x, y) = (1 - x^2)(1 - y^2) \sum_{i,j=1}^{N} a_{ij} P_{i-1}(x^2) P_{j-1}(y^2). \tag{5.25}$$

Using the matrices B to represent the second derivative, the collocation equations are

$$\sum_{i=1}^{N+1} B_{ji} u_{ik} + \varepsilon^2 \sum_{i=1}^{N+1} B_{ki} u_{ji} = -1, \tag{5.26}$$

where $u_{ij} = u(x_i, y_j)$. Of course, $u_{N+1,j} = u_{i,N+1} = 0$ since those are boundary points. If the region is square, $\varepsilon = 1$, the solution is symmetric about $x = y$ and the points u_{ij} with $i < j$ can be omitted since they are satisfied automatically by symmetry when the other equations are solved, along with $a_{ij} = a_{ji}$. It is a simple matter to find the solution to Eq. (5.26) using a computer.

Consider another problem: an entry-length problem in which the thermal conductivity depends on temperature:

$$\begin{gathered} c(T)u(r)\frac{\partial T}{\partial z} = \frac{1}{r}\frac{\partial}{\partial r}\left(k(T)r\frac{\partial T}{\partial r}\right), \\ T = T_0(r) \quad \text{at} \quad z = 0, \\ \frac{\partial T}{\partial r} = 0 \quad \text{at} \quad r = 0, \\ T = T_1 \quad \text{or} \quad -k(T)\frac{\partial T}{\partial r} = q \quad \text{at} \quad r = 1. \end{gathered} \tag{5.27}$$

Orthogonal collocation can be applied directly to Eqs. (5.27) to obtain the working equations

$$c(T_j)u(r_j)\frac{dT_j}{dz} = k(T_j)\sum_{i=1}^{N+1} B_{ji}T_i + \frac{dk}{dT}\bigg|_{T_j}\left(\sum_{i=1}^{N+1} A_{ji}T_i\right)^2,$$

$$T_j(0) = T_0(r_j), \tag{5.28}$$

$$T_{N+1} = T_1 \qquad \text{or} \qquad -k(T_{N+1})\sum_{i=1}^{N+1} A_{N+1,i}T_i = q.$$

These equations can be integrated numerically on a computer. Near the inlet, of course, it is necessary to take N very large to represent the step function change in the boundary conditions, just as in the case of the eigenfunction expansion for the linear problem with $k =$ constant. Using orthogonal collocation to go from the partial differential equation, Eqs. (5.27), to the set of ordinary differential equations, Eqs. (5.28), is straightforward. In this case the matrices A and B correspond to cylindrical geometry, $a = 2$, and either $w = 1$ or $w = 1 - x^2$ in Tables 5.2 and 5.3.

Next we construct additional polynomials for second-order equations which are orthogonal on the interval 0 to 1 but have no special symmetry properties, so that both even and odd powers of x are included. We choose a trial function of the form

$$y(x) = b + cx + x(1-x)\sum_{i=1}^{N} a_i P_{i-1}(x), \tag{5.29}$$

where the polynomials are defined by Eq. (5.7) with $a = 0$, $b = 1$. Note that this gives $N + 2$ constants: N conditions are provided by the residuals evaluated at the N collocation points, the N roots to $P_N(x) = 0$, and two conditions are provided by boundary conditions at $x = 0, 1$. The expressions for the matrices representing the first and second derivatives are given by the same formulas [Eqs. (5.15), (5.17)] but with different definitions for the matrices:

$$Q_{ji} = x_j^{i-1}, \qquad C_{ji} = (i-1)x_j^{i-2},$$

$$D_{ji} = (i-1)(i-2)x_j^{i-3}, \qquad f_i = \int_0^1 x^{i-1}\,dx = i^{-1}. \tag{5.30}$$

A column in the matrix is now $N + 2$ elements long. The integration of (5.16) is exact if $f(x)$ is a polynomial of degree $2N + 1$ in x. The polynomials satisfying (5.7) with $w = 1$ are shifted Legendre polynomials, and the roots and matrices are tabulated in Tables 5.4 and 5.5. These polynomials are useful

TABLE 5.4

ROOTS OF POLYNOMIALS DEFINED BY EQ. (5.7)[a]

$N=1$	0.50000 00000	$N=5$	0.04691 00771
			0.23076 53450
$N=2$	0.21132 48654		0.50000 00000
	0.78867 51346		0.76923 46551
			0.95308 99230
$N=3$	0.11270 16654		
	0.50000 00000	$N=6$	0.03376 52429
	0.88729 83346		0.16939 53068
$N=4$	0.06943 18442		0.38069 04070
	0.33000 94783		0.61930 95931
	0.66999 05218		0.83060 46933
	0.93056 81558		0.96623 47571

[a] In Eq. (5.7), $w=1, a=0, b=1$. For a given N the collocation points $x_2, \ldots, x_{N+1}$ are listed above; $x_1=0$, $x_{N+2}=1.0$.

TABLE 5.5

MATRICES FOR ORTHOGONAL COLLOCATION FOR POLYNOMIALS WITH ROOTS IN TABLE 5.4

$N=1$

$$W=\begin{pmatrix}\frac{1}{6}\\ \frac{2}{3}\\ \frac{1}{6}\end{pmatrix} \qquad A=\begin{pmatrix}-3 & 4 & -1\\ -1 & 0 & 1\\ 1 & -4 & 3\end{pmatrix} \qquad B=\begin{pmatrix}4 & -8 & 4\\ 4 & -8 & 4\\ 4 & -8 & 4\end{pmatrix}$$

$N=2$

$$W=\begin{pmatrix}0\\ \frac{1}{2}\\ \frac{1}{2}\\ 0\end{pmatrix} \qquad A=\begin{pmatrix}-7 & 8.196 & -2.196 & -1\\ -2.732 & 1.732 & 1.732 & -0.7321\\ 0.7321 & -1.732 & -1.732 & 2.732\\ -1 & 2.196 & -8.196 & 7\end{pmatrix}$$

$$B=\begin{pmatrix}24 & -37.18 & 25.18 & -12\\ 16.39 & -24 & 12 & -4.392\\ -4.392 & 12 & -24 & 16.39\\ -12 & 25.18 & -37.18 & 24\end{pmatrix}$$

for studying chemical reactors with axial diffusion (see Section 5.4). Note that if the polynomial (5.29) had been taken as $y(x)=\sum_{i=1}^{N} a_i P_{i-1}(x)$ we would have N adjustable constants. These could be chosen to satisfy 2 boundary conditions and only $N-2$ residuals. In this case we would either have to ignore 2 of the roots to P_N or else use the roots to P_{N-2}. By using the trial function in the form of Eq. (5.29) we avoid these problems.

Consider next the application of orthogonal collocation to an initial-value problem as developed by Villadsen and Sørensen (1969):

$$\frac{dy}{dx} = g(y, x), \qquad y(0) = y_0. \tag{5.31}$$

Since we have only one initial condition to satisfy we use a trial function in the form

$$y(x) = y_0 + x \sum_{i=1}^{N+1} a_i P_{i-1}(x). \tag{5.32}$$

Then y is a polynomial of degree $N + 1$ in x. The matrix operators are again given by Eqs. (5.15) and (5.17) with the definitions

$$Q_{ji} = x_j^{i-1}, \qquad C_{ji} = (i-1)x_j^{i-2}, \qquad f_i = i^{-1}. \tag{5.33}$$

Now the matrices have $N + 2$ elements in a column. The solution to (5.31) is written in the form

$$\sum_{i=1}^{N+2} A_{ji} y_i = g(y_j, x_j), \qquad j = 2, \ldots, N+2, \tag{5.34}$$

and y_1 is known from the initial conditions. This gives $N + 1$ equations to solve for y_2 through y_{N+2}. If the function g is nonlinear then (5.34) provides a set of nonlinear algebraic equations to solve. The roots of the polynomials are given in Table 5.4 and the matrix A is in Table 5.5. Applications of this technique are useful not so much for solving (5.31) on the interval from 0 to 1, but rather on the interval from 0 to Δx. Then the equations (5.34) are replaced by

$$\frac{1}{\Delta x} \sum_{i=1}^{N+2} A_{ji} y_i = g(y_j, x_j). \tag{5.35}$$

Equation (5.35) is used to predict $y(\Delta x)$, which is then used as the initial value for the next interval of integration. This technique is applied in Section 5.5 to systems of nonlinear equations like (5.31).

5.2 Unsteady Diffusion

An unsteady diffusion problem is solved using orthogonal collocation to illustrate the accuracy as a function of N. We also present a method of incorporating an asymptotic solution, valid for small time, into the approximate expansion in such a way that a good solution can be found for all time using $N = 1$.

Consider the unsteady diffusion in a sphere:

$$\frac{\partial c}{\partial t} = \frac{1}{x^2}\frac{\partial}{\partial x}\left(x^2 \frac{\partial c}{\partial x}\right),$$
$$c = 0 \quad \text{at} \quad t = 0, \qquad c = 1 \quad \text{at} \quad x = 1, t > 0, \tag{5.36}$$
$$\frac{\partial c}{\partial x} = 0 \quad \text{at} \quad x = 0, t > 0.$$

Villadsen and Stewart (1967) illustrate the application of orthogonal collocation to this problem. Following the procedure outlined in Section 5.1, we replace the Laplacian operator by a matrix operator, giving a set of ordinary differential equations

$$\frac{dc_j}{dt} = \sum_{i=1}^{N+1} B_{ji} c_i, \qquad c_j(0) = 0, \quad c_{N+1} = 1 \qquad (j = 1, \ldots, N). \tag{5.37}$$

The matrix B corresponds to $a = 3$, $w = 1$ or $1 - x^2$ in Tables 5.2 and 5.3. This equation can be solved explicitly (Amundson, 1966, p. 123). Let the $N \times N$ matrix $\bar{\bar{B}}$ be composed of the coefficients of the matrix B_{ij} with $i, j \le N$. The $N \times 1$ matrix $\bar{f}$ is defined as a matrix with elements from $B_{i,N+1}$, $i = 1, \ldots, N$. Then (5.37) is written as

$$\frac{d\bar{c}}{dt} = \bar{\bar{B}}\bar{c} + \bar{f} \tag{5.38}$$

and has the solution

$$\bar{c} = \sum_{r=1}^{N} \frac{\bar{w}_r^T \bar{\bar{B}}^{-1} \bar{f}}{\bar{w}_r^T \bar{Z}_r} \bar{Z}_r \exp(\lambda_r t) - \bar{\bar{B}}^{-1}\bar{f}, \tag{5.39}$$

where the λ_r are the eigenvalues of B, $\bar{Z}_r$ is an eigenvector satisfying $\bar{\bar{B}}\bar{Z}_r = \lambda_r \bar{Z}_r$, and $\bar{w}_r^T$ is a biorthogonal set of eigenrows of $\bar{\bar{B}}$ and $\bar{w}_r^T \bar{Z}_s = 0$, $r \ne s$. The equations (5.37) can also be integrated numerically. Using the polynomials with $w = 1$ and the Runge–Kutta method of integration, the solution for the flux at the surface of the sphere is shown in Fig. 5.1 and Table 5.6. For small times it is necessary to use a large number of terms to get good accuracy. Ferguson and Finlayson (1970) have shown that for diffusion in a slab the concentration at the collocation points is accurate to six significant figures for $t \ge 0.1$ and $N = 6$.

To find a good solution at small times using fewer coefficients we can make use of an idea used by Professor Abbott and his students (Koob and Abbott, 1968). This approach uses as the first term of the approximate solution an

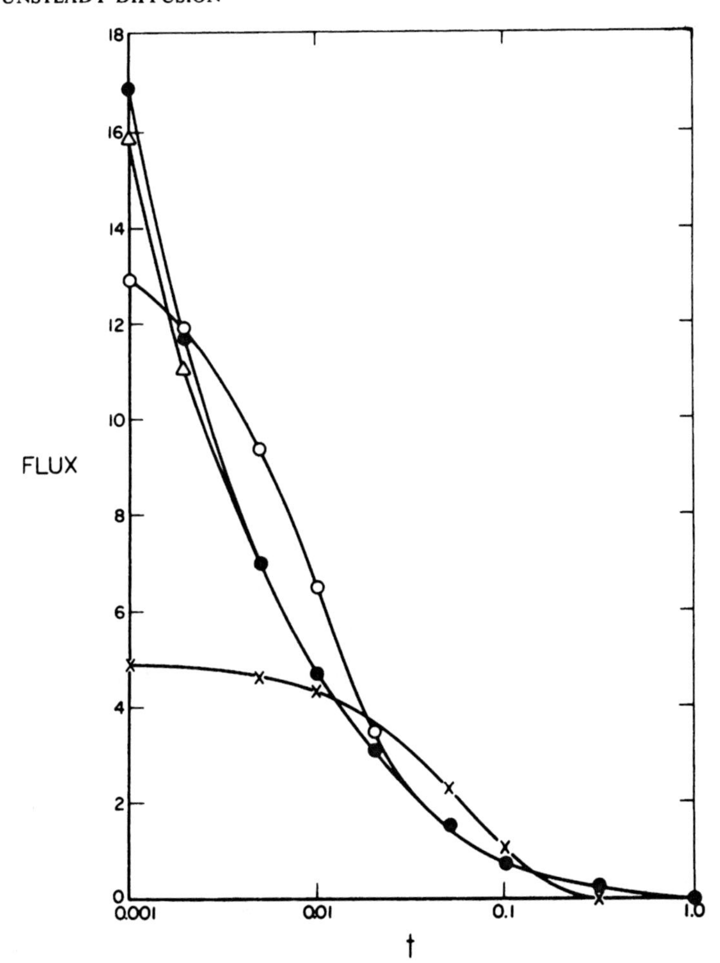

Fig. 5.1. Mass flux as a function of time (● exact and collocation–asymptotic, $N = 1, 2, \ldots$; collocation: × $N = 1$; ○ $N = 2$; △ $N = 6$).

TABLE 5.6

FLUX IN DIFFUSION PROBLEM

	Collocation				Asymptotic–collocation		
t	$N = 1$	$N = 2$	$N = 3$	$N = 6$	$N = 1$	$N = 2$	Exact
0.001	4.93	12.90	20.64	15.76	16.841	16.841	16.841
0.005	4.64	9.39	8.40	7.06	6.9788	6.9788	6.9788
0.01	4.30	6.48	4.38	4.64	4.6420	4.6419	4.6420
0.05	2.36	1.36			1.5231	1.5231	1.5231
0.10	1.11	0.77			0.7842	0.7843	0.7843

exact asymptotic solution for small time. We add to this a series of polynomials. For small time this part of the solution is zero, but as time proceeds it becomes more and more important. The Laplace transform solution converges rapidly for small time and is (Crank, 1956, p. 86)

$$c(x, t) = (1/x) \sum_{n=0}^{\infty} \{\mathrm{erfc}[(2n + 1 - x)/2t^{1/2}] - \mathrm{erfc}[(2n + 1 + x)/2t^{1/2}]\}. \quad (5.40)$$

We use as the first term in the approximate solution the first term of this infinite series, thereby taking

$$\begin{aligned} c(x, t) &= \theta(x, t) + u(1, t) + (1 - x^2) \sum_{i=1}^{N} a_i P_{i-1}(x^2), \\ \theta(x, t) &= (1/x)\{\mathrm{erfc}[(1 - x)/2t^{1/2}] - \mathrm{erfc}[(1 + x)/2t^{1/2}]\}. \end{aligned} \quad (5.41)$$

Notice that θ satisfies the differential equation, the initial condition, and the boundary condition at $x = 0$. It does not satisfy the condition $\theta(1, t) = 1$. Consequently we take

$$c(1, t) = 1 = \theta(1, t) + u(1, t). \quad (5.42)$$

Apply orthogonal collocation and the problem can be rearranged to give the equations for $u = c - \theta$ at the collocation points:

$$\frac{du_j}{dt} = \sum_{i=1}^{N+1} B_{ji} u_i, \quad u_j(0) = 0, \quad u_{N+1} = \mathrm{erfc}(t^{-1/2}) \quad (j = 1, \ldots, N). \quad (5.43)$$

These equations were integrated numerically, too, and for $N = 1$ the flux is correct to four significant figures for $0 \le t \le 0.1$. Thus for small times it is advantageous to use a trial function of the form (5.41). This technique is particularly useful when the diffusion problem must be solved in conjunction with other equations. An expansion function including an asymptotic term (5.41) can be used in nonlinear and entry-length problems, such as those discussed in Section 3.4. In these cases it is not necessary that the asymptotic term be an exact asymptotic solution, since the trial function will correct for any errors. The asymptotic term, however, makes the necessary correction smaller than if it were omitted.

5.3 Reaction and Diffusion in a Catalyst Particle

A catalyst particle is made from a porous material, such as alumina, impregnated with a catalytic material, such as platinum. When placed in a gas stream, the reactants diffuse into the particle, react on the active surface, and the products diffuse out. Since there is a heat of reaction, the temperature of the particle changes as well. The mathematical model of this process is highly nonlinear, due to the exponential nonlinearity of the reaction rate constant of Eq. (5.1). This leads to multiple steady-state solutions—that is the problem is not unique—and gives rise to interesting stability problems. The collocation method and Galerkin method have proved to be very useful tools for analyzing this problem. We first consider an isothermal case, then a nonisothermal case in steady state, and finally the time-dependent nonisothermal case. While we usually assume a spherical catalyst particle and a first-order irreversible reaction, the methods are not limited to these cases.

Isothermal Reaction

Consider a second-order, irreversible reaction in a catalyst slab with both sides exposed to reactant of concentration c_0. The reaction rate is

$$A \to B, \qquad R = -kc_A{}^2. \tag{5.44}$$

The dimensionless problem is a form of Eq. (1.10),

$$\begin{aligned} \nabla^2 y = \phi^2 y^2, &\qquad \phi^2 = kc_0 R^2/\mathscr{D}, \\ y'(0) = 0, &\qquad y(1) = 1, \end{aligned} \tag{5.45}$$

where $x_s = R$, the half-width, and $\mathscr{D}$ is the diffusivity. The orthogonal collocation method is applied using $N = 1$ and the polynomials from Tables 5.2 and 5.3 with $w = 1 - x^2$, $a = 1$:

$$-2.5y_1 + 2.5y_2 = \phi^2 y_1{}^2, \qquad y_2 = 1. \tag{5.46}$$

The equations can be rearranged to solve for y_1, the value of concentration at the collocation point, $x_1 = 0.447$:

$$y_1 = [-2.5 + (6.25 + 10\phi^2)^{1/2}]/2\phi^2. \tag{5.47}$$

Diffusion limits the reaction rate, since the concentration is not everywhere equal to one. This effect is conveniently expressed by the effectiveness factor, which is defined as the ratio of the amount reacted with diffusion to the amount that would be reacted if the concentration were everywhere the same, and equal to the value at the boundary:

$$\begin{aligned} \eta &\equiv \frac{\int_0^1 \phi^2 y^2\, dx}{\int_0^1 \phi^2 1\, dx} = \int_0^1 y^2\, dx = W_1 y_1^2 + W_2 y_2^2, \\ \eta &= \tfrac{1}{6} + \tfrac{5}{6} y_1{}^2, \end{aligned} \tag{5.48}$$

where the quadrature formula (5.16) is used to evaluate the integral accurately. The effectiveness factor η is plotted versus the Thiele modulus ϕ in Fig. 5.2.

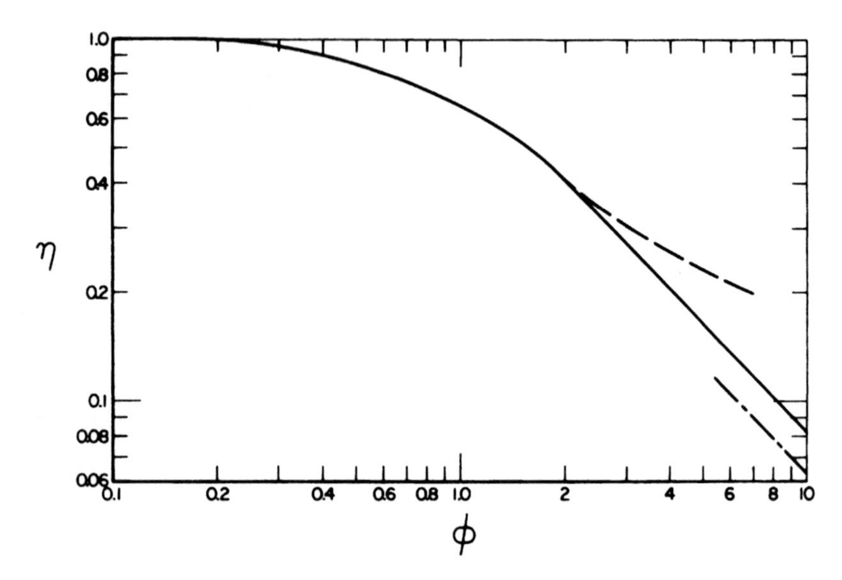

Fig. 5.2. Effectiveness factor as a function of Thiele modulus (— exact solution; - - collocation method; - - - integral method).

The approximation is accurate for $\phi \leq 2$. For larger ϕ a higher approximation is required. The reason for the poor approximation at large ϕ can be deduced from the approximate profile. Detailed examination of the solution, in the form $y = 1 + a(1 - x^2)$, reveals that the concentration becomes negative for $\phi^2 \geq 50$. This is clearly unrealistic and the approximation can be improved by using more terms in the expansion.

Rather than doing that here, let us consider another type of approximation. A large ϕ corresponds to a large rate of reaction or small diffusion coefficient. The process is diffusion limited and the reaction occurs only near the boundary. Further from the boundary the concentration is effectively zero, since all the reactant has been depleted. Consider a penetration-depth type of solution which satisfies $y = 1$ at $Z = 1 - x = 0$, $y' = y = 0$ at $Z = \delta$:

$$y = \begin{cases} [1 - (Z/\delta)]^2, & Z \leq \delta, \\ 0, & Z \geq \delta; \end{cases} \qquad Z \equiv 1 - x. \tag{5.49}$$

Application of the integral method gives

$$\int_0^1 (y'' - \phi^2 y^2)\, dZ = 0, \quad \delta^2 = 10/\phi^2, \qquad \eta = \sqrt{10}/5\phi. \tag{5.50}$$

This solution is valid for $\delta \leq 1$, or $\phi^2 \geq 10$ and is plotted on Fig. 5.2. It is a better approximation for large ϕ, but still somewhat inaccurate.

In Section 2.6 it was mentioned that if asymptotic solutions were available for some range of parameters, that information should surely influence the choice of trial function. In this case an asymptotic solution is available for large ϕ (Petersen, 1965, p. 70):

$$\eta = \frac{\sqrt{2}}{\hat{\phi}}\left[\int_0^1 R(y)\,dy\right]^{1/2}, \qquad \hat{\phi} = \frac{V_p}{A_p}\frac{\phi}{R}, \qquad R(1) = 1, \tag{5.51}$$

where V_p and A_p are the volume and external area of the catalyst particle. In this case the general formula becomes $\eta = (\frac{2}{3})^{1/2}/\phi$, and this is the best approximation for $\phi \geq 3$. The recommended procedure is then to use the collocation method, $N = 1$, for small ϕ and the asymptotic formula (5.51) for large ϕ. This gives adequate approximations for all ϕ. If desired the collocation method can be extended to higher N to improve the results.

Nonisothermal Reaction

Heat effects are next included. Take a spherical catalyst pellet with an irreversible first-order reaction, whose rate is

$$R = -ac_A \exp(-\Delta E/\hat{R}T), \tag{5.52}$$

where a is a constant, ΔE is the activation energy, $\hat{R}$ is the gas constant, and T is the absolute temperature. The dimensionless unsteady-state equations are

$$\begin{aligned}
&M_1 \frac{\partial T}{\partial t} = \frac{1}{x^2}\frac{\partial}{\partial x}\left(x^2 \frac{\partial T}{\partial x}\right) + \phi^2 \beta c \exp[\gamma(1 - (1/T))],\\
&M_2 \frac{\partial c}{\partial t} = \frac{1}{x^2}\frac{\partial}{\partial x}\left(x^2 \frac{\partial c}{\partial x}\right) - \phi^2 c \exp[\gamma(1 - (1/T))],\\
&\frac{\partial c}{\partial x} = \frac{\partial T}{\partial x} = 0 \quad \text{at} \quad x = 0\\
&-\frac{\partial T}{\partial x} = \frac{\text{Nu}}{2}(T - g_1(t)), \qquad -\frac{\partial c}{\partial x} = \frac{\text{Sh}}{2}(c - g_2(t)) \quad \text{at} \quad x = 1,\\
&T(x, 0) = h_1(x), \qquad c(x, 0) = h_2(x),
\end{aligned} \tag{5.53}$$

where $M_1 = \rho C R^2/kt_s$, $M_2 = \varepsilon R^2/\mathscr{D}t_s$, and the time standard t_s is not yet specified. The dimensionless variables are:

$$\begin{aligned}
&\phi^2 = \frac{k_0 R^2}{\mathscr{D}}, \qquad \beta = \frac{(-\Delta H_R)c_0 \mathscr{D}}{kT_0}, \qquad \gamma = \frac{\Delta E}{\hat{R}T_0},\\
&\text{Nu} = \frac{h2R}{k}, \qquad \text{Sh} = \frac{k_g 2R}{\mathscr{D}},
\end{aligned} \tag{5.54}$$

where $k_0 = a \exp(-\gamma)$, R is the radius of the sphere, $-\Delta H_R$ is the heat of reaction (a positive value of $-\Delta H_R$ indicates an exothermic reaction), and k is the thermal conductivity. In the Nusselt number and Sherwood number, h is the heat transfer coefficient and k_g the mass transfer coefficient. For many industrial reactions γ takes values from 6 to 40, β is usually less than 0.1 so that $\gamma\beta$ can take values up to 4 (Hlaváček *et al.*, 1969; McGreavy and Cresswell, 1969). The Thiele modulus ϕ can take all positive values.

We first solve for the steady state under conditions in which the Nusselt Sherwood numbers are very large, and $g_1 = g_2$ and $= 1$:

$$\begin{aligned} &\nabla^2 y = \phi^2 y \exp[\gamma(Z-1)/Z], \\ &\nabla^2 Z = -\beta\phi^2 y \exp[\gamma(Z-1)/Z], \\ &y(1) = Z(1) = 1, \qquad y'(0) = Z'(0) = 0. \end{aligned} \tag{5.55}$$

These equations can be combined, although this is not necessary to apply the collocation method. Multiply the first equation by β and add it to the second:

$$\nabla^2(\beta y + Z) = 0, \qquad \beta y(1) + Z(1) = \beta + 1. \tag{5.56}$$

The solution is

$$\beta y(x) + Z(x) = \beta + 1. \tag{5.57}$$

Thus we may define $Z(z) = \beta + 1 - \beta y(x)$ and solve

$$\begin{aligned} &\nabla^2 y = \phi^2 y \exp\{\gamma\beta(1-y)/[1+\beta(1-y)]\} = \phi^2 R(y), \\ &y(1) = 1, \qquad y'(0) = 0. \end{aligned} \tag{5.58}$$

Since the minimum value of y is zero, the maximum value of Z is $1 + \beta$.

First consider the question of uniqueness. The problem (5.58) admits multiple solutions for some parameters. The orthogonal collocation method can be used to predict what range of parameters gives rise to multiple solutions (Stewart and Villadsen, 1969). Apply a one-term solution, using the matrices from Table 5.3 for $a = 3$, $w = 1 - x^2$, and $N = 1$. The residual is satisfied at the collocation point:

$$(-10.5y_1 + 10.5)/\phi^2 = R(y_1). \tag{5.59}$$

Next examine (5.59) to see if it has multiple solutions. This could be done numerically, but it is more revealing to do it graphically. The right-hand side is plotted as a function of concentration (see Fig. 5.3) for $\beta = 0.6$, $\gamma = 20$. The left-hand side depends on the Thiele modulus, and is plotted on the same graph for various values of ϕ. For large particles, and hence large ϕ, the two curves intersect only once, as is the case for $\phi = 1.0$. This corresponds to a diffusion-controlled situation and gives a unique steady state.

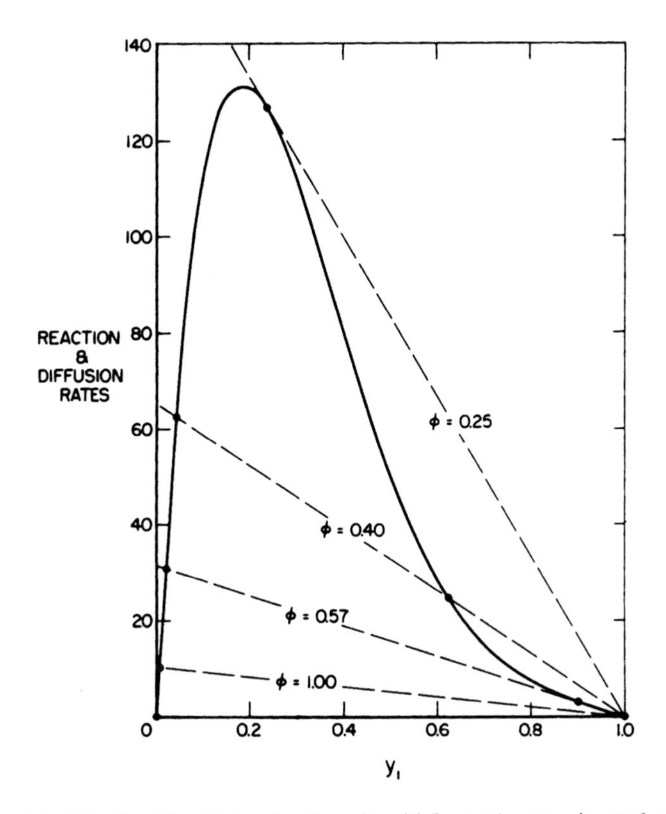

Fig. 5.3. Graphical determination of multiple steady states in catalyst.

For small particles only one intersection occurs ($\phi \leq 0.25$), which corresponds to the case when diffusion is very fast and the concentration gradients are small. For intermediate values of ϕ however, the two curves intersect at more than one place, as is the case for $\phi = 0.4$. This procedure predicts that multiple steady states occur for $0.25 \leq \phi \leq 0.57$. More accurate finite difference computations (using 1000 grid points) give the values $0.29 \leq \phi \leq 0.58$ (Weisz and Hicks, 1962). These can also be determined using orthogonal collocation with a higher N.

To obtain precise results the approximation is improved by taking a larger N. The problem (5.58) becomes

$$\sum_{i=1}^{N+1} B_{ji} y_i = \phi^2 R(y_i), \qquad j = 1, \ldots, N, \tag{5.60}$$

$$y_{N+1} = 1,$$

which gives a set of nonlinear algebraic equations. These are solved using a Newton–Raphson iterative procedure:

$$\sum_{i=1}^{N+1} B_{ji} y_i^{(n+1)} = \phi^2 \left[R(y_j^{(n)}) + \left.\frac{\partial R}{\partial y}\right|_{y_j^{(n)}} (y_j^{(n+1)} - y_j^{(n)}) \right]. \tag{5.61}$$

A low-order approximation, such as that given by (5.59) can be used as starting values. Such computations have been done for this problem by Ferguson (1971) and the steady-state profiles are shown in Fig. 5.4, calculated using

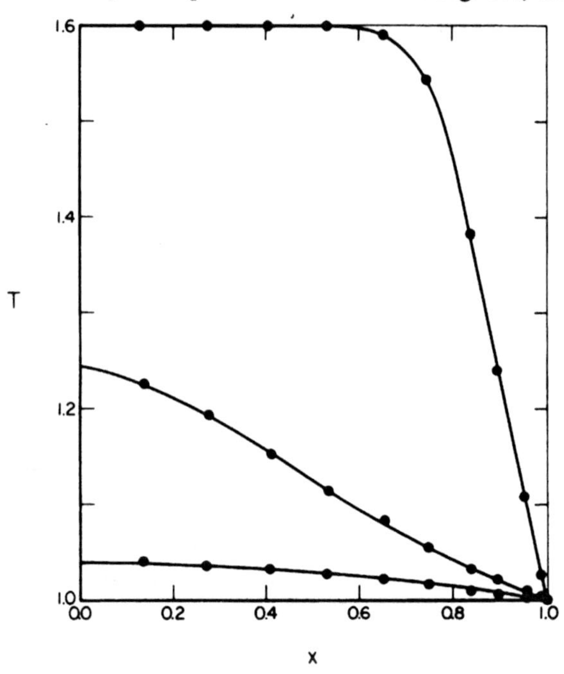

Fig. 5.4. Multiple steady states in catalyst.

$N = 10$, $w = 1 - x^2$, $\beta = 0.6$, $\gamma = 20$, and $\phi = 0.5$. The upper steady state is very flat over much of the region, which necessitates using a large number of terms to approximate it. Even so a finite difference calculation requires over 100 terms to achieve equivalent accuracy and takes about ten times as long to solve as the collocation method. Thus we see that the orthogonal collocation method is useful for providing qualitative information about the solution (such as a rough idea of the regions of multiple steady states) and can be used to obtain quantitative information, to any desired accuracy, merely by increasing the number of collocation points. Convergence is assured by Theorems 11.18, and 11.19. Error bounds for the case $\beta = 0.3$, which gives a unique solution, are discussed in Section 11.6.

Figure 5.4 displays the solution at the collocation points. If the trial function (5.10), (5.12) were evaluated at other positions, and the results plotted, it would exhibit small oscillations about the curve displayed. These oscillations disappear as N increases. For a small N, however, best graphical results are achieved by fairing a curve through the solution at the collocation points, where it is determined accurately.

When a catalyst particle can be in one of several steady states, the one it is actually in depends upon the time history. Some models of chemical reactors must account for the diffusional resistance in the catalyst, especially when studying the stability of the reactor (McGuire and Lapidus, 1965). In this case it is necessary to solve the transient equations (5.53) at many positions throughout the bed. The orthogonal collocation method is applied to a prototype problem, as discussed by Ferguson and Finlayson (1970).

Consider Eqs. (5.53) with $M_1 = 176$, $M_2 = 199$, where $t_s = 2R/u$, and u is the velocity of gas through the bed, exterior to the particle. Other numerical values are taken as $\mathrm{Nu} = 55.3$, $\mathrm{Sh} = 66.5$, $\gamma = 20$, $\beta = 0.6$, and $\phi^2 = 0.25$. The functions g and h are taken to be 1.1 and 1.0, respectively, for $t > 0$. The initial conditions are taken as the two-term approximation to the intermediate steady state for the problem with infinite Nu and Sh. The 10% temperature perturbation on the boundary is sufficient to drive the solution to the third steady state.

The collocation formulation of (5.53) is

$$\left.\begin{aligned} M_1 \frac{dT_j}{dt} &= \sum_{i=1}^{N+1} B_{ji} T_i + \phi^2 \beta c_j \exp[\gamma(1 - (1/T_j))] \\ M_2 \frac{dc_j}{dt} &= \sum_{i=1}^{N+1} B_{ji} c_i - \phi^2 c_j \exp[\gamma(1 - (1/T_j))] \end{aligned}\right\} \quad j = 1, \ldots, N, \tag{5.62}$$

$$-\sum_{i=1}^{N+1} A_{N+1,i} T_i = \frac{\mathrm{Nu}}{2}(T_{N+1} - g)$$

$$-\sum_{i=1}^{N+1} A_{N+1,i} c_i = \frac{\mathrm{Sh}}{2}(c_{N+1} - h).$$

These ordinary differential equations are integrated numerically, using a convenient predictor–corrector method. To integrate the equation

$$\frac{dy}{dt} = f(t, y) \tag{5.63}$$

take the difference scheme

$$\begin{aligned} y^{(1)}(t + \Delta t) &= y(t) + \Delta t f(t, y(t)) \\ y(t + \Delta t) &= y(t) + \frac{\Delta t}{2}\,[f(t, y(t)) + f(t + \Delta t, y^{(1)}(t + \Delta t))]. \end{aligned} \tag{5.64}$$

This is the first iteration of a modified Euler method. The stability of the method has been studied and stability is assured for the system of linear equations

$$\frac{dT_j}{dt} = \sum_{i=1}^{N} N_{ji} T_i, \qquad N_{ji} = B_{ji} - \frac{A_{N+1,i} B_{j,N+1}}{(\mathrm{Nu}/2) + A_{N+1,N+1}} \tag{5.65}$$

provided

$$\Delta t \|N\|_\infty \le 2. \tag{5.66}$$

The value of the matrix norm is given in the original article. The value of Δt deduced from (5.66) made Eqs. (5.62) stable and gave good accuracy as well. Smaller Δt gave no change in the solution. This conclusion does not necessarily hold for other ratios of M_1 to M_2. Calculations were performed using polynomials with w equal to 1 or $1 - x^2$ and either the predictor–corrector method of integration, or Hammings method, which is an explicit, four-point method with fifth-order truncation error. Comparison calculations were also done using two different finite difference methods. The implicit method, with nonlinear terms treated explicitly, and Liu's method (1969) were used, with the latter being more accurate and faster. Comparisons of these two finite difference methods are in Liu (1969). Calculations were done on a CDC 6400 computer for $0 \le t \le 35$.

Exploratory calculations, Fig. 5.5, indicated the solution had large spatial and time derivatives. Consequently, we expect that many terms are needed to approximate the solution. Even though the temperature was approximated within 2 to 3 digits using 6 to 10 terms and the polynomials with $w = 1 - x^2$, the convergence with N was rather slow (see Table 5.7). If polynomials

TABLE 5.7

HEAT FLUX FROM CATALYST[a]

			Flux		% Error
Polynomial	N	Δt	$t = 22$	$t = 35$	in flux at $t = 35$
$w = 1 - x^2$ [b]	6	0.05	2.348	4.241	16
$w = 1 - x^2$ [b]	8	0.025	2.857	4.594	9
$w = 1 - x^2$ [b]	10	0.025	2.489	5.078	1
$w = 1$ [c]	6	0.10	2.477	5.159	3
$w = 1$ [c]	8	0.05	2.539	4.956	1
$w = 1$ [c]	10	0.04	2.572	5.042	0.34
$w = 1$ [c]	12	0.025	2.584	5.024	0.02

[a] Ferguson and Finlayson (1970); used by permission of the copyright owner, Elsevier Publishing Company, Ltd.
[b] $w = 1 - x^2$ results using Hamming's method.
[c] $w = 1$ results using predictor–corrector method.

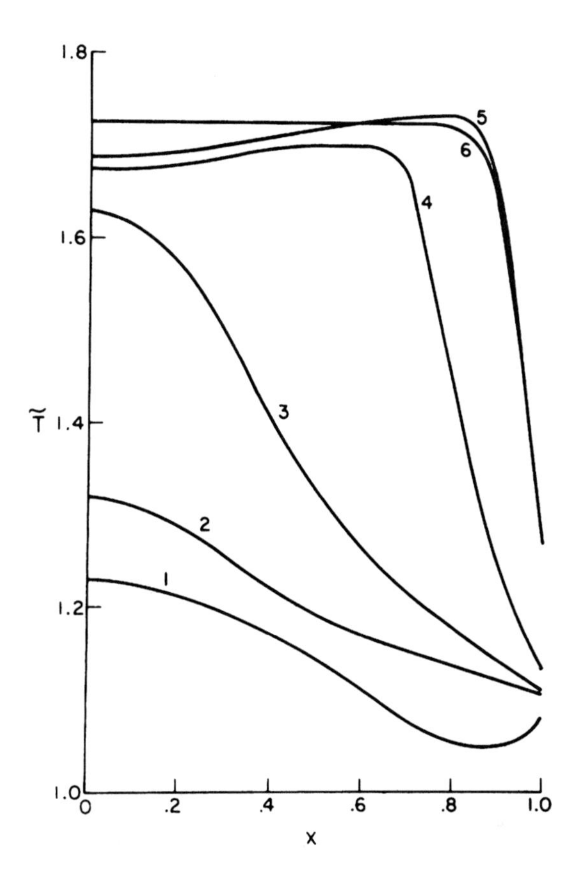

Fig. 5.5. Transient temperature distribution in catalyst (1 to 6 correspond to $t = 1$, 10, 15, 20, 25, 30; $\tilde{T}$ is the approximate solution; reprinted from the *Chemical Engineering Journal* with permission of the copyright owner).

with $w = 1$ are used, the convergence is much faster. This is probably due to the fact that with $w = 1$ the collocation points are closer to the boundary, where the flux is calculated.

The predictor–corrector method of integration gave accuracy comparable to Hamming's method, took about the same computation time for the same Δt, but allowed time steps about twice as large.

If we look at the difference between the surface derivative for a collocation solution and the best finite difference solution as a function of time, we find the errors can be characterized by three numbers:

$$\begin{aligned} &\text{(i)} \quad E_1 = \text{typical error for } 0 < t < 15, \\ &\text{(ii)} \quad E_2 = \text{maximum error for } 15 < t < 25, \\ &\text{(iii)} \quad E_3 = \text{typical error for } 25 < t, \end{aligned} \tag{5.67}$$

where E_1 represents the typical error during the initial transient, E_2 the maximum error during the time period when the temperature rises sharply and the flux increases rapidly, and E_3 is the typical error of the steady-state value. The effects of the integration scheme, the expansion functions, and the step size are summarized in Table 5.8.

TABLE 5.8

HEAT FLUX FROM CATALYST FOR DIFFERENT METHODS[a]

Entry	N or Δx	Δt	Method of integration	Expansion functions	E_1 (%)	E_2 (%)	E_3 (%)	Computation time (seconds)
			Collocation solutions					
(1)	6	0.05	Hamming	$w = 1$	<1	7.2	2.7	11.9
(2)	8	0.025	Hamming	$w = 1$	<1	2.4	1.4	30.6
(3)	6	0.10	Predictor–corrector	$w = 1$	<1	8.0	2.7	5.1
(4)	8	0.05	Predictor–corrector	$w = 1$	<1	2.4	1.4	13.3
(5)	10	0.04	Predictor–corrector	$w = 1$	<1	1.8	0.34	21.8
(6)	12	0.025	Predictor–corrector	$w = 1$	<1	1.4	0.02	45.6
			Finite difference solutions					
(7)	0.05	0.05	Implicit		6	15	20	11
(8)	0.05	0.05	Liu's method		2	7.4	1.5	21
(9)	0.025	0.0025	Liu's method		<1	2.8	0.1	830
(10)	0.01	0.005	Liu's method		—	—	—	1050

[a] Ferguson and Finlayson (1970), used by permission of the copyright owner, Elsevier Publishing Company, Ltd.

Based on results for another problem, Ferguson and Finlayson chose the following grid spacings to give representative results for the finite difference computations: $(\Delta x, \Delta t) = (0.05, 0.05)$, $(0.025, 0.0025)$, and $(0.01, 0.005)$. Additional values were not examined because of the excessive computation time necessary to use the finite difference methods.

The comparative errors for the classical implicit method are indicative of the unacceptable pointwise errors. Based on Liu's (1969) comparison of his own method to the implicit scheme and the errors found in Table 5.8, no additional computations were done using the implicit method.

Based on Table 5.7 the polynomials with $w = 1$ are recommended over those with $w = 1 - x^2$. Comparison of the collocation solution, entries (5) and (6) in Table 5.8, with a finite difference solution of comparable accuracy, entry (9), reveals that the collocation solution is from twenty to forty times faster. This advantage is due to the larger time step and the smaller number of terms in the collocation method (10–12 rather than 40–100). This problem clearly indicates that large computation times are necessary to model non-isothermal diffusion with reactions of this type, and the time savings made possible by the collocation method is especially welcome when several of these problems must be solved as is the case in the reactor model.

Stability

When multiple steady states occur, some of them may be unstable. Various arguments can be advanced to show that the middle steady state is unstable to small disturbances (Luss and Amudson, 1967; Gavalos, 1968, p. 91). We consider first the stability of the other two steady states to small disturbances and finally consider the complications encountered when the Lewis number is small.

Galerkin's method has been used to study the stability by Wei (1965), Kuo and Amundson (1967), and Lee and Luss (1970). Take the equations (5.53) with t_s chosen such that $M_2 = 1$. Then $M_1 = \rho C\mathscr{D}/\varepsilon k$ is the Lewis number Le. Let T^* and c^* be one of the steady-state solutions to (5.53). Define the new variables $y = T - T^*$ and $u = c - c^*$, where T and c satisfy (5.53) and subtract the equations for T^* and c^*. Then linearize the reaction rate expression about the steady-state value to obtain equations governing y and u for small disturbances:

$$\mathrm{Le}\frac{\partial y}{\partial t} = \nabla^2 y + \beta(R_T{}^* y + R_c{}^* u),$$

$$\frac{\partial u}{\partial t} = \nabla^2 u + (R_T{}^* y + R_c{}^* u), \tag{5.68}$$

$$y_x = u_x = 0 \qquad \text{at} \quad x = 0,$$

$$-y_x = \mathrm{Nu}\, y/2, \qquad -u_x = \mathrm{Sh}\, u/2 \qquad \text{at} \quad x = 1,$$

where $R_T{}^* = \partial R/\partial T|_{T^*,c^*}$, for example. The solution to these equations can be expressed as a sum of exponentials in time, since they are linear, and the disturbances die out if all eigenvalues are negative. Approximations to the eigenvalues are obtained using the Galerkin method. Expand the disturbances in terms of trial functions. If Nu or Sh are infinite, they must satisfy boundary conditions of the first kind. Otherwise they do not satisfy any boundary conditions:

$$y = \sum_{j=1}^{N} A_j(t) y_j(x), \qquad u = \sum_{j=1}^{N} B_j(t) u_j(x). \tag{5.69}$$

Form the residuals, make them orthogonal to y_i and u_i, respectively, and add in the weighted boundary residual as illustrated in Section 2.4. Then we have the equations (summation of repeated indices is assumed)

$$\begin{aligned} \text{Le}\,(y_j, y_i)\frac{dA_i}{dt} &= -A_i(\nabla y_j \cdot \nabla y_i) - \frac{\text{Nu}}{2} A_i y_j(1) y_i(1) \\ &\quad + \beta A_j(y_i, R_T{}^* y_j) + \beta B_j(y_i, R_c{}^* u_j), \\ (u_j, u_i)\frac{dB_i}{dt} &= -B_i(\nabla u_j \cdot \nabla u_i) - \frac{\text{Sh}}{2} B_i u_j(1) u_i(1) \\ &\quad + A_j(u_i, R_T{}^* y_j) + B_j(u_i, R_c{}^* u_j). \end{aligned} \tag{5.70}$$

The stability of this system of equations is usually determined by substituting

$$A_i = a_i e^{\lambda t}, \qquad B_i = b_i e^{\lambda t}, \tag{5.71}$$

and obtaining a set of homogeneous equations for the constants. For a nontrivial solution the determinant of the coefficients must vanish, and this happens only for certain eigenvalues. When all eigenvalues are negative, the steady state is stable. The stability of the set of ordinary differential equations can also be determined using methods described in detail in Chapter 6.

The Galerkin method has been applied to study the stability of steady-state solutions of Eqs. (5.68) by Wei (1965),† Kuo and Amundson (1967) (slab geometry), and Lee and Luss (1970) (spherical geometry). The last two authors did calculations for large Nusselt and Sherwood numbers, so that the boundary conditions are of the first kind. Then the functions y_i and u_i vanish at $x = 1$ and the boundary terms in Eqs. (5.70) vanish. The trial functions are taken as eigenfunctions of the operator $\nabla^2 y + \lambda y = 0$. For spherical geometry these are

$$y_i = u_i = (\sin i\pi x)/x. \tag{5.72}$$

Lee and Luss found that seven terms in the expansions (5.69) were necessary to obtain good values for the eigenvalues. Another approach is that of McGowin and Perlmutter (1971). They apply the collocation method to (5.53) to reduce the problem to a large set of ordinary differential equations. The stability of these is not determined directly. Instead a Liapunov criterion is used to verify the stability of the set of ordinary differential equations.

† Wei applied the Rayleigh–Ritz method for the case $\text{Le} = 1$. Then the equations can be combined and a variational principle exists. See Chapters 7–9.

For equations (5.58) which have three solutions, the second solution is unstable. The upper steady state (see Fig. 5.4) is stable for large Lewis number, but unstable for small Lewis number. For small Lewis number a limit cycle develops and the concentration and temperature in the catalyst change in a periodic fashion in time. Lee and Luss (1970) integrated the unsteady-state equations (5.53) using a finite difference method, but needed 1000 spatial grid points and 40 hours of computer time for the calculations. The orthogonal collocation method has not been applied to this problem, but it possibly would reduce the computation time. See Garabedian and Lynch (1965) for applications of MWR to the stability of nuclear reactors.

Next we show that the one-term collocation solution predicts that the upper steady state becomes unstable for small Lewis number. Consider the first approximation to (5.53) using the matrices from Table 5.3 appropriate to spherical geometry ($a = 3$), and with weighting function $w = 1 - x^2$:

$$\begin{aligned}
\frac{dy_1}{dt} &= -\frac{10.5}{\text{Le}}(y_1 - y_2) + \frac{\beta}{\text{Le}} R_1, \\
\frac{du_1}{dt} &= -10.5(u_1 - u_2) - R_1, \\
-(-3.5y_1 + 3.5y_2) &= \frac{\text{Nu}}{2}(y_2 - 1), \\
-(-3.5u_1 + 3.5u_2) &= \frac{\text{Sh}}{2}(u_2 - 1).
\end{aligned} \tag{5.73}$$

The boundary conditions are combined with the differential equations to give

$$\begin{aligned}
\frac{dy_1}{dt} &= a(1 - y_1) + \frac{\beta}{\text{Le}} R_1, \qquad a = \frac{10.5}{\text{Le}} \frac{\text{Nu}/2}{3.5 + (\text{Nu}/2)}, \\
\frac{du_1}{dt} &= b(1 - u_1) - R_1, \qquad b = 10.5 \frac{\text{Sh}/2}{3.5 + (\text{Sh}/2)}.
\end{aligned} \tag{5.74}$$

Call the steady-state solution, y_0, u_0. With $y = y_1 - y_0$, $u = u_1 - u_0$, the linearized equations are

$$\begin{aligned}
\frac{dy}{dt} &= -ay + \frac{\beta}{\text{Le}}(R_y y + R_u u), \\
\frac{du}{dt} &= -bu - (R_y y + R_u u).
\end{aligned} \tag{5.75}$$

An exponential solution is assumed, $y = A\exp(\mu t)$, $u = B\exp(\mu t)$, giving rise to a set of homogeneous equations in A and B. A solution exists only for certain values of μ, which are roots of the determinant

$$\begin{vmatrix} \mu + a - \dfrac{\beta}{\text{Le}} R_y & -\dfrac{\beta}{\text{Le}} R_u \\ R_y & \mu + b + R_u \end{vmatrix} = 0. \tag{5.76}$$

This gives a quadratic in μ:

$$\begin{gathered} \mu^2 + \mu Q + c = 0, \\ Q = a + b - \frac{\beta}{\text{Le}} R_y + R_u, \qquad c = ab + aR_u - \beta b R_y/\text{Le}. \end{gathered} \tag{5.77}$$

The quadratic has the solution $2\mu = -Q \pm (Q^2 - 4c)^{1/2}$. If Q is negative then one eigenvalue has a positive real part and the solution is unstable. By expressing the terms R_u and R_y in terms of R_1, the reaction rate at steady state, we can write Q in the form

$$\begin{gathered} Q = a\left[1 + \gamma\frac{1 - y_1}{y_1^{\,2}} + \text{Le}\,\kappa/u_1\right], \\ \kappa = \frac{b}{a\,\text{Le}} = \frac{\text{Sh}}{\text{Nu}}\frac{3.5 + (\text{Nu}/2)}{3.5 + (\text{Sh}/2)}. \end{gathered} \tag{5.78}$$

Clearly if $y_1 < 1$ then Q is positive. If $y_1 > 1$ then (5.78) gives the following condition for Q negative

$$\text{Le}\,\kappa \le u_1\left(-1 - \gamma\frac{1 - y_1}{y_1^{\,2}}\right). \tag{5.79}$$

The right-hand side is a numerical value, which depends on the steady-state solution. If the Lewis number is below a critical value, the steady state is unstable. For the example in Fig. 5.3 the upper steady state is $u_1 = 0.026$, $y_1 = 1.58$. Equation (5.79) then gives the following sufficient condition for instability of this steady state:

$$\text{Le} \le \text{Le}^* = 0.15/\kappa. \tag{5.80}$$

When both Sherwood and Nusselt numbers are large this becomes $\text{Le} \le 0.15$. As pointed out by Hlaváček *et al.* (1969) however, Lewis numbers are greater than one for industrial reactions so that such a phenomena is unlikely to be observed. It is interesting that a very simple approximation leads to this qualitatively correct result. More precise results can be obtained using higher approximations. When the Nusselt and Sherwood numbers are not infinite, the critical Lewis number depends on them. Villadsen (1970) gives possible ranges of Sh/Nu as between 8 and 400 and Hlaváček and Kubiček (1970) list values between 3 and 4300. Then $\kappa \geq 1$ and (5.80) can apparently never be satisfied for industrial reactions.

The analysis just presented is similar to that given in Hlaváček *et al.* (1969). See also Hellinckx *et al.* (1971). It is of interest to show that Hlaváček's results can be interpreted as an application of the collocation method. Hlaváček *et al.* use as a first approximation in place of (5.74) the equations (in the present notation)

$$\frac{dy_1}{dt} = \frac{\rho_1^2}{\text{Le}}(1 - y_1) + \frac{\beta}{\text{Le}} R_1, \qquad \frac{du_1}{dt} = \beta_1^2(1 - u_1) - R_1, \tag{5.81}$$

where ρ_1 and β_1 are the first roots to

$$\rho \cot \rho + (\text{Nu}/2) = 1, \qquad \beta \cot \beta + (\text{Sh}/2) = 1. \tag{5.82}$$

Let us apply MWR to obtain the equations (5.81). Expand the solutions in the form

$$y = 1 + \sum_{i=1}^{N} A_i f_i(x), \qquad u = 1 + \sum_{i=1}^{N} B_i g_i(x), \tag{5.83}$$

where f_i and g_i satisfy the problems

$$\begin{gathered} \nabla^2 f + \rho^2 f = 0, \qquad \nabla^2 g + \beta^2 g = 0, \\ -\left.\frac{\partial f}{\partial x}\right|_{x=1} = \left.\frac{\text{Nu}}{2}(f - 1)\right|_{x=1}, \qquad -\left.\frac{\partial g}{\partial x}\right|_{x=1} = \left.\frac{\text{Sh}}{2}(g - 1)\right|_{x=1}. \end{gathered} \tag{5.84}$$

Then the Laplacian operator is

$$\nabla^2 y = -\sum_{i=1}^{N} A_i \rho_i^2 f_i, \qquad \nabla^2 u = -\sum_{i=1}^{N} B_i \beta_i^2 g_i. \tag{5.85}$$

Now consider the first approximation ($N = 1$). Substitute (5.83) into the equations, use (5.85) and evaluate the equations at some point x_1. Define $y_1 = 1 + A_1 f_1(x_1)$ and $u_1 = 1 + B_1 g_1(x_1)$ to obtain

$$\frac{dy_1}{dt} = -\rho_1^2(y_1 - 1) + \frac{\beta}{\text{Le}} R_1, \qquad \frac{du_1}{dt} = -\beta_1^2(u_1 - 1) - R_1. \tag{5.86}$$

Comparison to Eqs. (5.74) reveals that the only difference is the value of the numerical coefficient representing the Laplacian operator. These are compared in Table 5.9 for several Nusselt numbers (the Sherwood number results

TABLE 5.9

COMPARISON OF APPROXIMATIONS FOR DIFFUSION OPERATOR

Nu/2	$\dfrac{10.5\ \mathrm{Nu}/2}{3.5 + (\mathrm{Nu}/2)}$	$\rho_1^{\,2}$
0	0	0
0.1	0.29	0.29
1.0	2.3	2.5
10.	7.8	8.0
100.	10.1	9.7

are similar). The small difference is insignificant compared to the gross approximations used when representing the problem by the first approximation alone. We conclude then the method advanced by Hlaváček *et al.* (1969) can be interpreted as a form of the collocation method using the trial functions (5.84). Higher approximations can be calculated by retaining more terms in (5.83) and evaluating the residual at more points. The orthogonal collocation method, based on orthogonal polynomials as expansion functions, is easily extended to higher approximations.

5.4 Tubular Reactor with Axial Dispersion

We next consider chemical reactors which are cylindrical tubes packed with catalyst. The reactants flow through the packed bed, react on the surface of the catalyst, and the products continue through the bed. Dispersion in the radial and axial directions occurs because of the flow around the particles. We consider problems with only axial dispersion here, and treat radial diffusion by itself in the next section. Dispersion in both directions could be accounted for by combining the techniques of the two sections.

Consider first an isothermal situation with an nth order irreversible reaction. The mass balance is, in dimensionless form,

$$\frac{1}{\mathrm{Pe}}\frac{d^2y}{dx^2} - \frac{dy}{dx} - Ry^n = 0, \tag{5.87}$$

where Pe is the axial Peclet number VL/D (based on the reactor length), and R is the reaction rate group, $R = kLc_0^{n-1}/V$, with V the axial velocity, L the length of reactor, D the axial dispersion coefficient, k the reaction rate constant, c_0 the inlet concentration, and n the reaction order, which can take fractional as well as integer values. The boundary conditions are (Danckwerts, 1953; Wehner and Wilhelm, 1956)

$$1 = y - \frac{1}{\text{Pe}}\frac{dy}{dx}, \qquad x = 0, \tag{5.88a}$$

$$\frac{dy}{dx} = 0, \qquad x = 1. \tag{5.88b}$$

There are several techniques applicable to this problem. See Villadsen (1970) and Hlaváček and Hofmann (1970) for a detailed discussion of them. A shooting technique can be used, in which an outlet concentration is assumed, the equations are integrated backwards as an initial-value problem to the inlet. There the inlet condition (5.88a) is checked, and if not satisfied, the iteration is repeated. The equations cannot be integrated in the reverse direction because they are unstable. A change of 10^{-9} in a guess at the inlet may cause the outlet concentration to change from -20 to $+20$. Quasilinearization can also be applied. A finite difference formulation of the equations is written and the nonlinear algebraic equations are solved using some iterative procedure. A time derivative can be included in (5.87) and the equations integrated to steady state. Finally a variational technique is applicable (see Section 9.5) although it is time consuming. We present here two other techniques: one a Galerkin method, useful for a first approximation, and the other based on orthogonal collocation for very accurate results.

Grotch (1969) takes a trial function in the form

$$y = \sum_{i=1}^{N} a_i(x-1)^{i-1} \tag{5.89}$$

and satisfies the boundary conditions by requiring

$$a_2 = 0, \qquad \sum_{i=1}^{N} (-1)^{i-1} a_i \left[1 + \frac{i-1}{\text{Pe}}\right] = 1. \tag{5.90}$$

For a three-term solution we get

$$a_3 = (1 - a_1)/\lambda, \qquad \lambda = 1 + 2/\text{Pe} \tag{5.91}$$

and a_1 is determined from the first Galerkin conditions (weighting function = 1)

$$1 - y(1) = R\int_0^1 y^n\, dx. \tag{5.92}$$

Grotch presents algebraic results for $n = 0$, 1, 2, and 3. Higher approximations are obtained using the weighting functions

$$1, \quad (x-1)^2, \quad (x-1)^3, \ldots \tag{5.93}$$

For $n = 1$ the results with four terms are accurate to 2% for nearly the full range of R and Pe. For general reaction order the integrals must be evaluated numerically.

The convergence theorems (Chapter 11) for the Galerkin method applied to problems with boundary conditions of the second or third kind say that the weighting functions must be complete for the class of functions in C^2. This means the term $(x - 1)$ should be included in the sequence of weighting functions, even though it need not be included as a trial function.

The orthogonal collocation method is simpler to apply, especially if higher approximations are desired. We apply it here to the case of mass and energy transfer governed by the equations

$$\begin{gathered}
\frac{d^2c}{dz^2} - \mathrm{Pe}_M \frac{dc}{dz} - \mathrm{Pe}_M R(c, T) = 0, \\
\frac{d^2T}{dz^2} - \mathrm{Pe}_H \frac{dT}{dz} - \mathrm{Pe}_H \beta R(c, T) = 0, \\
\frac{dc}{dz} = \mathrm{Pe}_M(c-1), \qquad \frac{dT}{dz} = \mathrm{Pe}_H(T-1) \qquad \text{at} \quad z = 0, \\
\frac{dc}{dz} = \frac{dT}{dz} = 0 \qquad \text{at} \quad z = 1.
\end{gathered} \tag{5.94}$$

The collocation method is applied using trial functions (5.29) and the roots and matrices from Tables 5.4 and 5.5:

$$\begin{gathered}
\sum_{i=1}^{N+2} B_{ji} c_i - \mathrm{Pe}_M \sum_{i=1}^{N+2} A_{ji} c_i - \mathrm{Pe}_M R(c_j, T_j) = 0, \\
\sum_{i=1}^{N+2} A_{1i} c_i = \mathrm{Pe}_M(c_1 - 1), \qquad \sum_{i=1}^{N+2} A_{N+2,i} c_i = 0,
\end{gathered} \tag{5.95}$$

with similar equations for T_i. These equations represent a set of nonlinear algebraic equations for the concentration and temperature at the collocation points. They can be represented as

$$F_i(\bar{y}) = 0, \qquad \bar{y}^T = (c_1, \ldots, c_{N+2}, T_1, \ldots, T_{N+2}). \tag{5.96}$$

One method for solving (5.96) is the Newton–Raphson procedure:

$$y_j^{(m+1)} - y_j^{(m)} = -\sum_{i=1}^{N+2} \left[\frac{\partial F}{\partial y^{(m)}}\right]_{ji}^{-1} F_i(\bar{y}^{(m)}), \tag{5.97}$$

where y^m represents the value at the mth iteration. Successive application of (5.97) gives the solution provided the initial guess is within the radius of convergence. Since this is not known initially it may be necessary to use another method to get close to the answer before applying (5.97). Table 5.10 compares the orthogonal collocation results to exact or finite difference solutions.

TABLE 5.10

AXIAL DIFFUSION IN REACTOR

N	$c(0)$	$T(0)$	$c(1)$	$T(1)$
	Isothermal, $Pe_M = 1$, $R = 2c^2$			
3	0.636809		0.457600	
6	0.636784		0.457589	
Finite difference[a]	0.63678		0.45759	
	Isothermal, $Pe_M = 15$, $R = 8c$			
3	0.740151		0.0044352	
6	0.722085		0.0028608	
Exact	0.721990		0.0028616	
	Nonisothermal, $Pe_M = Pe_H = 2$, $R = 3.36c^2 \exp(\gamma - (\gamma/T)), \gamma = 17.6, \beta = -0.056$			
1	0.62609	1.02094	0.25217	1.04188
3	0.58031	1.02350	0.23537	1.04282
6	0.58006	1.02352	0.23528	1.04282
	Nonisothermal, $Pe_M = Pe_H = 96$, $R = 3.817037\, c^2 \exp(\gamma - (\gamma/T)), \gamma = 17.6, \beta = -0.056$			
3	0.96510	1.00195	0.12564	1.04896
6	0.96333	1.00205	0.12410	1.04905
Finite difference[b]			0.12370	1.04904

[a] Reported by Pakes and Storey (1967); more than 100 grid points.
[b] Reported by Lee (1968, p. 67); 481 grid points.

5.5 Packed Bed Reactor with Radial Dispersion

Radial dispersion is important in packed bed reactors because the reactor is cooled at the wall, the energy must be transported to the wall by dispersion, and this induces temperature, and hence concentration, gradients in the radial direction. The equations for temperature and conversion are

$$\frac{\partial T}{\partial z} = \frac{1}{r}\frac{\partial}{\partial r}\left(\alpha' r \frac{\partial T}{\partial r}\right) + \beta'(1-c)\exp\frac{\gamma(T-1)}{T},$$

$$\frac{\partial c}{\partial z} = \frac{1}{r}\frac{\partial}{\partial r}\left(\alpha r \frac{\partial c}{\partial r}\right) + \beta(1-c)\exp\frac{\gamma(T-1)}{T},$$

$$\frac{\partial c}{\partial r} = 0, \qquad -\frac{\partial T}{\partial r} = \mathrm{Bi}(T - T_w) \quad \text{at} \quad r = 1, \tag{5.98}$$

$$\frac{\partial c}{\partial r} = \frac{\partial T}{\partial r} = 0 \quad \text{at} \quad r = 0,$$

$$c(r, 0) = c_0, \qquad T(r, 0) = T_0.$$

We apply orthogonal collocation for the parameter values $\alpha = \alpha' = 1$, $\beta = 0.3$, $\beta' = 0.2$, $\mathrm{Bi} = 20$, $T_w = 1$, $T_0 = 1$, and $c_0 = 0$ [see Finlayson (1971) for additional calculations]:

$$\frac{dT_j}{dz} = \alpha \sum_{i=1}^{N+1} B_{ji} T_i + \beta'(1-c_j)\exp\frac{\gamma(T_j - 1)}{T_j},$$

$$\frac{dc_j}{dz} = \alpha' \sum_{i=1}^{N+1} B_{ji} c_i + \beta(1-c_j)\exp\frac{\gamma(T_j - 1)}{T_j}$$

$$\sum_{i=1}^{N+1} A_{N+1,i} c_i = 0, \qquad -\sum_{i=1}^{N+1} A_{N+1,i} T_i = \mathrm{Bi}(T_{N+1} - 1), \tag{5.99}$$

$$c_j(0) = 0, \qquad T_j(0) = 1.0.$$

These are integrated using the Runge–Kutta method. The solutions of (5.99), as $N \to \infty$, converge to the solution of (5.98) as a result of Theorems 11.18 and 11.19.

The collocation method gives a first approximation which can be examined to gain physical insight. Consider the first approximation using the matrices from Table 5.3 appropriate to cylindrical geometry ($a = 2$) and with $w = 1 - x^2$. The equations can be rearranged as was done in Eqs. (5.73) and (5.74):

$$\frac{dT_1}{dz} = -\frac{6\alpha'\mathrm{Bi}}{3 + \mathrm{Bi}}(T_1 - T_w) + \beta' R_1, \tag{5.100a}$$

$$\frac{dc_1}{dz} = \beta R_1 \tag{5.100b}$$

These equations take the same form as the lumped parameter model, in which case no gradients are allowed and the equations are

$$\frac{dT}{dz} = -\mathrm{Nu}_w(T - T_w) + \beta' R, \tag{5.101a}$$

$$\frac{dc}{dz} = \beta R, \tag{5.101b}$$

$$\mathrm{Nu}_w = 2UL/c_p GR. \tag{5.101c}$$

The equations (5.100a) and (5.101a) are the same provided we take

$$\mathrm{Nu}_w = \mathrm{Bi}\,\frac{6\alpha'}{3 + \mathrm{Bi}}. \tag{5.102}$$

In dimensional notation this is

$$\frac{1}{U} = \frac{1}{h_w} + \frac{R}{3k_e} = \frac{R}{k}\left(\frac{1}{3} + \frac{1}{\mathrm{Bi}}\right). \tag{5.103}$$

The relative importance of the wall resistance to heat transfer is evident in (5.103). The ratio of $1/\mathrm{Bi}$ to $(1/\mathrm{Bi}) + (1/3)$ gives the fraction of the thermal resistance occurring at the wall (e.g., Bi = 1, 75%; Bi = 20, 13%). For small Biot numbers most of the resistance occurs at the wall and only a one-dimensional treatment (5.100), is necessary. As the Biot number increases, and radial dispersion becomes more important, the full two-dimensional treatment, Eqs. (5.98) and (5.99) is necessary, and N is increased as Bi increases. Note that the one-dimensional treatment (5.101) refers to a temperature which is constant across the radius, whereas the first approximation (5.100) refers to a temperature which is parabolic in radius and hence is a better approximation. The reaction rate expression in (5.100) is evaluated at the collocation point, and the temperature there is above the average temperature, so that this approximates more closely the average rate of reaction.

Results using polynomials with roots in Table 5.2, $a = 2$, are shown in Figs. 5.6, 5.7, and 5.8. The average temperature is shown in Fig. 5.6, and it is seen that three terms must be used to obtain good results. Near the hot spot at $z = 0.6$ the center temperature rises rapidly, as indicated by the temperature profiles in Fig. 5.7. The convergence with N is shown in Fig. 5.8, which illustrates why the polynomials with $w = 1$ are preferred for faster convergence. Also illustrated is the fact that the solution at the collocation points is accurate, whereas the error is greater for the solution at some

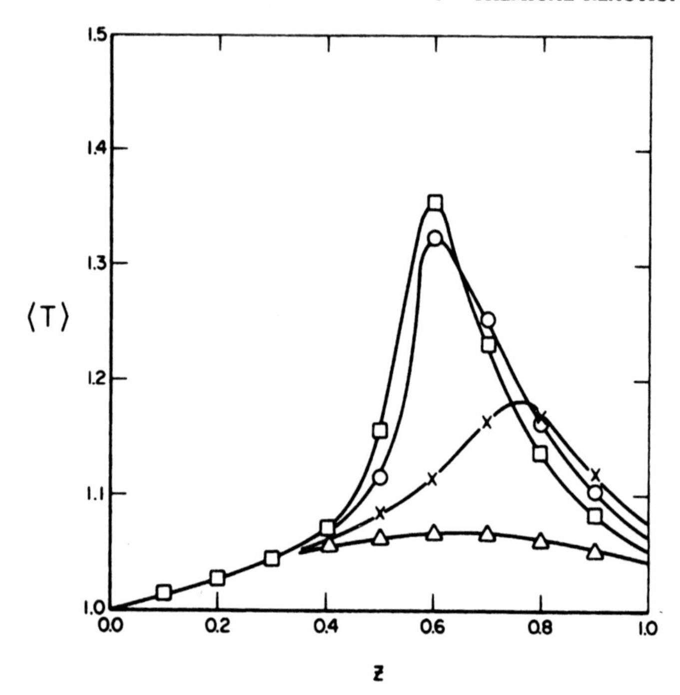

Fig. 5.6. Average temperature in reactor (polynomials with $w = 1$: △ $N = 1$, ○ $N = 2$, □ $N = 3, 4, \ldots$; $w = 1 - x^2$: × $N = 1$; reprinted from *Chemical Engineering Science* with permission of the copyright owner).

other point, such as at the center, for the same N. Finlayson (1971) compares accuracy and computation time for the collocation method to those for an implicit finite difference method. In some cases the collocation solution was twice as fast and ten times as accurate or four times as fast and five times as accurate.

Collocation can also be applied in the axial direction in place of the Runge–Kutta method. The set of equations (5.99) is written in the form

$$\frac{dy_j}{dz} = F_j(y_1, \ldots, y_{2N+2}), \qquad j = 1, \ldots, 2N,$$
$$\sum_{i=1}^{N+1} A_{N+1,i}\, y_i = 0, \qquad -\sum_{i=N+2}^{2N+2} A_{N+1,i-N-1}\, y_i = \mathrm{Bi}(y_{2N+2} - 1). \tag{5.104}$$

The collocation in the axial direction uses the polynomials with roots in Table 5.4 and matrix A in Table 5.5. For a position step of Δz the equations become

$$\frac{1}{\Delta z} \sum_{i=1}^{M+2} A_{ki}\, y_{ji} = F_j(y_{1k}, \ldots, y_{2N+2,k}), \qquad j = 1, \ldots, 2N, \quad k = 2, \ldots, M+2,$$

(*equation continues*)

$$\left.\begin{aligned} \sum_{i=1}^{N+1} A_{N+1,i}\, y_{ik} &= 0 \\ -\sum_{i=N+2}^{2N+2} A_{N+1,i-N-1}\, y_{ik} &= \mathrm{Bi}(y_{2N+2,k} - 1) \end{aligned}\right\} \quad k = 2, \ldots, M+2. \tag{5.105}$$

Here $y_j(z) = y(x_j, z)$ and $y_{ji} = y(x_j, z_i)$. This is a set of $(2N + 2)(M + 1)$ equations to advance from the position z to $z + \Delta z$. The Newton–Raphson method (5.97) can be used to solve them. Villadsen (1970) proposes an iterative procedure in which the linear terms are placed on the left-hand side of (5.105), which is written in the form

$$\bar{\bar{L}}\bar{\theta} = \bar{N}(\theta). \tag{5.106}$$

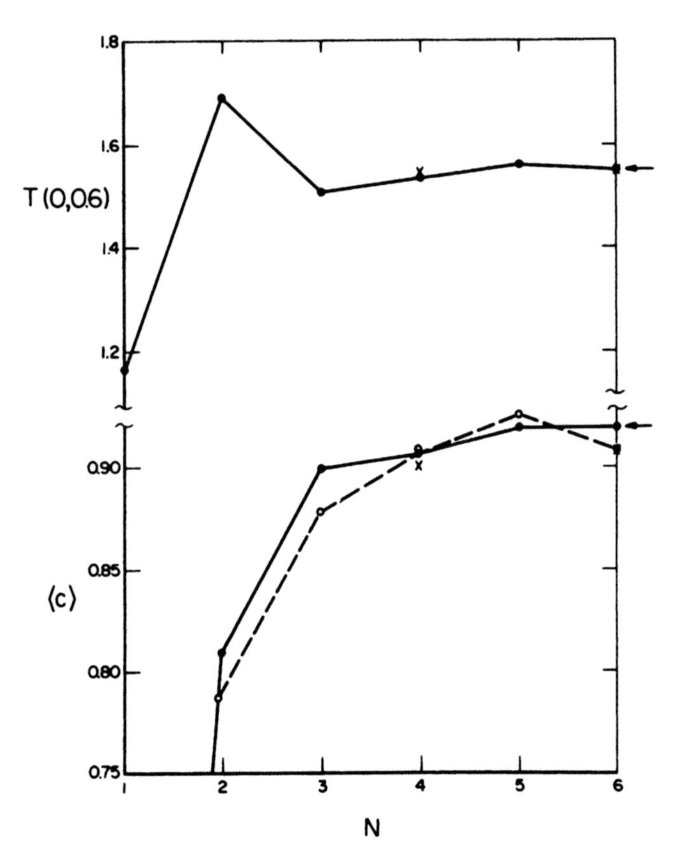

Fig. 5.7. Rate of convergence (● polynomials with $w = 1$, ○ polynomials with $w = 1 - x^2$, × finite difference; reprinted from *Chemical Engineering Science* with permission of the copyright owner).

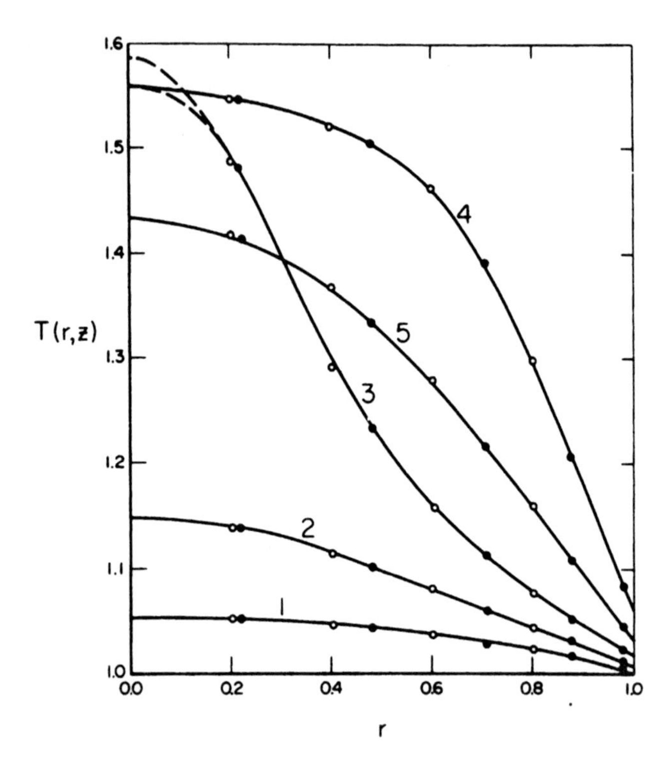

Fig. 5.8. Temperature distribution in reactor (1 to 5 correspond to $z = 0.2$, 0.4, 0.5, 0.6, 0.7; ● collocation, ○ finite difference; reprinted from *Chemical Engineering Science* with permission of the copyright owner).

The matrix $\bar{L}$ is inverted once and the iterations follow

$$\bar{\theta}^{(m+1)} = \bar{L}^{-1}\bar{N}(\bar{\theta}^{(m)}), \tag{5.107}$$

where $\bar{N}$ represents the nonlinear terms.

This procedure was applied to (5.99) with $\text{Bi} = 1$, $T_w = 0.92$. Before the hot spot the double collocation solution (5.105) with $M = 2$, $\Delta z = 0.025$ was as accurate and three times as fast as using the Runge–Kutta method to integrate (5.99) with $\Delta z = 0.001$. At the hot spot the iterative procedure (5.107) tended to oscillate, although these oscillations were reduced as Δz decreased. For a problem without a hot spot (5.99 with $\alpha = \alpha' = 1$, $\beta = \beta' = 0.2$, $\text{Bi} = 1$, $T_w = 0.8$, and $\gamma = 10$) the double collocation was about three times faster than the Runge–Kutta method for equivalent accuracy. The computer codes for the double collocation method are more complicated than for the Runge–Kutta method of integrating (5.99), but if a great many calculations are to be done the extra programming effort may be worthwhile.

5.6 Relation to Other Techniques: Galerkin, Least Squares, Finite Difference, Finite Element Methods

In the usual finite difference computations the difference expression corresponds to a linear interpolation between the grid points. When the solution at the grid points is known, the solution at other positions is found by linear interpolation. Other more sophisticated schemes are possible. With spline functions the region between two grid points is interpolated with second- or higher-degree polynomials. The derivatives at the grid points can then be made continuous. For example, for a linear interpolation, the first derivative is discontinuous at the grid points. If a second-degree polynomial is used the first derivatives can be made to be continuous. Examples of these types of approximations are given by Varga (1966), using variational methods, Loscalzo and Talbot (1967) for integrating nonlinear ordinary differential equations as initial-value problems, and Chu (1970) for applying cubic spline functions for the same type of problem. These methods differ in approach from the orthogonal collocation method in that with spline functions a low-order polynomial is used in each region from x_i to $x_i + \Delta x$. In the orthogonal collocation method, however, a single higher-order polynomial is used over the whole range of interest $x_1 \leq x \leq x_N$ and this polynomial is arranged to satisfy the boundary conditions. The obvious exception to this statement is application of orthogonal collocation to solve initial-value problems as in the previous section. There a Legendre polynomial was used on the range t to $t + \Delta t$. Thus this orthogonal collocation method can be thought of as a spline function method using Legendre polynomials.

Galerkin, Least Squares, and Orthogonal Collocation Methods

Consider a linear equation, $Ly = 0$, and the Galerkin method. Expand the solution in orthogonal polynomials, such as (5.10), and determine the constants a_i by the Galerkin method, which uses weighting functions $\partial y/\partial a_i = (1 - x^2)P_{i-1}(x^2)$:

$$\int_0^1 (1 - x^2)P_{i-1}(x^2)Lyx^{a-1}\,dx = \sum_{l=1}^{N+1} W_l(1 - x_l^2)P_{i-1}(x_l^2)Ly\big|_{x_l} = 0. \quad (5.108)$$

The evaluation of the integral is exact provided the linear operator is a polynomial of degree N or less in x^2. In that case the collocation method, which makes

$$Ly\big|_{x_l} = 0, \qquad l = 1, \ldots, N, \quad (5.109)$$

clearly implies the Galerkin method equations are satisfied also. When a variational principle exists the Galerkin method is equivalent to the Rayleigh–Ritz method (see Section 7.4). Thus the orthogonal collocation method is also equivalent to the Rayleigh–Ritz method when the quadrature (5.108) is exact.

The least squares method requires

$$\frac{\partial}{\partial a_i}\int_0^1 (Ly)^2 x^{a-1}\,dx = 0. \tag{5.110}$$

Provided the integrand is a polynomial of degree $2N$ in x^2 we can evaluate the integral exactly using the quadrature formula:

$$2\int_0^1 Ly\,\frac{\partial Ly}{\partial a_i}\,x^{a-1}\,dx = 2\sum_{l=1}^{N+1} W_l\,Ly\Big|_{x_l}\,L[(1-x^2)P_{i-1}(x^2)]\Big|_{x_l} = 0,$$

$$i = 1, \ldots, N. \tag{5.111}$$

Thus if (5.109) is satisfied and $W_{N+1} = 0$, the collocation method also satisfies the least squares criterion. This is true for $w = 1$ in (5.11) but not for $w = 1 - x^2$. If $w = 1 - x^2$, a different type of least squares criterion holds.

Suppose Ly is a polynomial of degree $2N$ in x^2 for an Nth order approximation. It can then be written as

$$Ly = a(x^2 - x_1^{\,2})(x^2 - x_2^{\,2})\cdots(x^2 - x_N^{\,2}) = a\prod_{k=1}^{N}(x^2 - x_k^{\,2}). \tag{5.112}$$

Consider the least squares criterion

$$\frac{\partial}{\partial x_i^{\,2}}\int_0^1 w(x^2)(Ly)^2 x^{a-1}\,dx = 0. \tag{5.113}$$

Now the collocation points are chosen such that

$$\int_0^1 w(x^2)P_{i-1}(x^2)\prod_{k=1}^{N}(x^2 - x_k^{\,2})x^{a-1}\,dx = 0, \tag{5.114}$$

since P_N can be written as $\prod_{k=1}^{N}(x^2 - x_k^{\,2})$ and is orthogonal to the lower-order polynomials. Thus $w(x^2)\prod_{k=1}^{N}(x^2 - x_k^{\,2})$ is orthogonal to every polynomial of order $N - 1$ in x^2, since every such polynomial can be expressed exactly as a linear combination of the P_i, $i \le N - 1$. In Eq. (5.113) $\partial Ly/\partial x_i^{\,2}$ is a polynomial of degree $N - 1$ in x^2. Consequently (5.113) is satisfied by the choice of the collocation points as the roots to $P_N(x^2) = 0$, as is done in the orthogonal collocation method. In summary, for linear equations whose residual is a polynomial of degree N or less in x^2, when the trial function is of degree N in x^2, the orthogonal collocation method satisfies both a Galerkin and least-square criterion. Equations (5.108) and (5.113) were first shown by Villadsen and Stewart (1967).

For the nonlinear problem

$$\frac{\partial \phi}{\partial t} = \nabla^2\phi + R(\phi) \tag{5.115}$$

let us apply the Galerkin method and use an M-point quadrature to evaluate the integrals, $M \geq N$. The first two terms can be evaluated exactly using $M = N$:

$$\int_0^1 (1 - x^2) P_{j-1}(x^2) \frac{\partial \phi}{\partial t} x^{a-1}\, dx$$

$$= \int_0^1 (1 - x^2) P_{j-1}(x^2)[\nabla^2 \phi + R(\phi)] x^{a-1}\, dx, \qquad j = 1, \ldots, N, \quad (5.116a)$$

$$\sum_{k+1}^{N+1} W_k P_{j-1}(x_k)(1 - x_k{}^2) \frac{d\phi_k}{dt}$$

$$= \sum_{k=1}^{N+1} W_k P_{j-1}(x_k)(1 - x_k{}^2) \sum_{l=1}^{N+1} B_{kl} \phi_l + \sum_{k=1}^{M+1} W_k^{(M)} P_{j-1}(x_k)(1 - x_k{}^2) R_k.$$

$$(5.116b)$$

The matrix

$$T_{jk} = W_k P_{j-1}(x_k)(1 - x_k{}^2), \qquad j, k = 1, \ldots, N \quad (5.117)$$

has an inverse so that the equations (5.116b) can be written as

$$\frac{d\phi_k}{dt} = \sum_{l=1}^{N+1} B_{kl} \phi_l + \sum_{j=1}^{N} (T)_{kj}^{-1} \sum_{l=1}^{M+1} W_l^{(M)} P_{j-1}(x_l)(1 - x_l{}^2) R_l. \quad (5.118)$$

These are the collocation equations if we choose $M = N$. Thus the collocation method is equivalent to applying the Galerkin method and evaluating the integral approximately using an N-point quadrature.

Finite Difference Methods

Next consider trial functions which are piecewise continuous and derive a relationship with finite difference methods. Consider the triangular functions

$$f_j(x) = \begin{cases} 1 - (|x - x_j|/\Delta x), & |x - x_j| \leq \Delta x, \\ 0, & \text{otherwise.} \end{cases} \quad (5.119)$$

Harrington (1968) shows that the Galerkin method using these trial functions is related to a finite difference method. The trial function is

$$T(x) = \sum_{i=1}^{n} a_i f_i(x),$$

$$(5.120)$$

$$T_j \equiv T(x_j) = a_j, \qquad \frac{da_j}{dt} = \frac{dT_j}{dt}.$$

Consider the linear equation

$$\frac{\partial T}{\partial t} = \nabla^2 T. \tag{5.121}$$

The Galerkin method applied to (5.121) yields

$$\sum_{i=1}^{n} \frac{da_i}{dt}(f_j, f_i) = \sum_{i=1}^{n} a_i(f_j, \nabla^2 f_i). \tag{5.122}$$

In order to calculate the right-hand side we must introduce the concept of a generalized inner product. Integrate the term by parts and define

$$(f_j, \nabla^2 f_i) \equiv -\int \nabla f_j \cdot \nabla f_i + [f_j \mathbf{n} \cdot \nabla f_i]. \tag{5.123}$$

The right-hand side of (5.122) can now be evaluated. The complete result is

$$\begin{gathered} \frac{1}{6}\frac{dT_{j-1}}{dt} + \frac{2}{3}\frac{dT_j}{dt} + \frac{1}{6}\frac{dT_{j+1}}{dt} = T_{xx,j}, \\ T_{xx,j} = (T_{j+1} - 2T_j + T_{j-1})/\Delta x^2. \end{gathered} \tag{5.124}$$

This set of equations is not exactly equivalent in form to finite difference equations because the left-hand side is tri-diagonal. However, if we apply the implicit method of solving (5.124) and use a first-order difference expression to approximate the time derivatives, the result is

$$\begin{aligned} &\frac{1}{6}\frac{T_{j-1}^{n+1} - T_{j-1}^{n}}{\Delta t} + \frac{4}{6}\frac{T_j^{n+1} - T_j^n}{\Delta t} + \frac{1}{6}\frac{T_{j+1}^{n+1} - T_{j+1}^{n}}{\Delta t} \\ &\quad = \lambda T_{xx,j}^{n+1} + (1-\lambda)T_{xx,j}^{n}. \end{aligned} \tag{5.125}$$

Swartz and Wendroff (1969) show that this can be arranged to give

$$\frac{T_j^{n+1} - T_j^n}{\Delta t} = \left(\lambda - \frac{\Delta x^2}{6\,\Delta t}\right)T_{xx,j}^{n+1} + \left(1 - \lambda + \frac{\Delta x^2}{6\,\Delta t}\right)T_{xx,j}^{n}. \tag{5.126}$$

With the choice $\theta = \lambda - (\Delta x^2/6\,\Delta t)$, Eq. (5.126) becomes the standard implicit equation for solving (5.121). Consequently the finite difference expressions for solving (5.121) can be regarded as a Galerkin method using triangular trial functions of Eq. (5.119). With this interpretation, the proper method to interpolate to find values of T at $x \neq x_j$ is a straight line interpolation. The results of applying the Galerkin method in this way are the same as a finite difference calculation.

We can also apply a more general method of weighted residuals. Use as weighting functions the step functions

$$P_j(x) = \begin{cases} 1, & x_j - (\Delta x/2) \le x \le x_j + (\Delta x/2), \\ 0, & \text{otherwise.} \end{cases} \tag{5.127}$$

For the trial function (5.119) the MWR equations reduce to

$$\frac{1}{8}\frac{dT_{j-1}}{dt} + \frac{3}{4}\frac{dT_j}{dt} + \frac{1}{8}\frac{dT_{j+1}}{dt} = T_{xx,\,j}, \tag{5.128}$$

which are also equivalent to an implicit finite difference scheme.

For nonlinear problems, such as Eq. (5.115), the correspondence does not hold since in general

$$\int f_j R(\phi)\, dx \neq \Delta x f_j R(\phi_j). \tag{5.129}$$

In fact the integrals on the left-hand side of (5.129) may be difficult to evaluate analytically. This was precisely the motivation for applying the orthogonal collocation method.

Finite Element Method

For two dimensions, comparisons between MWR and finite difference calculations are best revealed through the finite element method. In the finite element method the domain is divided into small elements, and a trial function of some specified shape is taken in each element. For example, in one dimension a triangular shape (5.119), may be used, whereas in two dimensions a pyramid is one possible choice. The trial function is then taken in the form (5.120) and the finite element method determines the coefficients a_i. In the historical development of the method, the criterion for the a_i was a variational principle, since the method was developed for problems in structural mechanics which are governed by variational principles. More recently, however, the Galerkin method has been used as the criterion, making the methods applicable to problems for which no variational principle exists. The details of the method are given in Zienkiewicz and Cheung (1967) and a review is available (Zienkiewicz, 1970). We limit consideration to two-dimensional problems for simplicity, and illustrate the application to a heat transfer problem.

Consider the triangular finite element shown in Fig. 5.9a. Within this triangle the temperature is taken as a linear function of x and y:

$$T = \alpha_1 + \alpha_2 x + \alpha_3 y. \tag{5.130}$$

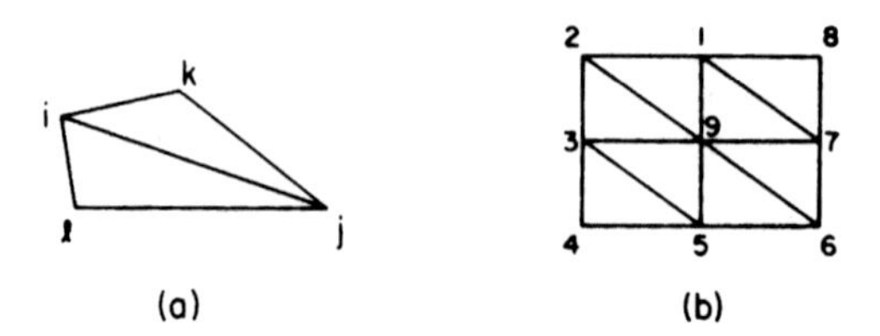

Fig. 5.9. Finite elements: (a) general element, (b) regular array.

This can be evaluated at each of the nodes, e.g.,

$$T_j = \alpha_1 + \alpha_2 x_j + \alpha_3 y_j \tag{5.131}$$

to provide a set of three equations to solve for the constants in (5.131). The solution is written in terms of the nodal temperatures, T_j:

$$T = N_i(x, y)T_i + N_j(x, y)T_j + N_k(x, y)T_k, \tag{5.132a}$$

$$N_i = (a_i + b_i x + c_i y)/2\Delta, \tag{5.132b}$$

$$\left.\begin{aligned} &a_i = x_j y_k - x_k y_j, \\ &b_i = y_j - y_k, \qquad c_i = x_k - x_j, \end{aligned}\right\}\text{plus permutations on } i, j, k \tag{5.132c}$$

$$2\Delta = \det \begin{vmatrix} 1 & x_i & y_i \\ 1 & x_j & y_j \\ 1 & x_k & y_k \end{vmatrix} = 2\text{ (area of }\Delta), \tag{5.132d}$$

$$a_i + a_j + a_k = 2\Delta, \tag{5.132e}$$

$$b_i + b_j + b_k = c_i + c_j + c_k = 0, \tag{5.132f}$$

$$N_i + N_j + N_k = 1. \tag{5.132g}$$

This provides the form of the trial function within the finite element.

Outside the element, e.g., when the point (x, y) does not fall within a triangle having a vertex at (x_i, y_i), then $N_i(x, y) = 0$. The expansion function then consists of a series of these trial functions, each of which is a pyramid on some triangle and zero elsewhere. Apply the Galerkin method for two-dimensional steady-state heat conduction:

$$\begin{aligned} \nabla \cdot k\, \nabla T &= Q \qquad \text{in } A, \\ T &= T_c \qquad \text{on } C. \end{aligned} \tag{5.133}$$

The weighting function is taken as N_m and the Galerkin method gives

$$\int_A N_m(-\nabla \cdot k\, \nabla T + Q)\, dx\, dy = 0. \tag{5.134}$$

Within any triangular element having a vertex at (x_m, y_m), N_m is given by Eq. (5.132b); thus N_m is different in different elements because the a, b, and c are different. One difficulty is immediately apparent in (5.134). The trial function of Eqs. (5.132) does not have a second derivative. Define a generalized derivative within any triangular element as follows:

$$\int_\Delta N_m \nabla \cdot k \nabla T \, dA = -\int k \nabla N_m \cdot \nabla T \, dA + [N_m k\mathbf{n} \cdot \nabla T]_{s_\Delta}, \quad (5.135)$$

where s_Δ is the boundary at the element Δ. Since both T and N_m have first derivatives, the right-hand side can be calculated. There is another difficulty, however. Equation (5.135) is correct within the finite element, but to evaluate (5.134) we must account for the discontinuity in the first derivative at the boundary, and hence the infinite value of the second derivative there. This problem is resolved (in one dimension) by defining the integral along the discontinuity in the following fashion:

$$\int_{a-\varepsilon}^{a+\varepsilon} p(x)q''(x)\, dx = p(a)[q'(a+) - q'(a-)], \quad (5.136)$$

when $p(x)$ is continuous. We employ a similar definition in two dimensions. Consider the boundary $i-j$. There are six contributions to Eq. (5.134) coming from the discontinuity along $i-j$: one each from the weighting functions N_i, N_j, and N_k, as well as N_i', N_j', and N_l'. (The prime is used to remind the reader that N_i is different in ijk and ilj.) However, the value of the flux is constant in element ijk and element ilj. Thus the discontinuity gives rise to the following contribution:

$$\int (N_i + N_j + N_k)\, \nabla^2 T \, dA + \int (N_i' + N_j' + N_l')\, \nabla^2 T \, dA$$
$$= k\mathbf{n} \cdot \nabla T \Big|_{\substack{\text{along } ij \\ \text{inside } ilj}} - k\mathbf{n} \cdot \nabla T \Big|_{\substack{\text{along } ij \\ \text{inside } ijk}}, \quad (5.137)$$

when $\mathbf{n}$ is the normal pointing outward from ijk. The boundary term in (5.137) cancels the sum of the boundary terms in (5.135) at each interior boundary, so that the boundary terms give no net contribution and can be ignored.

Along the boundary of the domain we can show that $N = 0$ and the boundary term vanishes. Consider the boundary $l-j$ as an external boundary, on which the temperature is specified. No weighting functions N_l or N_j are used in (5.134). The only contribution to (5.134) is with N_m of Eq. (5.135). However, along the boundary y and x are related,

$$\frac{y - y_l}{y_j - y_l} = \frac{x - x_l}{x_j - x_l}. \quad (5.138)$$

Calculation of $N_i(x, y)$ using this relation shows that

$$N_i = 0 \tag{5.139}$$

along an external boundary. Thus the boundary term in Eq. (5.135) vanishes along an external boundary, too.

The next task is to evaluate the weighted residual for the finite element ijk:

$$\begin{aligned} \int_\Delta N_m \nabla \cdot k \nabla T \, dA &= -\int_\Delta k \nabla N_m \cdot \nabla T \, dA \\ &= -\frac{1}{(2\Delta)^2} \int_\Delta k[(b_i T_i + b_j T_j + b_k T_k) b_m \\ &\qquad + (c_i T_i + c_j T_j + c_k T_k) c_m] \, dx \, dy. \end{aligned} \tag{5.140}$$

If k and Q are constant this can be simplified. The terms in the integrand are then constant and the value of the integral is their value times the area of the element. The last integral in (5.134) is

$$\begin{gathered} Q \int_\Delta N_m \, dx \, dy = \frac{Q}{2\Delta} (a_m + b_m \bar{x} + c_m \bar{y}) \, \Delta \\ \bar{x} = \frac{1}{\Delta} \int_\Delta x \, dx \, dy, \qquad \bar{y} = \frac{1}{\Delta} \int_\Delta y \, dx \, dy. \end{gathered} \tag{5.141}$$

Calculation of the centroid of the element gives

$$\begin{aligned} \bar{x} &= (x_i + x_j + x_k)/3, \\ \bar{y} &= (y_i + y_j + y_k)/3. \\ a_m + b_m \bar{x} + c_m \bar{y} &= \tfrac{2}{3}\Delta, \end{aligned} \tag{5.142}$$

The total contribution due to the weighting function N_m is then

$$\begin{aligned} \text{contribution from } N_m &= \sum_{q=i,j,k} h_{mq} T_q + F_m, \\ h_{mq} &= (+k/4\Delta)[b_m b_q + c_m c_q], \\ F_m &= +Q(\Delta/3), \end{aligned} \tag{5.143}$$

and the final equations are the sum of (5.143) at all nodes:

$$\sum_{\text{all nodes}} \left[\sum h_{mq} T_q + F_m\right] = 0. \tag{5.144}$$

Notice that the temperature at a node appears once for each triangle that has a vertex at that node. The derivation is simpler if a variational method is used (Section 7.4) since it avoids the necessity to define generalized derivatives. More general boundary conditions are also treated in Section 7.4.

Consider next a comparison to finite difference methods. For the regular array in Fig. 5.9b, the usual finite difference method would evaluate the differential equation at the point 9 by

$$\frac{k}{h^2}(T_1 + T_3 + T_5 + T_7 - 4T_9) - Q = 0. \tag{5.145}$$

Compare this result to that derived using the finite element method. Consider the contribution due to the weighting function N_9 in the triangle 1–2–9. Equations (5.132) and (5.143) give $4\Delta = 2h^2$ and

$$\begin{array}{llll} b_1 = h, & c_1 = h, & h_{11} = k, & h_{22} = \tfrac{1}{2}k, \\ b_2 = -h, & c_2 = 0, & h_{12} = -\tfrac{1}{2}k, & h_{29} = 0, \\ b_9 = 0, & c_9 = -h, & h_{19} = -\tfrac{1}{2}k, & h_{99} = \tfrac{1}{2}k. \end{array} \tag{5.146}$$

The contribution to the weighted integral is then

$$129\colon \quad \text{contribution from } N_9 = \tfrac{1}{2}(-T_1 + T_9)k. \tag{5.147}$$

Similar operations on the other triangles give the results

$$\begin{array}{ll} 569\colon\ \tfrac{1}{2}(-T_5 + T_9)k, & 197\colon\ \tfrac{1}{2}(-T_1 - T_7 + 2T_9)k, \\ 239\colon\ \tfrac{1}{2}(-T_3 + T_9)k, & 359\colon\ \tfrac{1}{2}(-T_3 - T_5 + 2T_9)k, \\ \multicolumn{2}{c}{679\colon\ \tfrac{1}{2}(-T_7 + T_9)k.} \end{array} \tag{5.148}$$

Combination of the results gives

$$k[4T_9 - T_1 - T_3 - T_5 - T_7] + h^2 Q = 0, \tag{5.149}$$

which is equivalent to (5.145). Thus with the regular array (Fig. 5.9b) the finite element method gives results which are identical to the finite difference method. Other arrays are not necessarily equivalent (see Zienkiewicz and Cheung, 1967).

When time derivatives are included it is necessary to add

$$\int_A N_m \frac{\partial T}{\partial t}\, dx\, dy = \frac{d}{dt} \int_\Delta N_m(N_i T_i + N_j T_j + N_k T_k)\, dx\, dy \tag{5.150}$$

to Eq. (5.134). Transient computations are outlined by Zienkiewicz and Cheung (1967, p. 166) and Zienkiewicz and Parekh (1970) using Galerkin's method, and by Wilson and Nickell (1966) using Gurtin's variational principle, which is shown in Section 9.4 to be equivalent to Galerkin's method. For nonlinear problems, such as Eq. (5.115), the integrals become difficult or impossible to evaluate. Then a quadrature formula or a very simple approximation like Eq. (5.129) may be used. Elements can take other shapes, as well. Zienkiewicz *et al.* (1967) give the formulas for rectangular, tetrahedral, and eight-cornered elements in three-dimensional problems, while

Zienkiewicz and Parekh (1970) use curved elements in two and three dimensions. Zlámal (1969) uses piecewise cubic polynomials and gives an algorithm to compute the matrices. Reference to applications in heat transfer are given by Zienkiewicz (1970); see also the application to the Boltzmann equation by Bramlette and Mallett (1970).

We see that in some cases the finite element method is identical to the finite difference method. The advantage of the finite element method is that the elements can be changed in size and shape to follow arbitrary boundaries and rapid variations of the expected solution. Furthermore, the equations can be solved using standard programs which are available to solve structural problems. These advantages are particularly welcome in those practical problems which have very irregular geometries. From the standpoint of MWR the finite element method replaces the choice of a trial function valid over the whole region by the choice of the shape of the finite element. At the same time a large number of terms is necessary, just as in finite difference calculations, so that this step gives both advantages and disadvantages. The literature on applications of the finite element method indicates that this is one method of MWR which is becoming increasingly important in engineering fields, particularly in structural analysis.

EXERCISES

5.1. Which polynomials must be used to apply orthogonal collocation to the following problems? Define the polynomials by identifying w, a, b, and the table listing the roots.

(a)–(d) $y'' + 2y = x^2$ in $0 < x < 1$ subject to the boundary conditions:

(a) $y'(0) = 0, \quad y(1) = 1,$
(b) $y'(0) = 0, \quad y'(1) + 3y(1) = 1,$
(c) $y(0) = 0, \quad y(1) = 1,$
(d) $-y'(0) + y(0) = 0, \quad y(1) = 1.$

(e)–(h) The same boundary conditions as (a)–(d) but the differential equation $y'' + 2y = x$.

(i)–(j) The differential equation $y'' - 2y' + 3y^2 = 0$ subject to the boundary conditions:

(i) $y'(0) = 0, \quad y(1) = 1,$
(j) $-y(0) + y(0) = 1, \quad y'(1) = 0.$

Answer: Polynomials in Table 5.2 with $w = 1 - x^2$, $a = 1$ (plane geometry) for a, b; polynomials in Table 5.4 for c, f, i.

5.2. Evaluate the integral $I = \int_0^1 f(x^2)\,dx$ for $f(x^2) = 1,\ x^2,\ x^4$, and x^6 using the quadrature formula for $N = 1$ and 2. Note that the results should be exact for $f(x^2)$ a polynomial of degree $2N$ in x^2.
Answer: $N = 1, f = x^4, I = 0.2$; $N = 2, f = x^4, I = 0.2$.

5.3. Apply orthogonal collocation to the problem solved in Section 2.1, Eq. (2.3). How does the first approximation compare to the collocation solution found in Chapter 2?
Answer: Results should be the same. Why?

5.4. Rework the problem in Eqs. (5.44)–(5.45) for spherical geometry. Compare the solutions η versus $\hat{\phi}$ for the planar and spherical geometry.

5.5. For spherical geometry, illustrate the effect of Sherwood number on the η versus $\hat{\phi}$ curve for Eqs. (5.44)–(5.45). The boundary condition is then

$$y'(1) + \frac{\mathrm{Sh}'}{2} y(1) = \frac{\mathrm{Sh}'}{2}.$$

The asymptotic dependence is then

$$\eta = \frac{\sqrt{2}}{\hat{\phi}} \left[\int_0^{y(1)} R(y)\,dy \right]^{1/2}, \qquad R(y) = y^2 \quad \text{here,}$$

where $y(1)$ satisfies

$$\frac{\mathrm{Sh}'}{2} [y(1) - 1] = -\sqrt{2}\phi \left[\int_0^{y(1)} R(y)\,dy \right]^{1/2}, \qquad \phi = 3\hat{\phi} \quad \text{for sphere}$$

in place of Eqs. (5.51). This expression for η is the effectiveness factor based on the reaction rate for the external concentration and includes internal and external (film) diffusion effects. Also it is assumed that $R(1) = 1$. (Hint: Try a graphical solution.)

5.6. Find the effectiveness factor for a cylindrical pellet of radius R and length $2R$:

$$\frac{1}{r}\frac{\partial}{\partial r}\left(r \frac{\partial y}{\partial r} \right) + \frac{\partial^2 y}{\partial z^2} = \phi^2 y,$$

$$\frac{\partial y}{\partial r}(0, z) = \frac{\partial y}{\partial z}(r, 0) = 0, \qquad y(1, z) = y(r, 1) = 1.$$

This two-dimensional problem can be solved by applying orthogonal collocation as in Eq. (5.26), but with different matrices to represent the derivatives in the r and z directions. Find η for $\phi = 1$. What is the trial function?
Answer: For $N = 1$, $\eta = 0.93$, $y_{11} = 8.5/(8.5 + \phi^2)$, $y = 1 + A(1 - r^2)(1 - z^2)$; $A = -0.197$.

5.7. Determine if the following equation can have multiple solutions:

$$\frac{1}{x^2}\frac{d}{dx}\left(x^2\frac{dc}{dx}\right) = \phi^2 R(c),$$

$$\frac{dc}{dx}(0) = 0, \qquad c(1) = 1, \qquad R(c) = c(E + c)/(1 + Kc)^2.$$

The reaction rate occurs with Langmuir–Hinshelwood kinetics and one of the reactants is in stoichiometric excess (fraction excess represented by E). Solve for:

	a	b	c	d	e
K	1	10	100	100	100
E	0	0	0	1	10

Answer: Case d has multiple solutions for $4300 < \phi^2 < 19{,}000$.

5.8. Consider Eqs. (5.99) for $N = 1$. Write out the equations and combine them to obtain Eqs. (5.100). Derive an expression for T_2 as a function of T_1, T_w, and Bi, and for $T_2 - T_w$ as a function of $T_1 - T_w$ and Bi. At what r values do T_2 and T_1 represent the temperature? How does $T_2 - T_w$ (the temperature drop at the wall) compare to $T_1 - T_w$ as Bi is changed? What is $T(0, z)$ in terms of $T_1(z)$, $T_2(z)$, and $T_w(z)$? What is the temperature profile corresponding to $N = 1$?
Answer: $(T_2 - T_w)/(T_1 - T_w) = 6/(\mathrm{Bi} + 6)$; $T(r, z) = T_2(z) + a_1(z)(1 - r^2)$.

5.9. Consider the entry-length problem, Eqs. (3.81). Formulate the equations, similar to Eqs. (5.43), which govern an approximate solution incorporating an asymptotic term. Note that for small z the mass penetrates only to small x, and the fact that the velocity profile is $(1 - x^2)$ is immaterial. Thus the asymptotic term can be the solution to

$$\frac{3}{2}\frac{\partial c}{\partial z} = \frac{\partial^2 c}{\partial x^2}, \qquad c(0, z) = 1, \quad c(\infty, z) = 0, \quad c(x, 0) = 0.$$

A similar approach can be used for nonlinear problems: the asymptotic solution can be for a related, linear problem. How large N must be for numerical convergence at small z depends on how closely the correct behavior is modeled by the asymptotic term.

REFERENCES

Amundson, N. R. (1966). "Mathematical Methods in Chemical Engineering." Prentice-Hall, Englewood Cliffs, New Jersey.

Bramlette, T. T., and Mallett, R. H. (1970). A Finite Element Solution Technique for the Boltzmann Equation, *J. Fluid Mech.* **42**, 177–191.

Chu, S.-C. (1970). Piecewise Polynomials and the Partition Method for Nonlinear Ordinary Differential Equations, *J. Eng. Math.* **4**, 65–76.

Clenshaw, C. W., and Norton, H. J. (1963). The Solution of Non-Linear Ordinary Differential Equations in Chebyshev Series, *Comp. J.* **6**, 88–92.

Crank, J. (1956). "The Mathematics of Diffusion." Oxford Univ. Press (Clarendon), London and New York.

Danckwerts, P. V. (1953). Continuous Flow Systems, *Chem. Eng. Sci.* **2**, 1–13.

Finlayson, B. A. (1971). Packed Bed Reactor Analysis by Orthogonal Collocation, *Chem. Eng. Sci.* **26**, 1081–1091.

Ferguson, N. B. (1971). Orthogonal Collocation as a Method of Analysis in Chemical Reaction Engineering, Ph.D. Thesis, Univ. of Washington, Seattle, Washington.

Ferguson, N. B., and Finlayson, B. A. (1970). Transient Chemical Reaction Analysis by Orthogonal Collocation, *Chem. Eng. J.* **1**, 327–336.

Garabedian, H. L., and Lynch, R. E. (1965). Nonlinear Reactor Kinetics Analysis, *Nucl. Sci. Eng.* **21**, 550–564.

Gavalas, G. R. (1968). "Nonlinear Differential Equations of Chemically Reacting Systems." Springer-Verlag, Berlin.

Grotch, S. L. (1969). The Solution of a Boundary Value Problem in Reactor Design Using Galerkin's Method, *AIChE J.* **15**, 463–465.

Harrington, R. F. (1968). "Field Computation by Moment Methods." Macmillan, New York.

Hellinckx, L., Grootjans, J. and Van den Bosch, B. (1971). Stability Analysis of the Catalyst Particle through Orthogonal Collocation, private communication.

Hlaváček, V., and Hofmann, H. (1970). Modeling of Chemical Reactors—XVI—Steady-State Axial Heat and Mass Transfer in Tubular Reactors. An Analysis of the Uniqueness of Solutions, *Chem. Eng. Sci.* **25**, 173–185.

Hlaváček, V., and Kubíček, M. (1970). Modeling of Chemical Reactors—XXI—Effect of Simultaneous Heat and Mass Transfer inside and outside of a Pellet on Reaction Rate, *Chem. Eng. Sci.* **25**, 1537–1547.

Hlaváček, V., Kubíček, M. and Marek, M. (1969). Analysis of Nonstationary Heat and Mass Transfer in a Porous Catalyst Particle I, *J. Catalysis* **15**, 17–30.

Horvay, G., and Spiess, F. N. (1954). Orthogonal Edge Polynomials in the Solution of Boundary Value Problems, *Quart. Appl. Math.* **12**, 57–69.

Koob, S. J., and Abbott, D. E. (1968). Investigation of a Method for the General Analysis of Time Dependent Two-Dimensional Laminar Boundary Layers, *J. Basic Eng., Trans. ASME, Ser. D* **90**, 563–571.

Kopal, Z. (1955). "Numerical Analysis." Wiley, New York.

Kuo, J. C. W., and Amundson, N. R. (1967). Catalytic Particle Stability Studies—III. Complex Distributed Resistances Model, *Chem. Eng. Sci.* **22**, 1185–1197.

Lanczos, C. (1938). Trigonometric Interpolation of Empirical and Analytical Functions, *J. Math. Phys.* **17**, 123–199.

Lanczos, C. (1956). "Applied Analysis." Prentice-Hall, Englewood Cliffs, New Jersey.

Lee, E. S. (1968). "Quasilinearization and Invariant Imbedding." Academic Press, New York.

Lee, J. C. M., and Luss, D. (1970). The Effect of Lewis Number on the Stability of a Catalytic Reaction, *AIChE J.* **16**, 620–625.

Liu, S. L. (1969). Stable Explicit Difference Approximations to Parabolic Partial Differential Equations, *AIChE J.* **15**, 334–338.

Loscalzo, F. R., and Talbot, T. D. (1967). Spline Function Approximations for Solutions of Ordinary Differential Equations, *SIAM J. Numer. Anal.* **4**, 433–445.

Luss, D., and Amundson, N. R. (1967). Uniqueness of the Steady State Solutions for Chemical Reaction Occurring in a Catalyst Particle or in a Tubular Reactor with Axial Diffusion, *Chem. Eng. Sci.* **22**, 253–266.

McGowin, C. R., and Perlmutter, D. D. (1971). Regions of Asymptotic Stability for Distributed Parameter Systems, *Chem. Eng. Sci.* **26**, 275–286.

McGreavy, C., and Cresswell, D. L. (1969). A Lumped Parameter Approximation to a General Model for Catalytic Reactors, *Can. J. Chem. Eng.* **47**, 583–589.

McGuire, M. L., and Lapidus, L. (1965). On the Stability of a Detailed Packed-Bed Reactor, *AIChE J.* **11**, 85–95.

Norton, H. J. (1964). The Iterative Solution of Nonlinear Ordinary Differential Equations in Chebyshev Series, *Comp. J.* **7**, 76–85.

Pakes, H. W., and Storey, C. (1967). Solution of the Equations for a Tubular Reactor with Axial Diffusion by a Variational Technique, *Trans. Inst. Chem. Eng.* **45**, CE96–98, 108.

Petersen, E. E. (1965). "Chemical Reaction Analysis." Prentice-Hall, Englewood Cliffs, New Jersey.

Stewart, W. E., and Villadsen, J. V. (1969). Graphical Calculation of Multiple Steady States and Effectiveness Factors for Porous Catalysts, *AIChE J.* **15**, 28–34.

Stroud, A. H., and Secrest, D. (1966). "Gaussian Quadrature Formulas." Prentice-Hall, Englewood Cliffs, New Jersey.

Swartz, B., and Wendroff, B. (1969). Generalized Finite-Difference Schemes, *Math. Comp.* **23**, 37–49.

Varga, R. S. (1966). Hermite Interpolation-Type Ritz Methods for Two-Point Boundary Value Problems, *in* "Numerical Solution of Partial Differential Equations" (J. H. Bramble, ed.), pp. 365–373. Academic Press, New York.

Vichnevetsky, R. (1969). Generalized Finite-Difference Approximations for the Parallel Solution of Initial Value Problems, *Simulation* **12**, 233–237.

Villadsen, J. (1970). "Selected Approximation Methods for Chemical Engineering Problems." Inst. for Kemiteknik Numer. Inst. Danmarks Tekniske Højskole.

Villadsen, J., and Sørensen, J. P. (1969). Solution of Parabolic Partial Differential Equations by a Double Collocation Method, *Chem. Eng. Sci.* **24**, 1337–1349.

Villadsen, J. V., and Stewart, W. E. (1967). Solution of Boundary-Value Problems by Orthogonal Collocation, *Chem. Eng. Sci.* **22**, 1483–1501.

Wehner, J. F., and Wilhelm, R. (1956). Boundary Conditions of Flow Reactor, *Chem. Eng. Sci.* **6**, 89–93.

Wei, J. (1965). The Stability of a Reaction with Intra-Particle Diffusion of Mass and Heat: The Liapunov Methods in a Metric Function Space, *Chem. Eng. Sci.* **20**, 729–736.

Weisz, P. B., and Hicks, J. S. (1962). The Behavior of Porous Catalyst Particles in View of Internal Mass and Heat Diffusion Effects, *Chem. Eng. Sci.* **17**, 265–275.

Wilson, E. L., and Nickell, R. E. (1966). Application of the Finite Element Method to Heat Conduction Analysis, *Nucl. Eng. Design* **4**, 276–286.

Wright, K. (1964). Chebyshev Collocation Methods for Ordinary Differential Equations, *Comp. J.* **6**, 358–365.

Zienkiewicz, O. C. (1970). The Finite Element Method: From Intuition to Generality, *Appl. Mech. Rev.* **23**, 249–256.

Zienkiewicz, O. C., Bahrani, A. K. and Arlett, P. L. (1967). Solution of Three-Dimensional Problems by the Finite Element Method, *Engineer* **224**, 547–550.

Zienkiewicz, O. C., and Cheung, Y. K. (1967). "The Finite Element Method in Structural and Continuum Mechanics." McGraw-Hill, New York.

Zienkiewicz, O. C., and Parekh, C. J. (1970). Transient Field Problems—Two-Dimensional and Three-Dimensional Analysis by Iso-Parametric Finite Elements, *Int. J. Num. Meth. Eng.* **2**, 61–71.

Zlámal, M. (1969). On Some Finite Element Procedures for Solving Second-Order Boundary-Value Problems, *Num. Math.* **14**, 42–48.

Chapter

6

Convective Instability Problems

The spontaneous generation of motion in a system previously at rest and the transition from a laminar to turbulent flow are two instability problems of great interest in science and engineering. These are eigenvalue problems for systems of equations, and their solution presents a formidable task. The Galerkin method and variational methods are standard techniques for solution of these eigenvalue problems. Here we present the entire topic in the framework of the Galerkin method, pointing out where results derived using some other method are applicable. The choice of trial functions is discussed, followed by applications of the Galerkin method to a variety of convective instability problems. The Galerkin method is contrasted to variational methods before proceeding to nonlinear convective instability and hydrodynamic stability problems. Application of Galerkin's method to elastic stability problems is discussed by Bolotin (1963), Beal (1965) and Ames (1965, p. 258), and combustion instability problems by Powell and Zinn (1969) and Zinn and Powell (1970). In the last two references the boundary residuals are combined with the differential equation residuals in order that the trial functions need not have to satisfy the boundary conditions.

6.1 Choice of Trial Functions

In many applications it is necessary to expand the solution in terms of functions which satisfy four boundary conditions. These trial functions can be taken as eigenfunctions of some simpler eigenvalue problem. The eigenfunctions corresponding to a fourth-order differential equation can be made to satisfy four boundary conditions, and when the eigenvalue problem is constructed properly certain orthogonality conditions are satisfied as well. Take the operator as

$$L^4 y = \lambda^4 N y, \tag{6.1}$$

where L^4 is a fourth-order ordinary differential operator and N is a general operator of order less than four. One set of boundary conditions of interest is

$$y = y' = 0 \qquad \text{at} \quad x = \pm\tfrac{1}{2}. \tag{6.2}$$

Under these boundary conditions the eigenfunctions of (6.1) are orthogonal for certain choices of L^4:

$$\int_{-1/2}^{1/2} y_n N y_m \, dx = 0, \qquad n \neq m. \tag{6.3}$$

Write Eq. (6.1) for y_m, set $L^4 y = y^{\text{IV}}$, multiply by y_n, and integrate the left-hand side by parts twice to obtain

$$\int_{-1/2}^{1/2} y_n'' y_m'' \, dx = \lambda_m \int_{-1/2}^{1/2} y_n N y_m \, dx \tag{6.4}$$

after application of the boundary conditions. Do the same thing for y_n in (6.1), multiply by y_m, and subtract from (6.4) to give (6.3). The operator N is chosen for convenience in particular problems.

Consider the approach advanced by Chandrasekhar and Reid (1957), Harris and Reid (1958) and Reid and Harris (1958). (See also Chandrasekhar, 1961, Appendix V.) Take the eigenvalue problem

$$y^{\text{IV}} = \lambda^4 y, \qquad y = y' = 0 \qquad \text{at} \quad x = \pm\tfrac{1}{2}. \tag{6.5}$$

The solution is obtained by trying $y = \exp(mx)$, giving $m = \pm\lambda$, $\pm i\lambda$. The general solution is then divided into even and odd functions

$$y = B_1 \cosh \lambda x + B_2 \cos \lambda x + B_3 \sinh \lambda x + B_4 \sin \lambda x. \tag{6.6}$$

The even functions satisfy the boundary conditions if

$$\begin{aligned} B_1 \cosh \tfrac{1}{2}\lambda + B_2 \cos \tfrac{1}{2}\lambda &= 0, \\ \lambda B_1 \sinh \tfrac{1}{2}\lambda - \lambda B_2 \sin \tfrac{1}{2}\lambda &= 0, \end{aligned} \tag{6.7}$$

or

$$\cosh \tfrac{1}{2}\lambda \sin \tfrac{1}{2}\lambda + \sinh \tfrac{1}{2}\lambda \cos \tfrac{1}{2}\lambda = 0.$$

Thus we take the even functions as

$$C_n = \frac{\cosh \lambda_n x}{\cosh \frac{1}{2}\lambda_n} - \frac{\cos \lambda_n x}{\cos \frac{1}{2}\lambda_n}, \qquad \tanh \tfrac{1}{2}\lambda_n + \tan \tfrac{1}{2}\lambda_n = 0. \tag{6.8}$$

Similarly the odd functions are

$$S_n = \frac{\sinh \mu_n x}{\sinh \frac{1}{2}\mu_n} - \frac{\sin \mu_n x}{\sin \frac{1}{2}\mu_n} \qquad \coth \tfrac{1}{2}\mu_n - \cot \tfrac{1}{2}\mu_n = 0. \tag{6.9}$$

It can be shown that the eigenfunctions satisfy the following orthogonality properties:

$$\langle C_n, C_m \rangle = \delta_{nm}, \qquad \langle C_n, S_m \rangle = 0, \qquad \langle S_n, S_m \rangle = \delta_{nm}. \tag{6.10}$$

Other integrals are evaluated by Reid and Harris (1958) and Chandrasekhar (1961, Appendix V). Chandrasekhar and Elbert (1958) and Chandrasekhar (1961) construct similar functions for cylindrical regions which satisfy the boundary conditions

$$y = Dy = 0 \qquad \text{at} \quad x = \kappa, 1. \tag{6.11}$$

Finlayson and Scriven (1969) use similar functions on the interval $0 \leq x \leq 1$, in which case the boundary condition at 0 is that the functions are finite. Bisshopp (1958) and Roberts (1969) use functions for a spherical domain.

For reasons which are apparent below it is convenient to have eigenfunctions which are orthogonal with the following inner product:

$$-\int_a^b y_n(D^2 - a^2)y_m \, dx = \int_a^b (Dy_n Dy_m + a^2 y_n y_m) \, dx = \delta_{nm}. \tag{6.12}$$

These can be generated from the eigenvalue problem

$$L^4 y = \lambda(a^2 - D^2)y, \tag{6.13}$$

where L^4 is a general fourth-order operator, possibly depending on a. Specific examples are due to Dolph and Lewis (1958) and Mahler *et al.* (1968) (see pp. 183, 197).

Trial functions like Eqs. (6.8) give accurate results but polynomials are more convenient to use. We illustrate the generation of polynomial trial functions and later (Section 6.4) compare results obtained with them and

(6.8). Consider the boundary condition (6.2) and the requirement of a set of functions which is symmetric about $z = 0$. It can be represented as a power series in z^2:

$$W_1 = \sum_{i=0}^{4} a_i z^i. \tag{6.14}$$

Application of the boundary conditions (6.2), along with $a_4 = 1$ as the normalization condition, gives

$$W_1 = (z^2 - \tfrac{1}{4})^2. \tag{6.15}$$

For higher approximations we then take

$$W_i = (z^2 - \tfrac{1}{4})^{i+1}. \tag{6.16}$$

Trial functions are also needed for temperature. If the problem has boundary conditions of the first kind, then the trial functions must satisfy $T = 0$ at $z = \pm\frac{1}{2}$. Following the same procedure for the velocity functions gives

$$T_i = (z^2 - \tfrac{1}{4})^i. \tag{6.17}$$

In applications of the Galerkin method it is necessary to compute various integrals involving these trial functions. For example,

$$\langle T_j T_i \rangle \equiv \int_{-1/2}^{1/2} T_i T_j \, dZ = \int_{-1/2}^{1/2} (z^2 - \tfrac{1}{4})^{i+j} \, dz. \tag{6.18}$$

It is thus convenient to define a function

$$\mathscr{E}(m, n) = \int_{-1/2}^{1/2} z^m (z^2 - \tfrac{1}{4})^n \, dz. \tag{6.19}$$

Successive integration yields the recursion formulas:

$$\begin{aligned} \mathscr{E}(m, n) &= -n\mathscr{E}(m, n-1)/2(2n + m + 1), \\ \mathscr{E}(m, 0) &= \begin{cases} 0, & m \text{ odd}, \\ (m+1)^{-1}(\tfrac{1}{2})^m, & m \text{ even}. \end{cases} \end{aligned} \tag{6.20}$$

These can be easily generated on a computer and a few values are tabulated in Table 6.1. Formulas for various integrals are given in Table 6.2.

TABLE 6.1

VALUES OF THE INTEGRAL $\mathscr{E}(m, n) = \int_{-1/2}^{1/2} z^m(z^2 - \frac{1}{4})^n \, dz$

m	$n=0$	$n=1$	$n=2$
0	1.000000	−0.166667	0.333333(−1)
2	0.833333(−1)	−0.833333(−2)	0.119048(−2)
4	0.125000(−1)	−0.892857(−3)	0.992063(−4)

m	$n=3$	$n=4$	$n=5$	$n=6$
0	−0.714286(−2)	0.158730(−2)	−0.360750(−3)	0.832501(−4)
2	−0.198413(−3)	0.360750(−4)	−0.693751(−5)	0.138750(−5)
4	−0.135281(−4)	0.208125(−5)	−0.346875(−6)	0.612133(−7)

TABLE 6.2

INTEGRALS ARISING IN GALERKIN METHOD

$$W_i = (z^2 - \tfrac{1}{4})^{i+1}, \quad T_i = (z^2 - \tfrac{1}{4})^i, \quad W_i = DW_i = T_i = 0 \quad \text{at} \quad z = \pm\tfrac{1}{2}$$

$$\langle uv \rangle = \int_{-1/2}^{1/2} uv \, dz$$

$$\langle W_j W_i \rangle = \mathscr{E}(0, i+j+2)$$

$$\langle DW_j DW_i \rangle = 4(i+1)(j+1)\mathscr{E}(2, i+j)$$

$$\langle D^2 W_j D^2 W_i \rangle = 16ij(i+1)(j+1)\mathscr{E}(4, i+j-2) + 8(i+1)(j+1)(i+j)\mathscr{E}(2, i+j-1) + 4(i+1)(j+1)\mathscr{E}(0, i+j)$$

$$\langle T_j T_i \rangle = \mathscr{E}(0, i+j)$$

$$\langle DT_j DT_i \rangle = 4ji\,\mathscr{E}(2, i+j-2)$$

$$\langle W_j T_i \rangle = \mathscr{E}(0, j+i+1)$$

$$\langle T_j W_i F(z^2) \rangle = \int_{-1/2}^{1/2} f(z^2)(z^2 - \tfrac{1}{4})^{j+i+1} \, dz$$

$$\Xi_i = z(z^2 - \tfrac{1}{4})^i, \quad \Xi = 0 \quad \text{at} \quad z = \pm\tfrac{1}{2}$$

$$\langle \Xi_i \Xi_j \rangle = \mathscr{E}(2, i+j)$$

$$\langle D\Xi_i D\Xi_j \rangle = \mathscr{E}(0, i+j) + 2(i+j)\mathscr{E}(2, i+j-1) + 4ij\mathscr{E}(4, i+j-2)$$

$$\langle W_j D\Xi_i \rangle = -\langle DW_j \Xi_i \rangle = \mathscr{E}(0, i+j+1) + 2i\mathscr{E}(2, i+j)$$

If the solution is not known to possess a symmetry about $z = 0$, both even and odd powers of z must be included in the trial function. The first approximation is the same as Eq. (6.15) but higher approximations are different for both velocity and temperature:

$$W_i = (z^2 - \tfrac{1}{4})^2 z^{i-1}, \qquad T_i = (z^2 - \tfrac{1}{4}) z^{i-1}. \tag{6.21}$$

The integrals needed with these trial functions are given in Table 6.3. Using

TABLE 6.3

INTEGRALS ARISING IN GALERKIN METHOD

$$W_i = (z^2 - \tfrac{1}{4})^2 z^{i-1}, \quad T_i = (z^2 - \tfrac{1}{4}) z^{i-1}, \quad W_i = DW_i = T_i = 0 \text{ at } z = \pm\tfrac{1}{2}$$

$$\langle uv \rangle = \int_{-1/2}^{1/2} uv\, dz$$
$$\langle W_j W_i \rangle = \mathscr{E}(i+j-2, 4)$$
$$\langle DW_j DW_i \rangle = 16\mathscr{E}(i+j, 2) + (i-1)(j-1)\mathscr{E}(i+j-4, 4) + 4(i+j-2)\mathscr{E}(i+j-2, 3)$$
$$\begin{aligned}\langle D^2W_j D^2W_i \rangle = {} & (i-1)(i-2)(j-1)(j-2)\mathscr{E}(i+j-6, 4) \\ & + 16(2i-1)(2j-1)\mathscr{E}(i+j-2, 2) + 64\mathscr{E}(i+j+2, 0) \\ & + 4[(i-1)(i-2)(2j-1) + (j-1)(j-2)(2i-1)]\mathscr{E}(i+j-4, 3) \\ & + 8[(i-1)(i-2) + (j-1)(j-2)]\mathscr{E}(i+j-2, 2) \\ & + 64(i+j-1)\mathscr{E}(i+j, 1)\end{aligned}$$

$$\langle T_j T_i \rangle = \mathscr{E}(i+j-2, 2)$$
$$\langle DT_j DT_i \rangle = (i-1)(j-1)\mathscr{E}(i+j-4, 2) + 2(i+j-2)\mathscr{E}(i+j-2, 1) + 4\mathscr{E}(i+j, 0)$$
$$\langle W_j T_i \rangle = \mathscr{E}(i+j-2, 3)$$
$$\langle T_j W_i f(z) \rangle = \int_{-1/2}^{1/2} f(z) z^{i+j-2}(z^2 - \tfrac{1}{4})^3\, dz$$

$$T_i = z^{i-1}$$

$$\langle T_j T_i \rangle = \mathscr{E}(i+j-2, 0)$$
$$\langle DT_j DT_i \rangle = (i-1)(j-1)\mathscr{E}(i+j-4, 0)$$
$$\langle W_j T_i \rangle = \mathscr{E}(i+j-2, 2)$$
$$\langle T_j W_i f(z) \rangle = \int_{-1/2}^{1/2} f(z) z^{i+j-2}(z^2 - \tfrac{1}{4})^2\, dz$$

these tables it is possible to compute a first approximation on a slide rule or desk calculator to assess the importance of various complicating factors, such as temperature-dependent properties, heat generation, different boundary conditions, etc. The formulas are also useful for computing higher approximations on a computer.

If the boundary condition is the third kind another trial function must be chosen for temperature. It is simpler to choose trial function which are complete, but do not satisfy the boundary conditions. These are satisfied using boundary residuals in a manner described in Eqs. (6.34) and (6.59). Thus we take

$$T_i = z^{i-1}, \tag{6.22}$$

which gives rise to the additional integrals in Table 6.3.

In treating surface-tension-driven flows we need trial functions satisfying the boundary conditions

$$W(0) = DW(0) = W(1) = 0. \tag{6.23}$$

These are provided by

$$W_i = (1 - z)z^{i+1}. \tag{6.24}$$

The integrals then involve the function

$$\mathscr{F}(m, n) = \int_0^1 z^m(1 - z)^n \, dz = \frac{m!\,n!}{(m + n + 1)!}, \tag{6.25}$$

which is tabulated in Table 6.4.

TABLE 6.4

VALUES OF THE INTEGRAL $\mathscr{F}(m, n) = \int_0^1 z^m(1 - z)^n \, dz = m!n!/(m + n + 1)!$

m	$n = 0$	$n = 1$	$n = 2$	$n = 3$	$n = 4$
0	1.000000	0.500000	0.333333	0.250000	0.200000
1	0.500000	0.166667	0.833333(−1)	0.500000(−1)	0.333333(−1)
2	0.333333	0.833333(−1)	0.333333(−1)	0.166667(−1)	0.952381(−2)
3	0.250000	0.500000(−1)	0.166667(−1)	0.714286(−2)	0.357143(−2)
4	0.200000	0.333333(−1)	0.952381(−2)	0.357143(−2)	0.158730(−2)
5	0.166667	0.238095(−1)	0.595238(−2)	0.198413(−2)	0.793651(−3)

A trial function is needed for vorticity in Section 6.3. The requirements are

$$\Xi = 0 \quad \text{at} \quad z = \pm\tfrac{1}{2}, \qquad \Xi(z) = -\Xi(-z), \tag{6.26}$$

and a trial function is

$$\Xi_i = z(z^2 - \tfrac{1}{4})^i. \tag{6.27}$$

Integrals for this function are given in Table 6.2.

If the problem is nonlinear, as in Section 6.5, the trial function must be easy to use. Trial functions which satisfy the boundary conditions (6.2) (at $z = 0, 1$) and are convenient to use in nonlinear problems are

$$W_i = \cos(i - 1)\pi z - \cos(i + 1)\pi z. \tag{6.28}$$

Other trial functions are discussed in the context of particular applications. The definition of polynomials in Eq. (6.16) is arbitrary. It would also be possible to generate orthogonal polynomials by writing the velocity function as

$$W = \sum_{i=1}^{N} a_i(z^2 - \tfrac{1}{4})^2 P_{i-1}(z^2), \tag{6.29}$$

where the P_i are polynomials defined by the orthogonality conditions

$$\int_{-1/2}^{1/2} (z^2 - \tfrac{1}{4})^2 P_j(z^2) P_i(z^2) \, dz = \delta_{ji}, \qquad i = 1, \ldots, j. \tag{6.30}$$

It is not as convenient to use the orthogonal collocation method for the problems treated below because the velocity and temperature trial functions are different. They satisfy different boundary conditions, so that the residuals cannot be evaluated at the collocation points as conveniently. For example, the temperature equation would be evaluated at the roots of the Nth order temperature polynomials. The velocity function is known at the roots of the velocity polynomial. The method is applicable in principle, but it is not as convenient. Furthermore, the calculations using Eqs. (6.16)–(6.28) are usually accurate with a small number of terms (say 3 or 4) so that the added advantage of having orthogonal polynomials is not realized.

6.2 Application of the Galerkin Method

Consider a fluid layer heated from below, infinite in horizontal extent, which is heated from below (see Fig. 6.1). For small temperature differences

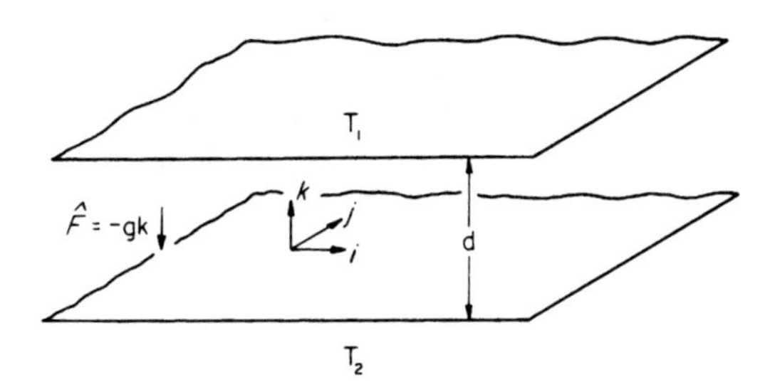

Fig. 6.1. Fluid layer heated from below.

the fluid remains motionless because the buoyancy forces cannot overcome the viscous dissipation. Heat transfer occurs only by conduction. The temperature difference across the layer is increased, such that the buoyancy forces are increased. A point is reached in which the buoyancy forces counterbalance the viscous and thermal dissipation and motion occurs. The onset of instability is predicted by the solution to an eigenvalue problem, derived from the Navier–Stokes and energy equations (Chandrasekhar, 1961, Chapter 2):

$$\frac{\partial}{\partial t}(D^2 - a^2)W = -R^{1/2}aT + (D^2 - a^2)^2 W,$$

$$\Pr \frac{\partial T}{\partial t} = (D^2 - a^2)T + R^{1/2}aWf(z), \tag{6.31}$$

$$f(z) = -\frac{1}{\beta}\frac{dT_0}{dz},$$

where $W = DW = T = 0$ on a rigid, perfectly conducting boundary, and $W = D^2W = T = 0$ on a free, flat, perfectly conducting boundary. Here a is the wave number, $R = \alpha g \beta d^4/\nu\kappa$ is the Rayleigh number, α is the coefficient of thermal expansion, g the acceleration of gravity, $\beta = \Delta T/d$, where ΔT is the temperature drop across the layer of thickness d, ν is the kinematic viscosity, κ is the thermal diffusivity, and $t_s = d^2/\nu$. W is the vertical component of the velocity and T the perturbation temperature (total temperature minus temperature in quiescent state, T_0). These equations govern the onset of instability; after the disturbance grows the full nonlinear equations must be used (Section 6.5). For $R < R_c$, the critical Rayleigh number, the system is stable. For $R = R_c$ the amplitude of the motion can remain at a constant value, termed neutral stationary instability, or can oscillate about an average value, termed neutral oscillatory instability. Oscillatory instability can sometimes be ruled out by proving the principle of exchange of stabilities, as done by Pellew and Southwell (1940), Sani (1963), or Davis (1969). An exact solution of Eq. (6.31) for $f(z) = 1$ was derived by Pellew and Southwell (1940) and for various $f(z)$ by Sparrow *et al.* (1964). In principle, the same methods are applicable to the more complicated problems treated below, although they lead to considerable numerical work which must be done on a computer. The Galerkin method is useful for these problems because the first approximation often gives the desired information, and if precise results are desired these can be generated with higher approximations calculated on a computer.

Next apply the Galerkin method to Eq. (6.31). Expand the velocity and temperature in terms of functions which satisfy the boundary conditions

$$W(z, t) = \sum_{i=1}^{N} A_i(t) W_i(z), \qquad T(z, t) = \sum_{i=1}^{N} B_i(t) T_i(z). \tag{6.32}$$

Substitute the trial functions into the differential equations to form the residuals which are made orthogonal to the respective trial functions:

$$C_{ji} \frac{dA_i}{dt} = -D_{ji} A_i + R^{1/2} E_{ji} B_i, \tag{6.33a}$$

$$\Pr F_{ji} \frac{dB_i}{dt} = R^{1/2} H_{ji} A_i - G_{ji} B_i, \tag{6.33b}$$

or

$$\bar{K} \frac{d\bar{A}}{dt} = \bar{J}\bar{A}, \qquad \frac{d\bar{A}}{dt} = \bar{K}^{-1}\bar{J}\bar{A} = \bar{L}\bar{A}. \tag{6.33c}$$

The matrices are defined explicitly for future reference:

$$
\begin{aligned}
C_{ji} &= \langle DW_j\, DW_i + a^2 W_j\, W_i \rangle, \\
D_{ji} &= \langle D^2 W_j\, D^2 W_i + 2a^2 DW_i\, DW_j + a^4 W_i\, W_j \rangle, \\
E_{ji} &= a\langle W_j T_i \rangle, \qquad H_{ji} = a\langle T_j\, W_i f(z) \rangle, \\
F_{ji} &= \langle T_j\, T_i \rangle, \\
G_{ji} &= \langle DT_j\, DT_i + a^2 T_j T_i \rangle + [\mathrm{Nu} T_j T_i]_{z=+1/2} + [\mathrm{Nu} T_j\, T_i]_{z=-1/2}, \\
\langle uv \rangle &= \int_{-1/2}^{1/2} u(z) v(z)\, dz.
\end{aligned}
\tag{6.34}
$$

When the boundary conditions are of the first kind, $T_j = 0$ at $z = \pm\frac{1}{2}$ so that the boundary terms vanish in G_{ji}. When the boundary conditions are of the second kind, $DT_j = 0$, then $\mathrm{Nu} = 0$ so that the boundary terms vanish. The formula for G is derived by combining the differential equation and boundary residuals as in Eq. (2.54). The inverse of the matrix K exists if its determinant is nonzero, which is usually the case and is assumed to be true here. We study the stability for successively larger sets of equations, Eqs. (6.33). When the results converge, the approximation represents the stability of the original system (6.31). The solution to (6.33) is well known (Amundson, 1966, p. 123):

$$
\bar{A} = \sum_{j=1}^{2N} \bar{C}_j \exp(\lambda_j t), \tag{6.35}
$$

where the $\bar{C}_j$ are constants if the eigenvalues λ_j are distinct. The system of equations is asymptotically stable if

$$
\lim_{t \to \infty} \bar{A} = \bar{0}. \tag{6.36}
$$

The necessary and sufficient condition that the system be asymptotically stable is that all the eigenvalues of L have negative real parts (Gantmacher, 1959, p. 144):

$$
\det(L - \lambda I) = 0, \qquad I_{ij} = \delta_{ij}. \tag{6.37}
$$

Since $\det(J - \lambda K) = \det K \det\,(L - \lambda I)$ the eigenvalues of L are just the exponential time factor in (6.35). The eigenvalues can be calculated, but whether or not the eigenvalues have negative real parts can be decided by methods which are much simpler and shorter than the methods used to actually determine the eigenvalues. In particular the Routh–Hurwitz criterion is especially useful.

The Routh–Hurwitz criterion gives necessary and sufficient conditions for all the roots of a real polynomial to have negative real parts. Equation (6.37) can be rewritten as the polynomial

$$
\lambda^n + a_1 \lambda^{n-1} + \cdots + a_{n-2} \lambda^2 + a_{n-1} \lambda + a_n = 0. \tag{6.38}
$$

The criterion (cf. Gantmacher, 1959, p. 231) specifies that the roots all have negative real parts if

$$T_i > 0, \qquad i = 1, 2, \ldots, n, \tag{6.39}$$

where T_i are the successive determinants formed from the matrix

$$\begin{bmatrix} a_1 & a_3 & a_5 & \cdots & 0 \\ 1 & a_2 & a_4 & \cdots & 0 \\ 0 & a_1 & a_3 & \cdots & 0 \\ \vdots & \vdots & \vdots & \vdots & \vdots \\ 0 & 0 & 0 & \cdots & a_n \end{bmatrix} \tag{6.40}$$

that is, $T_1 = a_1$, $T_2 = a_1 a_2 - a_3$, and so on. The coefficients a_i depend on the matrix L (Aris, 1962, p. 270):

$$a_i = (-1)^i \operatorname{tr}_i L, \tag{6.41}$$

where $\operatorname{tr}_p L$ is the sum of all the $p \times p$ determinants that can be formed with diagonal elements that are diagonal elements of L. For example,

$$a_1 = -\operatorname{tr} L, \qquad a_n = (-1)^n \det L. \tag{6.42}$$

Additional information is given by Orlando's formula (Gantmacher, 1959, p. 234): $T_{n-1} = 0$ if and only if the sum of at least one pair of roots of the polynomials is zero. Neutral oscillatory instability is characterized by $\lambda = i\omega$. When the coefficients are real, the complex roots of Eq. (6.38) can occur only in pairs, including complex conjugates $\pm i\omega$. The sum of these two roots is zero, and $T_{n-1} = 0$ for neutral oscillatory instability. The case of neutral stationary instability is characterized by $\lambda = 0$, in which case $a_n = 0$.

The following procedure can be used to study the stability of the system of equations (6.33) as a function of Prandtl and Rayleigh numbers. Duncan (1952, p. 117) first developed this procedure for the study of dynamical systems governed by sets of ordinary differential equations. For a specified value of Pr, consider the system as a function of R. The system is stable for $R = 0$ since only dissipation mechanisms occur; all the Hurwitz determinants are positive. Increase R monotonically until a critical condition is reached such that any further change would result in instability. If one looks at the path of eigenvalues in the complex plan as R is changed, the critical condition corresponds to an eigenvalue passing from the left half-plane, in which the real part is negative, to the right half-plane, in which the real part is positive. The eigenvalue can pass from one side to the other in two ways: either it will pass through zero, corresponding to neutral stationary instability and $a_n = 0$, or a pair of purely imaginary complex roots will exist, corresponding to neutral oscillatory instability and $T_{n-1} = 0$. Consequently, the stability of any system can be settled by examining a_n and T_{n-1} as functions of increasing R. Whichever condition

$$a_n = 0, \qquad T_{n-1} = 0 \tag{6.43}$$

occurs first determines the type of instability, and the value of R there is the critical stability parameter. For that value of R the other Hurwitz determinants must be positive, of course.

The set of ordinary differential equations (6.33) is just an approximation to the system of partial differential equations (6.31). Successive approximations must be compared to ensure the approximation is a good one. For approximations beyond the first, the computations can conveniently be done on a computer, and the program code is suitable for many different stability problems.

To illustrate the application consider the first approximations,

$$C\frac{dA}{dt} = -DA + R^{1/2}EB, \qquad \Pr F\frac{dB}{dt} = -GB + R^{1/2}HA, \tag{6.44}$$

where all matrices have a subscript 11 or 1 understood. In the form of (6.33c) this is

$$\bar{K} = \begin{pmatrix} C & 0 \\ 0 & \Pr F \end{pmatrix}, \qquad \bar{J} = \begin{pmatrix} -D & R^{1/2}E \\ R^{1/2}H & -G \end{pmatrix}. \tag{6.45}$$

The Hurwitz determinants are then

$$a_1 = -(D/C + G/\Pr F), \qquad a_2 = \frac{DG - RHE}{\Pr CF}. \tag{6.46}$$

It is clear that $T_1 = a_1 \neq 0$ for all values of Rayleigh number. This is due to the viscous and thermal dissipation. Oscillatory instability cannot occur since $T_{n-1} \neq 0$. Stationary instability occurs when $a_2 = 0$:

$$R = DG/EH. \tag{6.47}$$

The right-hand side depends on the wave number a and must be minimized to determine the critical Rayleigh number, or lowest value of R satisfying (6.47).

The growth rate is determined by the eigenvalues λ:

$$\lambda = \frac{-(CG + DF\Pr) \pm ((CG - DF\Pr)^2 + 4\mathrm{CFEH}\Pr R)^{1/2}}{2\Pr CF}. \tag{6.48}$$

To fully solve the initial-value problem the initial conditions must of course be given. In the physical problem the random thermal perturbations provide the source of perturbations to trigger the instability. The form of the solution in Eq. (6.48) indicates how the various parameters affect the behavior of the system in time.

The analysis thus far applies to the first approximation. Higher approximations must be calculated on a computer. Since oscillatory instability does not occur for this problem we consider only neutral stationary instability. The determinant, a_n, becomes

$$J_{ji} = \left| \begin{array}{c|c} -D_{ji} & R^{1/2}E_{ji} \\ \hline R^{1/2}H_{ji} & -G_{ji} \end{array} \right| = 0, \tag{6.49}$$

$$J_{ji} = \begin{cases} -D_{ji}, & i, j \leq N, \\ R^{1/2}E_{j-N,\,i}, & i \leq N, \quad N+1 \leq j \leq 2N, \\ R^{1/2}H_{j,\,i-N}, & j \leq N, \quad N+1 \leq i \leq 2N, \\ -G_{j-N,\,i-N}, & N+1 \leq i, \quad j \leq 2N. \end{cases}$$

Numerical results for this problem are given in Table 6.5. The first approximation is within 2.5% of the exact answer. The third approximation is the exact result, indicating that three trial functions are adequate for the velocity and temperature. This gives rise to a 6×6 determinant for this problem. Convergence as $N \to \infty$ is assured by Theorem 11.28 and even the first approximation is adequate for many engineering purposes. This is indeed fortunate because the first approximation can be easily calculated using Tables 6.1 to 6.4. Thus it is possible to assess the importance of complicating effects by examining the first approximation alone.

TABLE 6.5

CRITICAL RAYLEIGH NUMBERS FOR BUOYANCY-INDUCED CONVECTION WITH TWO RIGID BOUNDARIES AND TEMPERATURE SPECIFIED

N	R_c	a_c
1	1750	3.117
2	1708.5	3.116
3	1707.762	3.116
Exact	1707.762	3.117

Consider some numerical values for the growth rate. Take two rigid boundaries at $z = \pm\frac{1}{2}$ with temperature specified and use Tables 6.1 and 6.2 to evaluate the integrals (6.33). Then (6.47) becomes

$$R = \frac{(0.8 + 0.0381a^2 + 0.00159a^4)(0.333 + 0.0333a^2)}{(0.00714)^2 a^2}, \qquad (6.50)$$

which can be minimized with respect to wave number. The minimum is $R_c = 1750$ for $a = 3.12$. Then Eq. (6.48) can be evaluated for this a, and can be written in terms of $R - R_c$:

$$\lambda = \left(\frac{9.836}{\text{Pr}} + 19.14\right)\left[-1 \pm \left(1 + \frac{1.721\,\text{Pr}\,(R - R_c)}{(19.67 + 38.28\,\text{Pr})^2}\right)^{1/2}\right]. \qquad (6.51)$$

The effect of Prandtl number is apparent in Table 6.6. The effect of Nusselt number on the growth rate can be determined in a similar fashion.

TABLE 6.6

GROWTH RATE AT ONSET OF MOTION FOR TWO RIGID BOUNDARIES

	$R - R_c$		
Pr	10	50	100
1000	0.00011	0.00056	0.0011
7	0.015	0.075	0.19
0.7	0.093	0.46	0.91

Nonlinear Initial Temperature Profile

When the initial temperature profile is a quadratic in z, the term $f(z)$ in (6.31) becomes $f(z) = Qz + 1$. Since f is not symmetric about $z = 0$ we must include all powers of z in the expansion for W and T, Eqs. (6.21). Oscillatory instability can be ruled out for free boundaries but not for rigid boundaries. It does not occur for rigid boundaries in the first approximation (6.46) so that we limit consideration to stationary instability. The definitions for D, E, and G matrices are the same but H differs:

$$H_{11} = a\langle T_1 W_1(1 + Qz)\rangle = a[\mathscr{E}(0, 3) + Q\mathscr{E}(1, 3)]. \qquad (6.52)$$

Since $\mathscr{E}(m, n) = 0$ for m odd it is clear that the parameter Q cannot influence the first approximation. Consider the second approximation:

$$\begin{vmatrix} -D_{11} & -D_{21} & R^{1/2}E_{11} & R^{1/2}E_{12} \\ -D_{21} & -D_{22} & R^{1/2}E_{21} & R^{1/2}E_{22} \\ R^{1/2}H_{11} & R^{1/2}H_{12} & -G_{11} & -G_{12} \\ R^{1/2}H_{21} & R^{1/2}H_{22} & -G_{21} & -G_{22} \end{vmatrix} = 0. \qquad (6.53)$$

The properties of the function $\mathscr{E}(m, n)$ make the off-diagonal elements zero, except for those involving H:

$$\begin{aligned} E_{12} &= E_{21} = D_{12} = D_{21} = G_{12} = G_{21} = 0, \\ H_{12} &= H_{21} = \mathscr{E}(1, 3) + Q\mathscr{E}(2, 3) = Q\mathscr{E}(2, 3). \end{aligned} \tag{6.54}$$

The matrix (6.53) can be expanded to give

$$\begin{aligned} R^2 E_{11} E_{22} [H_{11} H_{22} - Q^2 \mathscr{E}^2(2, 3)] - R(D_{11} E_{22} G_{11} H_{22} \\ + D_{22} E_{11} G_{22} H_{11}) + D_{11} D_{22} G_{11} G_{22} = 0. \end{aligned} \tag{6.55}$$

Note that R depends on Q^2 so that the sign of Q is immaterial. When the parameter Q is small the asymptotic expression (for $Q \to 0$) is

$$R = \frac{D_{11} G_{11}}{E_{11} H_{11}} \left[1 + Q^2 \mathscr{E}^2(2,3) \frac{D_{11} E_{22} G_{11}/H_{11}}{D_{11} E_{22} G_{11} H_{22} - D_{22} E_{11} G_{22} H_{11}} \right]. \tag{6.56}$$

This means that the parameter Q has only a second-order influence on the Rayleigh number.

This problem was solved numerically by Sparrow *et al.* (1964) using a power series expansion. While they obtained accurate numerical results, they did not obtain the insight that the parameter influences the Rayleigh number only in second order, Eq. (6.56), when Q is small. This fact can also be determined from the asymptotic analyses developed by Busse (1967b) for the problem allowing property variations with temperature.

For large values of Q computations have been performed by Roja (1969) and are shown in Table 6.7. The results are for the fourth approximation

TABLE 6.7

CRITICAL RAYLEIGH NUMBERS FOR NONLINEAR TEMPERATURE PROFILES

Q	N_s	Galerkin ($N = 4$)	Exact
0	0	1708.5	1707.76
−0.5	0.25	1707.0	1706.95
−20	10	1118.6	1118.43
−40	20	717.44	717.20

using the trial functions (6.21). The third and fourth approximation differ by only 0.01% and the results compare favorably with those obtained by Sparrow *et al.* (1964) which should be the same when $Q = -2N_s$, $N_s > 0$. N_s is a parameter defined in Sparrow's article.

Combined Surface Tension and Buoyancy Flow

A fluid layer with a free surface at the top can have motion induced by surface tension variations on the surface. When hot fluid rises to the surface it cools as it moves along the surface. Thus the temperature can vary along the upper surface. Since the surface tension depends on temperature, there is also a surface-tension gradient along the surface. This induces a surface traction that either tends to pull the fluid along, leading to instability, or to restrain the fluid motion, leading to stability. We include these effects and compute numerical results in the absence of buoyancy forces.

The differential equations remain the same, but the boundary conditions are different. The velocity condition is that the surface traction vanishes (Pearson, 1958). The dimensionless equations are

$$\frac{\partial}{\partial t}(D^2 - a^2)W = -RM^{-1/2}aT + (D^2 - a^2)^2 W,$$

$$\Pr \frac{\partial T}{\partial t} = (D^2 - a^2)T + aM^{1/2}W,$$

$$D^2W + M^{1/2}aT = 0, \quad W = 0, \quad DT = -\mathrm{Nu}_1 T \qquad \text{at} \quad z = 1, \tag{6.57}$$

$$W = DW = 0, \quad DT = \mathrm{Nu}_0 T \qquad \text{at} \quad z = 0.$$

To apply the Galerkin method, choose trial functions which satisfy the boundary conditions

$$W(0) = DW(0) = W(1) = 0. \tag{6.58}$$

If the lower boundary condition is $T = 0$, that must be satisfied as well. The trial functions are substituted into the differential equations to form the residuals. The velocity equation is made orthogonal to the velocity expansion functions, and the temperature equation is made orthogonal to the temperature expansion functions. After performing several integrations by parts some boundary terms remain since the trial functions do not satisfy all the boundary conditions:

$$\begin{aligned} \langle W_j D^4 W_i \rangle &= \langle D^2 W_j D^2 W_i \rangle - [DW_j D^2 W_i]_0^1 \\ &= \langle D^2 W_j D^2 W_i \rangle + aM^{1/2} DW_j(1) T_i(1). \end{aligned} \tag{6.59}$$

We have substituted into this equation the appropriate boundary condition (this is equivalent to combining the differential equation and boundary residual) to obtain the final results,

$$\begin{aligned} C_{ji}\frac{dA_i}{dt} &= -D_{ji}A_i + M^{1/2}\left(U_{ji} + \frac{R}{M}V_{ji}\right)B_i, \\ \Pr F_{ji}\frac{dB_i}{dt} &= M^{1/2}H_{ji}A_i - G_{ji}B_i. \end{aligned} \tag{6.60}$$

These equations are in the form of (6.33) with the same C, D, and F matrices and

$$\begin{aligned} E_{ji} &= -aDW_j(1)T_i(1) + a(R/M)\langle W_j T_i\rangle \equiv U_{ji} + (R/M)V_{ji}, \\ G_{ji} &= \langle DT_j\, DT_i + a^2 T_j T_i\rangle + \mathrm{Nu}(1)T_j(1)T_i(1) + \mathrm{Nu}(0)T_j(0)T_i(0), \\ H_{ji} &= a\langle T_j W_i\rangle. \end{aligned} \tag{6.61}$$

Vidal and Acrivos (1966) report numerical calculations which prove that oscillatory instability cannot occur. The first approximation of the Galerkin method gives the same result. Thus we look only at stationary instability:

$$\begin{vmatrix} -D & M^{1/2}(U + (R/M)V) \\ M^{1/2}H & -G \end{vmatrix} = 0, \tag{6.62a}$$

$$MUH + RVH = DG. \tag{6.62b}$$

When only a single mechanism is operative

$$R = DG/VH \tag{6.63a}$$

or

$$M = DG/UH. \tag{6.63b}$$

These must be minimized with respect to a to determine R_c or M_c. Equations (6.62) can then be written in the form

$$\frac{M}{M_c} + \frac{R}{R_c} = 1. \tag{6.64}$$

This formula is oversimplified since the wave number may be different in (6.63a,b). For the combined mechanisms Eq. (6.62b) must be minimized with respect to wave number; for example, minimize M with respect to a for a given R. Detailed calculations verify that (Eq. 6.64) is a good approximation for a wide range of conditions.

We next choose trial functions for use in the Galerkin method. We require the velocity to satisfy the boundary conditions (6.58). The lowest-order polynomial satisfying these requirements is

$$W = z^2(1 - z)\sum_{i=1}^{N} a_i z^{i-1}. \tag{6.65}$$

Thus we take as the trial functions

$$W_i = (1 - z)z^{i+1}, \qquad T_i = z^{i-1}. \tag{6.66}$$

The various scalar products are given in Table 6.8.

TABLE 6.8

MORE INTEGRALS FOR THE GALERKIN METHOD

$$W_i = (1 - z)z^{i+1}, \quad T_i = z^{i-1}, \quad W(0) = DW(0) = W(1) = 0$$

$$\langle W_j W_i \rangle = \mathscr{F}(i+j+2, 2)$$
$$\langle DW_j DW_i \rangle = (i+1)(j+1)\mathscr{F}(i+j, 2) - (i+j+2)\mathscr{F}(i+j+1, 1) + \mathscr{F}(i+j+2, 0)$$
$$\langle D^2 W_j D^2 W_i \rangle = ij(i+1)(j+1)\mathscr{F}(i+j-2, 2) - 2i(i+1)(j+1)\mathscr{F}(i+j-1, 1) - 2j(j+1)(i+1)\mathscr{F}(i+j-1, 1) + 4(i+1)(j+1)\mathscr{F}(i+j, 0)$$

$$\langle T_j T_i \rangle = \mathscr{F}(i+j-2, 0)$$
$$\langle DT_j DT_i \rangle = (i-1)(j-1)\mathscr{F}(i+j-4, 0)$$
$$\langle W_j T_i \rangle = \mathscr{F}(i+j, 1)$$
$$DW_j(1) = -1, \quad T_j(0) = \delta_{ij}, \quad T_j(1) = 1$$

We consider here numerical results when only the surface tension mechanism is operative and the boundary condition: $\mathrm{Nu}(1) = 0$. The zero Nusselt number case is referred to in the literature as an insulating boundary condition and the infinite case (giving $T = 0$) is referred to as the conducting boundary condition. For the first approximation the results would be better if the boundary conditions are satisfied. Thus we take temperature functions which satisfy the thermal boundary conditions exactly in place of Eq. (6.66b). For higher approximations, of course, the trial function (6.66b) would be used because of the ease in automating the computations.

Consider then the first case with

$$DT(0) = DT(1) = 0. \tag{6.67}$$

The trial function $T = 1$ satisfies these boundary conditions. With that choice and $W = (1 - z)z^2$

$$H = a\mathscr{F}(2, 1) = a/12, \qquad G = a^2, \qquad U = a. \tag{6.68}$$

The Marangoni number is then given by Eq. (6.63b),

$$M = 12(4 + 0.266a^2 + 0.00952a^4). \tag{6.69}$$

The minimum value of M occurs for $a = 0$ and is $M_c = 48$. This result is the exact answer even though the approximate eigenfunction for temperature is not exact. This success should not be expected in other problems, and if the exact result were unknown higher approximations would have to be calculated to confirm the result.

For the conducting case the trial functions must satisfy

$$DT(1) = T(0) = 0. \tag{6.70}$$

Assume a quadratic expression and apply the boundary conditions to give

$$T = z - \tfrac{1}{2}z^2, \tag{6.71}$$

which will be used as the temperature trial function. The appropriate integrals are

$$\begin{aligned} H &= a[\mathscr{F}(3, 1) - \tfrac{1}{2}\mathscr{F}(4, 1)] = a/30, \qquad U = a/2, \\ G &= \mathscr{F}(0, 2) + a^2[\mathscr{F}(2, 0) - \mathscr{F}(3, 0) + \tfrac{1}{4}\mathscr{F}(4, 0)] = \tfrac{1}{3} + \tfrac{2}{15}a^2, \end{aligned} \tag{6.72}$$

and the expression for Marangoni number is

$$M = (4 + 0.266a^2 + 0.00952a^4)(20 + 8a^2)/a^2. \tag{6.73}$$

Numerical minimizations of Eq. (6.73) with respect to wave number gives $M_c = 69$ and $a_c = 2.25$, compared to exact values of 79.6 and 1.99. The numerical values could of course be improved by computing higher approximations.

Next, suppose we are doing an experiment and wish to know the effect of Nu(0) on our results. We know from the above calculations that as Nu goes from zero to infinity the critical Marangoni number goes from 48 to approximately 69. We wish to know for example, whether Nu = 0.1 is close enough to zero for the insulating results to apply and when Nu = 10 do the conducting results apply. To answer those questions we solve the problem with a general thermal boundary condition at the bottom.†

The thermal boundary conditions are

$$DT = \mathrm{Nu}_0 T, \quad z = 0; \qquad DT = 0, \quad z = 1. \tag{6.74}$$

A quadratic expression which satisfies Eq. (6.74) is

$$T = 1 + \mathrm{Nu}_0\, z(1 - \tfrac{1}{2}z). \tag{6.75}$$

The inner products involving temperature are then

$$\begin{aligned} H &= a[0.0833 + 0.0333\ \mathrm{Nu}_0], \qquad U = a(1 + \mathrm{Nu}_0/2) \\ G &= \mathrm{Nu}_0{}^2\mathscr{F}(0, 2) + a^2[1 + 2\ \mathrm{Nu}_0\mathscr{F}(1, 0) + (\mathrm{Nu}_0{}^2 - \mathrm{Nu}_0)\mathscr{F}(2, 0) \\ &\quad - \mathrm{Nu}_0{}^2\mathscr{F}(3, 0) + \mathrm{Nu}_0{}^2\mathscr{F}(4, 0)/4] + \mathrm{Nu}_0. \end{aligned} \tag{6.76}$$

The Marangoni number is then $M = DG/UH$. Minimization for several Nusselt numbers gives the results listed in Table 6.9. Since exact results are known for $\mathrm{Nu}_0 \to \infty$ we can improve the results in Table 6.9 by calculating

$$f(\mathrm{Nu}_0) = \frac{M_c(\mathrm{Nu}_0) - M_c(0)}{M_c(\infty) - M_c(0)}. \tag{6.77}$$

† Pearson (1958) has solved the surface tension problem for a general thermal boundary condition at the top. Sparrow *et al.* (1964) have solved the buoyancy problem for general thermal boundary conditions at both surfaces, a rigid bottom surface, and either rigid or free top surface. Hurle *et al.* (1967) have solved the buoyancy problem with rigid boundaries of finite conductivity.

TABLE 6.9

CRITICAL MARANGONI NUMBER VERSUS NUSSELT NUMBER

Nu_0	M_c	$f(\text{Nu}_0)$	M_c (improved)
0	48	0	48.0
0.1	55	0.368	59.6
1.0	63	0.789	72.8
10.0	66	0.947	77.8
∞	67	1.0	79.5

using *approximate* values for $M_c(\text{Nu}_0)$ and $M_c(\infty)$. Then the same formula is used with exact values of $M_c(0)$, $M_c(\infty)$ and the $f(\text{Nu}_0)$ determined from Table 6.9 to obtain improved results. With this information we can decide whether any given experimental situation is close enough to the conducting or insulating case or whether the Nusselt number affects the results. If it does, higher approximations can be calculated.

One feature of the above results deserves further mention. Equations (6.60) for the first approximation do not admit the possibility of oscillatory instability. When the buoyancy and surface tension driving forces are opposed it seems possible that oscillatory instability might occur, since perturbations might grow due to one mechanism and then be damped out due to the opposing mechanism. Jakeman (1968) investigated this problem using an approximate analysis when both surfaces were insulating, that is, $DT = 0$. He concluded that the minimum Rayleigh number (for given Marangoni number) occurred at $a = 0$, and that for a limited ratio $-1/6 \leq M/R \leq -1/7$ oscillatory instability was possible. Thus we must conclude that the question of oscillatory instability cannot be decided conclusively on a basis of the first approximation of the Galerkin method.

Other Applications

We next discuss several applications to other problems where the analyses have been done using Galerkin's method. In most cases we give only a brief treatment, and emphasize the choice of trial functions for the problem in question. Gershuni and Zhukhovitskii (1968) examined the influence of rigid boundaries with finite thermal conductivity. In most experimental situations the fluid layer is bounded by two plates made of a good thermal conductor. Fixed temperatures are maintained at the outer boundary of these two plates, but not necessarily at the boundary of the fluid. Thus the boundary conditions are not as simple as those assumed above. To examine this effect, Gershuni and Zhukhovitskii used a one-term expansion for velocity and temperature, plus a trial function for the temperature in the solid, and applied the Galerkin method. Exact results are also available for a similar problem (Hurle *et al.*, 1967).

Kurzweg (1965) considered convective instability of a hydromagnetic fluid within a rectangular cavity. In this problem it is no longer possible to employ separation of variables, as is used to derive (6.31), since the boundary conditions on the vertical walls do not allow a separable solution. The eigenvalue problem is then a system of partial differential equations rather than a set of ordinary differential equations with an unspecified wave number:

$$\left(L^2 - (\gamma M)^2 \frac{\partial^2}{\partial \zeta^2}\right)\psi = R\frac{\partial \theta}{\partial \eta}, \qquad L\theta = \frac{\partial \psi}{\partial \eta}, \qquad L = \frac{\partial^2}{\partial \eta^2} + \gamma^2 \frac{\partial^2}{\partial \zeta^2}, \tag{6.78}$$

where η and ζ are the position coordinates and the rectangular cavity is infinitely long in the third direction. M is the Hartmann number $\mu H d(\sigma/\rho\nu)^{1/2}$, where μ is the magnetic permeability, H the magnetic field, d the height of the layer, σ the electrical conductivity, and $\gamma = b/d$ the aspect ratio, the width divided by the height. The boundary conditions are

$$\begin{aligned} \psi = \partial\psi/\partial\eta = \partial\theta/\partial\eta = 0, \qquad & \eta = \pm 1, \\ \psi = \partial\psi/\partial\zeta = \theta = 0, \qquad & \zeta = \pm 1. \end{aligned} \tag{6.79}$$

The side walls are insulating, the top walls are conducting, and the velocity vanishes on all walls. Kurzweg solves the problem exactly in two limiting cases: large aspect ratio and zero aspect ratio. For arbitrary aspect ratio he applies the Galerkin method and the two exact special solutions provide a check on the results. Kurzweg expands the solution in the form

$$\psi = \sum_{m,n=1}^{N} A_{mn} U_m(\eta) V_n(\zeta), \qquad \theta = \sum_{m,n=1}^{N} B_{mn} X_m(\eta) Y_n(\zeta). \tag{6.80}$$

We need to find U, V, X, and Y functions satisfying the boundary conditions. The velocity functions U_m and V_n are similar to Eq. (6.8), rearranged so that the boundaries are at ± 1 rather than $\pm \frac{1}{2}$. The temperature functions can be sines and cosines with a symmetry such that it interacts with the velocity function in the differential equation:

$$X_m(\eta) Y_n(\zeta) = \sin\left(\frac{2m-1}{2}\right)\pi\eta \cos\left(\frac{2n-1}{2}\right)\pi\zeta. \tag{6.81}$$

Kurzweg does calculations using $N = 3$ so that the results require solving an 18×18 determinant. We discuss only the results with $M = 0$, that is, in the absence of magnetic effects. In the limit of large aspect ratio Kurzweg derives $R^{1/4} = (1708/16)^{1/4}\gamma$, and the results for finite aspect ratio are always above this value, because of the extra dissipation resulting from the vertical walls. For $\gamma < 1.6$ the motion is in the form of a single roll cell, while for $1.6 < \gamma < 2.6$ the motion is in the form of two roll cells.

Davis (1967) solved the same problem in a three-dimensional box. He considers the equations in the form

$$\nabla^2 \mathbf{u} + R^{1/2}\theta\mathbf{k} + \nabla p = 0, \qquad \nabla^2\theta + R^{1/2}\mathbf{k}\cdot\mathbf{u} = 0, \tag{6.82}$$

and expands the velocity in terms of a vector function which is solenoidal and vanishes on the boundary. He chooses

$$\theta_n = \sum_{i=1}^{3}\sum_{j=1}^{J} g_{n,j}^{(i)} x_i^{\,j}, \qquad (x_1, x_2, x_3) = (x, y, z), \tag{6.83}$$

where the constants are further restricted to satisfy the boundary conditions as well as whatever symmetry of roll cell it is desired to investigate. Similar expressions are used for the velocity functions $\mathbf{u}_n = (\phi_n^{(1)}, \phi_n^{(2)}, \phi_n^{(3)})$ and in addition to the boundary conditions the condition $\nabla\cdot\mathbf{u}_n = 0$ is applied. Davis found that the solution was essentially three-dimensional and that it corresponded closely with experimental results in that the shape of the box determined the geometry of the cells. Davis (1968) then extended the calculations to the nonlinear problem and showed that the walls of the box cause the preferred wave number to decrease when the Rayleigh number increases, as is experimentally observed. The theory for an infinite layer gives the opposite result.

Catton (1970) considered the same problem using different trial functions. He comments that Davis's trial functions were constructed out of a set of equally spaced rolls, and this might be the reason that convergence was not obtained for aspect ratios less than one. Catton and Davis both construct three-dimensional motions by superposing two-dimensional motions which satisfy the continuity equation. For example, if

$$\frac{\partial u_1}{\partial x} + \frac{\partial w_1}{\partial z} = 0, \qquad \frac{\partial v_2}{\partial y} + \frac{\partial w_2}{\partial z} = 0, \tag{6.84}$$

then $u = u_1$, $v = v_1$, $w = w_1 + w_2$ gives a three-dimensional motion which satisfies the continuity requirement. The two-dimensional motion is constructed from components such as

$$u_{pqr} = -f_p(x)g_q(y)h_r'(z), \qquad w_{pqr} = f_p'(x)g_q(y)h_r(z). \tag{6.85}$$

Clearly this satisfies the continuity requirement since $u_x + w_z = 0$. To make it satisfy the boundary conditions at $x = \pm\frac{1}{2}$ it is necessary that $u = w = 0$

there. Thus f and f' must vanish at $x = \pm\frac{1}{2}$. A possible choice for f is then (6.8) or (6.9). Similarly, h can be the same functions. The temperature is chosen to interact with the velocity function and vanish at $x = \pm\frac{1}{2}$, $y = \pm\frac{1}{2}$, $z = \pm\frac{1}{2}$. One of the sets of trial functions is

$$u_{pqr} = -\frac{H_1\lambda_r}{\lambda_p} C_p\left(\frac{x}{H_1}\right) \cos\left[(2q-1)\pi\frac{y}{H_2}\right]\frac{C_r'(z)}{\lambda_r},$$

$$w_{pqr} = \frac{C_p'\left(\dfrac{x}{H_1}\right)}{\lambda_p} \cos\left[(2q-1)\pi\frac{y}{H_2}\right] C_r(z). \tag{6.86}$$

Other sets, including the odd functions (6.9), must be included as well. Catton (1970) used as many terms as were necessary to determine the critical Rayleigh number to six significant figures.

Another problem is penetrative convection, in which a fluid layer with a stable density profile is above a layer with an unstable density profile. The perturbations in velocity and temperature can propagate into the upper layer, where they are damped out. Rintel (1967) studies penetrative convection for both buoyancy driven flows and instability between two rotating cylinders. For the buoyancy problem the equations are

$$(D^2 - a^2)^2 W = -Rg(z)T, \qquad (D^2 - a^2)T = W, \tag{6.87}$$

with boundary conditions

$$\begin{aligned} W = DW = T = 0, &\qquad z = 0, \\ W = D^2W = T = 0, &\qquad z = 1. \end{aligned} \tag{6.88}$$

Due to the changing density profile the function $g(z)$ takes two different constant values in the stable and unstable portion of the layer. To solve the problem the temperature is expanded in a series of functions which satisfy a simpler eigenvalue problem:

$$\begin{gathered} T = \sum_{n=1}^{M} B_n T_n, \\ (D^2 - a^2)T_n = -\lambda_n T_n, \qquad T_n = 0 \quad \text{at} \quad z = 0, 1, \\ \lambda_n T_n = \sin n\pi x, \qquad \lambda_n = a^2 + n^2\pi^2. \end{gathered} \tag{6.89}$$

The velocity functions are then chosen to satisfy the equations:

$$\begin{gathered} (D^2 - a^2)W_n = \lambda_n T_n, \\ W_n = DW_n = 0, \qquad z = 0, \\ W_n = D^2W_n = 0, \qquad z = 1. \end{gathered} \tag{6.90}$$

The Galerkin method is then applied to (6.87). Due to the special nature of the trial functions it is possible to eliminate the B_n from the computations. The final results correspond to retaining 35 terms in the temperature expansion and 6 in the velocity expansion, but because of this elimination the determinant is only a 6×6 determinant. The results show that the stable layer on top is destabilizing—it lowers the critical Rayleigh number—and that as the density profile in the stable layer becomes vertical the results reduce to those appropriate to the standard problem with a free boundary at the top.

Roja and Finlayson (1970) solved the convective instability problem for gases including the perfect gas law as an equation of state and taking into account the variation of physical properties, such as viscosity, thermal conductivity, etc. with temperature. The case of heat capacity variation is similar to the case studied above with a nonlinear temperature profile. The problem then becomes an eigenvalue problem with variable coefficients, and the Galerkin method was used to solve it. The trial functions used were Eqs. (6.21) and four terms were used, giving an 8×8 determinant. The results were then expressed in a perturbation series for small property variation:

$$R/R_0 = 1 + \sum_{i=1}^{5} \sum_{j=1}^{i} \alpha_{ij} B_i B_j, \tag{6.91}$$

where the B_i denotes the variation of the property across the fluid layer divided by the average value. Numerical values were given for two rigid boundaries with $T = 0$, and the effect on Rayleigh number is extremely small. A viscosity variation of 10% across the layer gives a change in critical Rayleigh number of only 0.05%.

Finlayson (1970) used the Galerkin method to study the convective instability of a ferromagnetic fluid. Here the convection is caused by spatial variations of magnetic field and magnetization which are established by the linear temperature profile, since the magnetization depends on temperature as well as magnetic field. The solution depends on three parameters, in addition to R and a, so that the computational efficiency of the Galerkin method is particularly welcome. The computations provide an example in which the boundary conditions (of the third kind) on magnetic potential are not satisfied exactly by the trial function and the boundary residual is combined with the differential equation residual.

Couette Flow between Rotating Cylinders

We next consider several problems which are similar to the convective instability problems. Consider two long coaxial cylinders with a fluid contained between them. The two cylinders are rotated at different velocities, in either the same or opposite directions. We then wish to determine the conditions under which the basic flow thus established will be unstable to small perturbations and a secondary flow will be established. The basic equations are, after simplification for a narrow gap (Chandrasekhar, 1961, Section 69),

$$(D^2 - a^2)^2 u = (1 + \alpha z)v, \tag{6.92a}$$

$$(D^2 - a^2)v = -Ta^2 u, \tag{6.92b}$$

$$\alpha = -(1 - \mu), \tag{6.92c}$$

$$u = Du = v = 0 \quad \text{at} \quad z = 0, 1. \tag{6.92d}$$

The functions u and v are the radial and azimuthal velocity functions, μ is the ratio of angular velocity of the outer cylinder to that of the inner cylinder, and T is the Taylor number and depends on the gap thickness between the cylinders, the viscosity and the initial flow. Methods applicable to this problem include the Galerkin method, adjoint variational methods, and modified Galerkin methods, which are compared in Section 6.4.

Consider the complication introduced when azimuthal variations are allowed in the disturbances, that is the velocity functions can be periodic in angle, θ. The equations for nonrotationally symmetric disturbances in the small gap approximation are (DiPrima, 1961)

$$\begin{aligned} (DLD - a^2 L)u &= -\tfrac{1}{2}a^2 T(1 + \mu)v, \\ Lv &= u, \\ L &= (D^2 - a^2) - i\mathrm{Re}\{\beta + kf(x)\}, \end{aligned} \tag{6.93}$$

where k is the azimuthal wave number and Re, β, and $f(x)$ depend on the initial flow. These equations must be solved subject to the same boundary conditions (6.92d). DiPrima (1961) solves the problem for $\mu \to 0$ using the Galerkin method and trial functions (6.8) and (6.9). Both even and odd functions must be included:

$$\begin{aligned} u &= \sum_{i=1}^{J} a_i C_i(z) + \sum_{i=1}^{M} b_i S_i(z), \\ v &= \sum_{i=1}^{J} c_i E_i(z) + \sum_{i=1}^{M} d_i F_i(z), \end{aligned} \tag{6.94}$$

$$E_n = \sqrt{2}\cos(2n - 1)\pi z, \qquad F_n = \sqrt{2}\sin 2n\pi z.$$

The solution indicates that the minimum Taylor number occurs for $k = 0$, that is, for rotationally symmetric functions ($\mu \geq 0$).

Another complication which has been studied is the stability of Couette flow with an axial flow superimposed on top of it. The equations for rotationally symmetric disturbances and a narrow gap are (Chandrasekhar, 1961)

$$\begin{aligned} [D^2 - a^2 - i\sigma + ia\,\mathrm{Re}\, f(x)](D^2 - a^2)u + 12ia\,\mathrm{Re}\, u &= -a^2 Tg(x)v, \\ [D^2 - a^2 - i\sigma + ia\,\mathrm{Re}\, f(x)]v &= u, \end{aligned} \tag{6.95}$$

where σ is the time factor introduced when the time dependence is assumed to be $W(x, t) = w(x)\exp(\sigma t)$. Re is the Reynolds number for the axial flow. The same boundary conditions (6.92d) apply. The solution to this system is a formidable task. For special cases [such as $g(x) = 1$ and $f(x) = 1$] the exact solution is known (DiPrima, 1960; Chandrasekhar, 1960). For the more general case most solutions have been derived using some form of the Galerkin method. DiPrima (1960) used trial functions of the type (6.94). Later Krueger and DiPrima (1964) did additional calculations using the same trial functions as well as the following more complicated ones. Krueger and DiPrima derived the solutions to

$$\begin{aligned}[D^2 - a^2 - i\sigma + ia\,\mathrm{Re}](D^2 - a^2)u_m &= E_m,\\ [D^2 - a^2 - i\sigma + ia\,\mathrm{Re}]v_m &= C_m,\end{aligned} \tag{6.96}$$

and expanded the solution in terms of these functions. Note the similarity of (6.95)–(6.96). For this problem the results using (6.96) (two-term series) agreed very well with those obtained using (6.94). This indicates the simpler trial functions give satisfactory results. Krueger and DiPrima present results of calculations for Re from 1 to 60 and various values of μ between $+1$ and -1. The instability is oscillatory in nature, and for each set of parameters they determine the critical wave number, frequency of oscillation and Taylor number. In this problem the Galerkin method has proved of great value in demonstrating the effect of the parameters on the instability.

When the same authors extended the non-axisymmetric problem (6.93) to other values of μ, including negative ones, and for finite gaps, they resorted to a numerical procedure (Krueger *et al.*, 1966). For negative μ, more terms must be used in the expansion, and the algebra became tedious. These results confirmed that for $\mu \geq 0$ the axisymmetric disturbances are the most critical, whereas for $\mu \leq -0.78$ the critical disturbances are non-axisymmetric. As more and more terms are used in the expansion it is imperative to easily automate the computations. Otherwise the algebra becomes too tedious.

Other applications of the Galerkin method include the study of Couette flow in the presence of an axial magnetic field (Kurzweg, 1964), and a circular magnetic field (DiPrima and Pan, 1964). The effects of radial temperature gradients are determined without the small gap approximation (Walowit *et al.*, 1964) and with the small gap approximations but including viscosity and density variations due to the temperature gradient (Walowit, 1966). Ritchie (1968) has recently applied the Galerkin method to the study of Couette flow between rotating cylinders when they are eccentric. This application requires using bipolar coordinate systems, but illustrates the wide range of problems amenable by the Galerkin method.

It is clear that the Galerkin method can be used as a standard and powerful computational method for studying eigenvalue problems.

6.3 Time-Dependent Motion

When instability manifests itself the motion can be either stationary or oscillatory. The oscillatory instability is plainly more complicated and deserves special attention. We consider the physical mechanisms which give rise to oscillatory instability and then outline computations for a rotating fluid layer. We conclude by treating problems in which the quiescent state itself has a time-dependent temperature or concentration profile.

To illustrate the calculations for oscillatory instability we study a fluid layer heated from below which is rotated about an axis perpendicular to the heated boundaries. This introduces Coriolis forces into the equations. The equations given by Chandrasekhar (1961, pp. 89 and 90) can be rearranged to give in place of (6.31)

$$(D^2 - a^2)\frac{\partial W}{\partial t} = (D^2 - a^2)^2 W - R^{1/2} a\theta - T^{1/2} D\Xi,$$

$$\Pr \frac{\partial \theta}{\partial t} = (D^2 - a^2)^2 \theta + R^{1/2} a W, \tag{6.97}$$

$$\frac{\partial \Xi}{\partial t} = (D^2 - a^2)\Xi + T^{1/2} DW.$$

The vertical component of vorticity must be included since it interacts with the velocity equation through the Coriolis forces. T is the Taylor number, given by $4\Omega^2 d^4/\nu^2$.

Consider first the physical reason these equations might give rise to oscillatory instability, following Veronis (1966a) and McConaghy and Finlayson (1969). The momentum equation represents a balance of local acceleration, Coriolis, gravitational and viscous forces. In time-dependent motions of dynamical systems, the local acceleration partially offsets the constraining force of rotation. When the Prandtl number is small, viscous forces become small and the local acceleration becomes more important in the dynamical balance. Consequently we would expect time-dependent motions to be less stable, or more easily generated, since then the local acceleration can offset part of the constraining force of rotation. Time-dependent neutral stability states are of course oscillatory. We would then expect oscillatory instability to occur for buoyancy-driven convection with Coriolis forces. If the convection is driven by surface-tension variations rather than density gradients, the gravitational force does not appear in the momentum balance and the driving force arises in the momentum balance at the free surface. The argument for oscillatory instability did not depend on the source of the driving force, however, and the local acceleration in time-dependent motion should still partially offset the constraining force of rotation. We then expect the same conclusion: oscillatory instability driven by surface-tension variations should be possible in a rotating fluid layer, especially for fluids with small Prandtl numbers. Oscillatory instability has in fact been shown to occur (see Chandrasekhar, 1953, 1961; Chandrasekhar and Elbert, 1955; Nakagawa and Frenzen, 1955 for the buoyancy problem; McConaghy and Finlayson, 1969, for the surface tension problem).

Another situation giving rise to oscillatory instability is the instability caused by density gradients due to both energy and mass transport across a fluid layer. The energy and mass transport give rise to temperature and concentration gradients, which thereby induce density gradients. This problem has been studied in detail for two free boundaries by Sani (1963, 1965) and Veronis (1968b). They find that oscillatory instability can occur provided the two mechanisms work in opposition: one of the gradients must induce a stabilizing effect and the other must encourage instability. In this case the cause for oscillatory motion is due to the competing effects of the two mechanisms. As one mechanism causes motion to grow, the other causes it to diminish, and oscillatory motion occurs. These two cases of oscillatory instability: a rotating fluid layer and transfer of both energy and mass, suggest when oscillatory instability might occur. If we decide to study oscillatory instability the Galerkin method can be applied as we will show.

The Galerkin method was applied to (6.97) by Finlayson (1968) and gives for the first approximation in place of (6.44)

$$\begin{aligned} C\frac{dA}{dt} &= -DA + R^{1/2}EB + T^{1/2}UC', \\ \Pr F\frac{dB}{dt} &= R^{1/2}EA - GB, \\ V\frac{dC'}{dt} &= -T^{1/2}UA - XC', \end{aligned} \tag{6.98}$$

where, for any approximation,

$$\begin{aligned} U_{ji} &= \langle W_j D\Xi_i \rangle = -\langle \Xi_i DW_j \rangle, \\ V_{ji} &= \langle \Xi_j \Xi_i \rangle, \\ X_{ji} &= \langle D\Xi_j D\Xi_i + a^2 \Xi_j \Xi_i \rangle. \end{aligned} \tag{6.99}$$

These equations can be put into the form

$$\bar{K}\frac{d\bar{A}}{dt} = \bar{J}\bar{A} \tag{6.100}$$

with

$$K = \begin{bmatrix} C & 0 & 0 \\ 0 & \Pr F & 0 \\ 0 & 0 & V \end{bmatrix}, \qquad J = \begin{bmatrix} -D & R^{1/2}E & 0 \\ R^{1/2} & -G & 0 \\ 0 & 0 & -X \end{bmatrix} + T^{1/2}\begin{bmatrix} 0 & 0 & U \\ 0 & 0 & 0 \\ -U & 0 & 0 \end{bmatrix}. \tag{6.101}$$

The matrix J has been split into its symmetric and antisymmetric parts. When the Taylor number is zero, the matrix is symmetric and can have only real eigenvalues; hence oscillatory instability is impossible. If oscillatory instability occurs for $T \neq 0$ the matrix J must have a certain amount of asymmetry. To study the instability we put the equations into the form:

$$\frac{d\bar{A}}{dt} = \bar{L}\bar{A},$$

$$\bar{L} = \begin{vmatrix} -D/C & R^{1/2}E/C & T^{1/2}U/C \\ R^{1/2}E/\Pr F & -G/\Pr F & 0 \\ -T^{1/2}U/V & 0 & -X/V \end{vmatrix}. \tag{6.102}$$

The stability of the system is governed by either $a_3 = \det L = 0$ for stationary instability or $T_2 = a_1 a_2 - a_3 = 0$ for oscillatory instability. The results are

$$R^s = \frac{DG}{E^2} + T\,\frac{U^2G}{E^2X}, \tag{6.103a}$$

$$R^o = \frac{(D + XC/V)}{E^2}\left[G + \frac{X \Pr F}{V} + T \Pr^2 \frac{U^2F/V}{(D \Pr + GC/F)}\right], \tag{6.103b}$$

and when oscillatory instability occurs the frequency of oscillation is given by $\omega^2 = -a_3/a_1 = -a_2$,

$$\omega^2 = T\,\frac{U^2(G - XF \Pr/V)}{V(GC + DF \Pr)} - \frac{X^2}{V^2}. \tag{6.104}$$

The Rayleigh number must be minimized with respect to wave number to obtain the critical value. If $R_c^s \leq R_c^o$, then stationary instability occurs and vice versa.

It is instructive to establish the results which are valid for large Taylor number. For this purpose we assume the trial functions are normalized, so that $\langle W^2 \rangle = \langle T^2 \rangle = \langle \Xi^2 \rangle = 1$. If Eq. (6.103a) is differentiated with respect to a^2, the result set to zero and solved for a, we get for large $T \to \infty$,

$$2a^2 - T(\langle WD\Xi \rangle^2/a^4) = 0. \tag{6.105}$$

The root to this equation is

$$a_c = (\tfrac{1}{2}T)^{1/6}\langle WD\Xi \rangle^{1/3} \tag{6.106}$$

and the critical Rayleigh number is

$$R_c^{\,s} = \tfrac{3}{2}\, 2^{1/3} T^{2/3} \langle WD\Xi \rangle^{4/3} \langle W\theta \rangle^{-2}. \tag{6.107}$$

The same procedure applied to Eq. (6.103b) gives

$$a_c = (\tfrac{1}{2}T)^{1/6} \left(\frac{\mathrm{Pr}}{1+\mathrm{Pr}} \right)^{1/3} \langle WD\Xi \rangle^{1/3},$$

$$R_c^{\,o} = 6(1+\mathrm{Pr}) \left(\frac{T\,\mathrm{Pr}^2}{(1+\mathrm{Pr})^2 2} \right)^{2/3} \langle WD\Xi \rangle^{4/3} \langle W\theta \rangle^{-2}, \tag{6.108}$$

$$\omega_c^{\,2} = \left(\frac{T}{2} \right)^{2/3} \langle WD\Xi \rangle^{4/3} \frac{(2 - 3\,\mathrm{Pr}^2)}{\mathrm{Pr}^{2/3}(1+\mathrm{Pr})^{4/3}}.$$

These formulas give the critical conditions as $\mathrm{Pr}^2 T \to \infty$. The value of $R_c^{\,o}$ depends on Prandtl number. Compare Eqs. (6.107) and (6.108) to see if there is a Prandtl number above which $R_c^{\,s} \le R_c^{\,o}$. Setting $R^s = R^o$ gives

$$\frac{\mathrm{Pr}^{4/3}}{(1+\mathrm{Pr})^{1/3}} = \frac{1}{2}. \tag{6.109}$$

The root to this equation is $\mathrm{Pr} = 0.676605$. Thus for a fluid with $\mathrm{Pr} \ge 0.68$ stationary instability is the preferred made since $R_c^{\,s} < R_c^{\,o}$. These results apply to any boundary conditions, rigid or free, including any thermal condition, but only for the first approximation. The value of the integrals changes as we insert approximations which satisfy the different boundary conditions.

Trial functions must be assumed to satisfy the boundary conditions. Consider two rigid boundaries which are good conductors:

$$W = DW = \theta = \Xi = 0, \qquad z = \pm\tfrac{1}{2}. \tag{6.110}$$

Appropriate trial functions are (6.16) for velocity, (6.17) for temperature, and (6.27) for vorticity. Calculations which converged by the third approximation are shown in Fig. 6.2. The effect of the rotation is to stabilize the fluid layer, and fluids with smaller Prandtl numbers are less stable with respect to oscillatory instability. Results similar to Eqs. (6.107) and (6.108) were obtained by Chandrasekhar (1961, Ch. 3) for the case of two free boundaries which are good conductors. In that case the exact solution can be found. In the more general case, Chandrasekhar used an approximate method, outlined in Section 6.4, but obtained only numerical results without the qualitative features of Eqs. (6.107) and (6.108). Regardless of what method one uses to obtain precise numerical results, it is clear that the first Galerkin approxi-

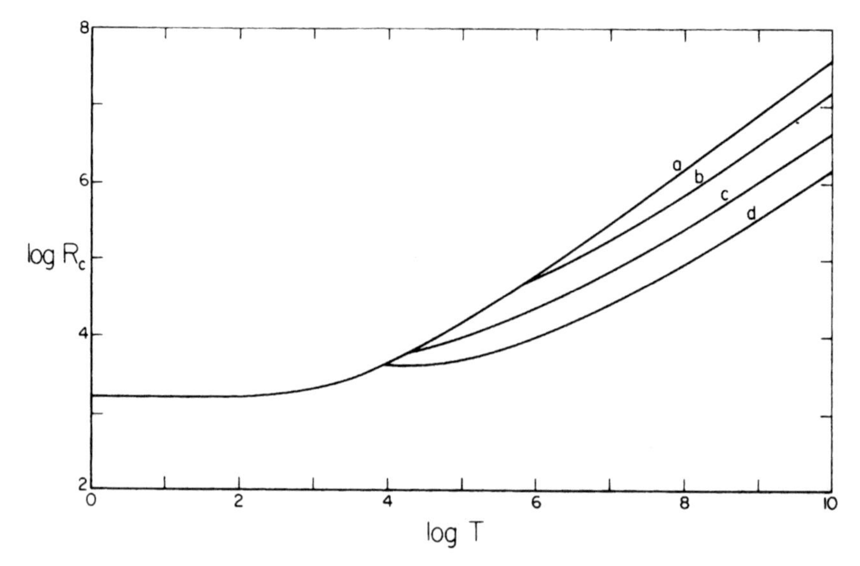

Fig. 6.2. Rayleigh number dependence on Taylor and Prandtl numbers. Curve a, stationary instability. Curves b–d, oscillatory instability, Pr: (b) 0.25, (c) 0.1, (d) 0.025.

mation gives considerable insight into the solution. This information can then be used to guide the more precise computations, whether done by the Galerkin method or some other scheme.

Stability of Time-Dependent States

In the above treatment of oscillatory instability by the Galerkin method, the time dependence was exponential and this could have been assumed. The ordinary differential equations were retained because they facilitated the solution. Situations exist where the exponential time dependence cannot be assumed and the full time-dependent equations must be retained. These are cases when the initial state itself is changing in time. Quasi-static approximations have been made: assume the perturbation grows much faster than the initial state, and freeze the initial state into some spatial distribution whose stability is studied. However, the results have been shown to be in error. We thus consider the full time-dependent problem here. Meister (1963) was apparently the first to apply the Galerkin method to a time-dependent problem. He considered the Couette flow between two coaxial cylinders. Initially the cylinders are rotating in a fashion that is stable. Then at time zero one of the cylinders is speeded up and the Couette flow begins to change to the new steady-state value. The new steady state is unstable, however, and somewhere in the transition an instability sets in. The state whose stability is to be determined is then a transient one—changing from one Couette flow to another. Instead of treating this case in detail we investigate natural convection phenomena giving rise to similar problems.

Consider the following heat transfer problem treated by Foster (1968). An infinitely deep fluid layer is at a constant temperature. At time zero the temperature of the cooling medium at the top surface is abruptly lowered to a new value and the temperature distribution in the fluid begins to change with time. With no velocity disturbance the temperature of the fluid is given by the solution to

$$\frac{\partial T_0}{\partial t} = \frac{\partial^2 T_0}{\partial z^2}. \tag{6.111}$$

The perturbation equations are derived as done above with the distance scale in units of $(\kappa\nu/\alpha g\ \Delta T)^{1/3}$ and the time scale is chosen as $(\nu^2/\alpha^2 g^2\ \Delta T^2\ \kappa)^{1/3}$. The resulting equations, after assuming a normal mode analysis, are

$$\begin{gathered}\frac{\partial T}{\partial t} = (D^2 - a^2)T + W\frac{\partial T_0}{\partial z}, \\ (1/\mathrm{Pr})(D^2 - a^2)\frac{\partial W}{\partial t} = (D^2 - a^2)^2 W - a^2 T.\end{gathered} \tag{6.112}$$

We consider the case of a rigid boundary with a general thermal boundary condition:

$$W = DW = DT - NT = 0 \qquad \text{at} \quad z = 0. \tag{6.113}$$

The initial state is easily determined using Laplace transforms to obtain the solution for small time. The result is

$$\frac{\partial T_0}{\partial z} = \frac{\exp[-z^2/4t]}{(\pi t)^{1/2}}. \tag{6.114}$$

The Galerkin method is then used to solve (6.112).

We must expand the solution in a series of functions with coefficients which are functions of time, Eqs. (6.32). Foster (1968) derives orthonormal functions in the following manner. He writes

$$W_n(z) = \sum_{m=1}^{n+1} C_{nm}(bz)^m e^{-bz}, \qquad T_n(z) = \sum_{m=1}^{n+1} D_{nm}(bz)^{m-1} e^{-bz}, \tag{6.115}$$

and determines the constants C_{nm} and D_{nm} by the conditions (6.113) and

$$\int_0^\infty W_n W_m\, dz = \delta_{nm}, \qquad \int_0^\infty T_n T_m\, dz = \delta_{nm}, \qquad m \le n. \tag{6.116}$$

The constant b was adjusted to give rapid convergence of the results, and from 3 to 6 terms give convergence. These computations can be done on a computer, after which the Galerkin integrals are calculated on a computer. The integrals involving $\partial T_0/\partial z$ are calculated numerically. This procedure reduces the set of partial differential equations (6.112), to a set of $2N$ coupled ordinary differential equations which are solved numerically. It is of course necessary to prescribe initial conditions and define what is meant by instability. The results are somewhat dependent upon the initial conditions assumed, but the qualitative behavior is not (whether the disturbance grows or diminishes, its growth rate after an initial transient period, etc.). Foster used many different combinations and then concentrated on $A_1(0) = 1$, and all other terms initially zero. The growth of the average velocity disturbance was calculated as a function of the horizontal wave number and the Prandtl number:

$$\langle W \rangle \equiv \left[\int_0^\infty W^2(z, t)\, dz \Big/ \int_0^\infty W^2(z, 0)\, dz\right]^{1/2}. \tag{6.117}$$

The results were expressed in terms of a critical time, t_c, which it took for (6.117) to reach various values, from 10 to 10^8. Various effects can then be examined: the effect of Nusselt number is small, and the critical time decreases as the Prandtl number increases. Asymptotic limits of small and large Prandtl number are also derived. Foster concludes that the effect of a free boundary is about the same as a rigid boundary insofar as the critical time is concerned, so that the free boundary can be used to investigate the variation of critical time on various parameters. Foster considered another case in which the surface temperature is decreased at a linear rate. Its effect was much greater than the effect of various boundary conditions on the perturbed velocity or temperature. Thus, the initial disturbance seems to be the most important variable. We note that while Foster (1968) used a general thermal boundary condition for the perturbed temperature, Eq. (6.113), he used the solution for $T_0(z, t)$ corresponding to a constant temperature at the surface, Eq. (6.114). Thus the results apply only to the case of N very large. In a later study Foster (1969b) compared the theory to experiment. While the comparison cannot be made precisely the general features of the analysis were verified. The amplification (6.117) usually had values between 10^3 and 10^5.

Another application of the Galerkin method to a time-dependent problem is the study of gas absorption into a liquid layer done by Mahler *et al.* (1968) and Mahler and Schechter (1970). The problem is very similar to that treated above, except for the initial concentration and boundary conditions. We consider here only the choice of trial functions for the following boundary conditions:

$$T(0) = DT(1) = 0, \qquad W(0) = W(1) = DW(1) = D^2 W(0) = 0. \tag{6.118}$$

We want trial functions which satisfy the orthogonality condition (6.12), since then the matrices on the left-hand side of (6.33) are tridiagonal and it is not necessary to invert a matrix before integrating the equations. The velocity functions are derived as the solution to the following eigenvalue problem [see Eq. (6.13)]:

$$D^4 W + \alpha(D^2 - a^2)W = 0, \qquad W(0) = W(1) = DW(1) = D^2 W(0) = 0. \tag{6.119}$$

The solution is

$$W_n = \frac{\sinh(\lambda_n z)}{\sinh \lambda_n} - \frac{\sin(\tau_n z)}{\sin \tau_n},$$
$$\tanh \lambda/\lambda = \tan\tau/\tau \tag{6.120}$$
$$\lambda^2 = [-\alpha + (\alpha^2 + 4\alpha a^2)^{1/2}]/2,$$
$$\tau^2 = [\alpha + (\alpha^2 + 4\alpha a^2)^{1/2}]/2.$$

The temperature trial function is taken as

$$T_n = \sin[(n - \tfrac{1}{2})\pi z]. \tag{6.121}$$

In most of the computations 15 temperature and 5 to 9 velocity functions were necessary to obtain convergence. For the results of the calculations the reader is referred to the original article.

In a later article Foster (1969) studied the effect of initial conditions and lateral boundaries on convection. He expanded the velocity and temperature in terms of sines and cosines appropriate to the assumed free, perfectly conducting boundaries. He retained nonlinear terms in the equations and made each residual orthogonal to a trial function, thus resulting in a set of ordinary differential equations. These were integrated numerically. The solution required more terms as the Rayleigh number was increased above the critical value, and for $R = 4R_c$ he needed 6 horizontal and 6 vertical terms in the velocity expansion, and 6 in the temperature expansion, giving rise to a set of $6 \times 6 + 6 = 42$ ordinary differential equations. Foster calls his method a Fourier series method, which it is, but we notice that the same results are achieved with the Galerkin method. In the case of free boundaries the trial functions can be sines and cosines, whereas in the case of rigid boundaries other functions must be used and the method is properly called a Galerkin method.

Gresho and Sani (1970) use the Galerkin method to study the effect on the stability of a fluid layer when the gravity field is a constant plus a part oscillating in time. A one-term solution is used to delineate the general features of the solution, and quantitative results are obtained using a five-term expansion. Both linear and nonlinear problems are tackled. The expansion functions are in terms of sine functions, which facilitates calculation of the integrals:

$$W_i(x, z) = \cos ax \sin \pi z \sin(2i - 1)\pi z, \qquad W = DW = 0 \quad \text{at} \quad z = 0, 1,$$
$$T_i(x, z) = \cos ax \sin(2i - 1)\pi z, \qquad T = 0 \quad \text{at} \quad z = 0, 1. \tag{6.122}$$

6.4 Variational Methods

Some of the problems already treated can be solved using variational principles, and all the Galerkin methods can be interpreted as either an adjoint variational principle or a regular variational principle. We examine those relationships here. The reader who is unfamiliar with variational principles should read Chapter 7 and Section 9.2 first.

Consider the set of equations governing the buoyancy problem for stationary instability with a nonlinear initial temperature profile (6.31):

$$(D^2 - a^2)^2 W = R^{1/2} aT, \tag{6.123a}$$

$$(D^2 - a^2)T = -R^{1/2} aWf(z). \tag{6.123b}$$

We first examine these equations to determine if they are self-adjoint (see Section 9.2). Multiply (Eq. 6.123a) by U and Eq. (6.123b) by V and integrate over the region:

$$\langle U, (D^2 - a^2)^2 W\rangle = R^{1/2} a\langle U, T\rangle,$$
$$\langle V, (D^2 - a^2)T\rangle = -R^{1/2} a\langle V, Wf(z)\rangle. \tag{6.124}$$

Integrate these terms by parts, and apply the boundary conditions $W = 0$ and either DW or $D^2W = 0$ (and the same boundary conditions for U):

$$\langle (D^2 - a^2)U, (D^2 - a^2)W\rangle = R^{1/2} a\langle U, T\rangle,$$
$$B(\tfrac{1}{2}) + B(-\tfrac{1}{2}) + \langle DVDT + a^2 VT\rangle = R^{1/2} a\langle V, Wf(z)\rangle, \tag{6.125}$$

where $B(z) = N(z)V(z)T(z)$ allows for a general thermal boundary condition. The system of equations is self-adjoint if the sum of these equations is symmetric in $U \leftrightarrow W$ and $V \leftrightarrow T$. Clearly the left-hand sides are symmetric whereas the right-hand side is symmetric only if

$$\langle U, T\rangle + \langle V, Wf(z)\rangle = \langle W, V\rangle + \langle T, Uf(z)\rangle. \tag{6.126}$$

This is true if $f(z) = 1$. The system is self-adjoint then only in the case $f(z) = 1$, that is, corresponding to a linear initial temperature distribution. For this case, then, we can formulate a variational principle by setting $U = W$, $V = T$ in Eq. (6.125) and adding:

$$\frac{1}{R^{1/2}} = \frac{2a\langle W, T\rangle}{\langle (D^2 - a^2)W, (D^2 - a^2)W\rangle + \langle DT^2 + a^2 T^2\rangle + B(\tfrac{1}{2}) + B(-\tfrac{1}{2})}. \tag{6.127}$$

The functional $R^{-1/2}$ is to be made stationary among all functions W, T that satisfy the following essential boundary conditions:

$$\begin{aligned} W = DW = 0, &\quad \text{rigid boundary,} \\ W = 0, &\quad \text{free boundary,} \\ T = 0, &\quad \text{if } N \text{ is infinite.} \end{aligned} \tag{6.128}$$

Note that the boundary condition $D^2 W = 0$ for a free boundary is a natural boundary condition and need not be satisfied by the trial function. The reader can easily verify that the Euler equations resulting from variations of W and T are just Eqs. (6.125). This variational principle was first presented by Sani (1963).

When the system of equations is self-adjoint we can formulate a variational principle. Let us examine the problems treated above to see which ones have variational principles associated with them. The symmetry needed in (6.126) for the system of equations to be self-adjoint can be easily related to the inner products derived in the Galerkin method. The symmetry of the left-hand side of (6.126) is clearly evident in the symmetry of the matrices D and G defined in (6.34). The symmetry of the right-hand side requires

$$E_{ji} = H_{ij}, \tag{6.129}$$

which is the case when $f(z) = 1$. For other problems then we merely need to examine the matrices E and H to see if they are transposes of each other. For the combined buoyancy and surface tension driven problem the matrices are given in (6.61) and they are not symmetric for either the combined mechanism or the surface-tension mechanism alone. Thus these problems are not self-adjoint. The buoyancy problem in a three-dimensional box, Eqs. (6.82), is self-adjoint and a variational principle exists. Note that the equations are written with $R^{1/2}$ in both equations rather than R and 1 in the two equations. The problem of penetrative convection in Eqs. (6.87) is non-self-adjoint, as is Couette flow between rotating cylinders, Eqs. (6.92) (except for $\alpha = 0$ and then only with a change of nondimensionalization). The problem (6.97) is non-self-adjoint, as is evident from the asymmetry in (6.101). This is true whether one studies stationary or oscillatory instability. We thus see that only the very simplest problem is self-adjoint: buoyancy driven convection in a horizontal infinite fluid layer or a three-dimensional box. Any other complication, such as a nonlinear initial temperature profile, surface tension forces, or rotation render the problem non-self-adjoint. The self-adjoint problems prove to be the exception, and we must study methods applicable to non-self-adjoint problems. The self-adjointness can be destroyed by combining the equations or by improper nondimensionalization.

Sani (1963) was the first to derive the variational principle (6.127). The ability to do so rests on the symmetry of (6.125) and this in turn rests on both terms on the right-hand side being multiplied by $R^{1/2}a$.

To derive (6.31) the dimensional standards for u_s and T_s were taken such that

$$\frac{\alpha g T_s\, d^2}{\nu u_s} = \frac{\beta u_s\, d^2}{\kappa T_s} = R^{1/2}, \tag{6.130}$$

$$\frac{T_s}{u_s} = \left(\frac{\beta \nu}{\alpha g \kappa}\right)^{1/2}, \qquad R = \frac{\alpha g \beta\, d^4}{\nu \kappa}. \tag{6.131}$$

Instead of choosing the T_s and u_s as in (6.130) let us choose u_s such that

$$\beta u_s\, d^2/\kappa T_s = 1. \tag{6.132}$$

Then in the first equation we get

$$\frac{\alpha g T_s\, d^2}{\nu u_s} = \frac{\alpha g\, d^2}{\nu} \cdot \frac{\beta\, d^2}{\kappa} = R. \tag{6.133}$$

After applying separation of variables we obtain in place of (6.123) and (6.126)

$$(D^2 - a^2)^2 W = Ra^2 T, \tag{6.134a}$$

$$(D^2 - a^2) T = -W f(z), \tag{6.134b}$$

$$Ra^2\langle U, T\rangle + \langle V, Wf(z)\rangle \neq Ra^2\langle W, V\rangle + \langle T, Uf(z)\rangle. \tag{6.134c}$$

Clearly Eq. (6.134c) is not symmetric even if $f = 1$. Thus by a different non-dimensionalization we have transformed a self-adjoint system of equations into a non-self-adjoint system. Thus the self-adjoint property depends on having $R^{1/2}a$ in both equations rather than Ra^2 in one and 1 in the other. In (6.134) replace T by $T/(R^{1/2}a)$. Then the equations can be rearranged to give (6.123).

Another way to solve (6.123) is to combine the equations to obtain the single equation

$$(D^2 - a^2)^3 W = -Ra^2 W f(z). \tag{6.135}$$

The boundary conditions must also be converted. We consider the temperature condition $T = 0$. Equation (6.123a) applied at the boundary gives $(D^2 - a^2)^2 W = 0$. Thus we need

$$\begin{aligned} &W = D^2 W = D^4 W = 0, && \text{free boundary,} \\ &W = DW = (D^2 - a^2)^2 W = 0, && \text{rigid boundary.} \end{aligned} \tag{6.136}$$

We examine (6.135) to see if it is self-adjoint. Multiplying (6.135) by U and integrating by parts (applying the boundary condition $W = U = 0$), we obtain

$$\begin{aligned}\langle U, (D^2 - a^2)^3 W\rangle &= -(\langle D^3U, D^3W\rangle + 3a^2\langle D^2U, D^2W\rangle \\ &\quad + 3a^4\langle DU, DW\rangle + a^6\langle U, W\rangle) - [DU, D^4W] \\ &\quad + [D^2U, D^3W] + 3a^2[DU, D^3W] \\ &= -Ra^2\langle U, Wf(z)\rangle,\end{aligned} \tag{6.137}$$

where [,] denotes a quantity to be evaluated at the boundaries. When the boundary is free, the boundary terms vanish and the result is symmetric in W and U. When the boundary is rigid, the boundary terms are not symmetric in W and U and the problem is not self-adjoint. Thus in this formulation the boundary conditions are crucial: free boundaries yield a self-adjoint problem whereas rigid boundaries yield a non-self-adjoint formulation. Note that the free boundaries give a self-adjoint problem even in the case where $f(z) \neq 1$.

We show next that a variational principle can be formed for the original problem and its adjoint, and we relate such a principle to several methods used to solve convective instability problems.

Consider a single linear, non-self-adjoint equation with boundary conditions

$$Lu - \lambda Mu = 0, \qquad B_i u = 0. \tag{6.138}$$

We define the adjoint operators L^* and M^*, after integrating by parts,

$$\begin{gathered}\langle v, Lu\rangle - \lambda\langle v, Mu\rangle = \langle u, L^*v\rangle - \lambda\langle u, M^*v\rangle + B(u, v) \\ B(u, v) \equiv \text{boundary terms.}\end{gathered} \tag{6.139}$$

The adjoint boundary conditions are those which make the boundary terms vanish when the function u satisfies the original boundary condition. We denote them by $B_j^*(v) = 0$:

$$B(u, v) = 0 \qquad \text{when} \quad B_i u = B_j^* v = 0. \tag{6.140}$$

Next consider the two eigenvalue problems:

$$Lu - \lambda Mu = 0, \qquad L^*v - \lambda^* M^* v = 0, \tag{6.141a}$$

$$B_i u = 0 \qquad\qquad B_j^* v = 0. \tag{6.141b}$$

Let u and v be eigenfunctions of the two problems, respectively, and multiply (6.141a) by v and u, respectively; integrate over the domain. Subtraction of the two equations gives

$$\begin{gathered}\lambda\langle v, Mu\rangle - \lambda^*\langle u, M^*v\rangle = \langle v, Lu\rangle - \langle u, L^*v\rangle, \\ (\lambda^* - \lambda)\langle u, M^*v\rangle = 0.\end{gathered} \tag{6.142}$$

The right-hand side is zero, however, because of the definition of the adjoint operator (6.139), and the boundary conditions which u and v satisfy. Thus either the eigenfunctions u and v are orthogonal, $\langle v, M^*u\rangle = 0$, or the eigenvalues are the same, $\lambda = \lambda^*$. For each eigenvalue there are two eigenfunctions: one for the original equation and one for the adjoint equation. Next formulate a variational principle for the combined problem (6.141).

Make the functional

$$\lambda = -\langle v, Lu\rangle / \langle v, Mu\rangle \tag{6.143}$$

stationary to variations in functions u and v which satisfy

$$B_i u = B_j^* v = 0.$$

The Euler equations are just Eqs. (6.141). Thus we have a variational principle for any linear, non-self-adjoint problem.

Examine the natural boundary conditions for this variational principle. Make the functional stationary among functions u satisfying $B_i u = 0$, but with no restrictions on the functions v, other than differentiability. The variation gives, using (6.139),

$$\delta\lambda = -\frac{1}{\langle v, Mu\rangle}[\langle \delta v, Lu - \lambda Mu\rangle + \langle \delta u, L^*v - \lambda M^*v\rangle + B(\delta u, v)]. \tag{6.144}$$

When this is true among all variations, δv δu, it follows that

$$Lu - \lambda Mu = 0, \qquad L^*v - \lambda M^*v = 0, \qquad B_j^*(v) = 0, \tag{6.145}$$

and we obtain the equation and boundary conditions for the adjoint. Thus the variational principle (6.143) applies even if the adjoint trial functions do not satisfy the adjoint boundary conditions: they are natural conditions. The reverse is also true: we can make the functional (6.143) stationary among functions u and v which satisfy $B_j^* v = 0$. The conditions $B_i u = 0$ then are natural boundary conditions.

Now apply the variational method. We expand the function u in terms of functions satisfying the boundary conditions $B_i u = 0$ and apply no boundary conditions to the v_i:

$$u = \sum c_i u_i, \qquad v = \sum d_i v_i \tag{6.146}$$

and substitute into the functional (6.143). Taking the partial derivatives with respect to c_i and d_j gives

$$\frac{\partial\lambda}{\partial c_k} = 0 = \sum_j c_j \langle v_k, Lu_j - \lambda Mu_j\rangle, \tag{6.147a}$$

$$\frac{\partial\lambda}{\partial d_k} = 0 = \sum_j d_j [\langle u_k, L^*v_j - \lambda M^*v_j\rangle + B(u_k, v_j)], \tag{6.147b}$$

or

$$\sum_j A_{kj} c_j = 0, \qquad \sum_j B_{kj} d_j = 0, \qquad A_{kj} = B_{jk}. \tag{6.148}$$

Thus the eigenvalues which satisfy Eq. (6.147a) automatically force Eq. (6.147b) to be satisfied as well. This is true even if the equations for (6.147a) are written down using MWR. For example, (6.147a) are the equations obtained using MWR with weighting functions v_k. Thus the best weighting functions are trial functions for the adjoint eigenfunctions, that is, functions satisfying the adjoint boundary conditions. Then the eigenvalue is stationary, $\delta\lambda = 0$. This was shown by Kaplan (1963) and Finlayson (1970). The unimportance of adjoint boundary conditions was proved in a less general way by Crouch *et al.* (1970). In any MWR the eigenvalue is stationary, $\delta\lambda = 0$, since (6.147a) implies (6.147b) even when v_k does not satisfy the adjoint boundary conditions. Better results would be expected, however, if the adjoint trial functions were used as weighting functions. It is not necessary that (6.147b) be calculated, since it follows from (6.147a). In most convective instability problems the adjoint and original boundary conditions are the same (the reader can verify that this applies to all linear problems in this chapter). In that case the Galerkin method is equivalent to an adjoint variational principle since $v_k = u_k$. We conclude that *in all Galerkin methods for linear problems in this chapter, the eigenvalue is stationary.*

Let us consider a specific case: either a nonlinear initial temperature profile (6.31) or penetrative convection (6.87). We examine both at the same time using the equations

$$\begin{gathered}(D^2 - a^2)^2 W = R^{1/2} a T g(z),\\ (D^2 - a^2) T = -R^{1/2} a W f(z),\\ T = W = 0, \qquad \text{at } z = 0, 1, \qquad DW(0) = D^2 W(1) = 0.\end{gathered} \tag{6.149}$$

The boundary conditions are appropriate to one rigid and one free boundary, both of which are good thermal conductors. Find the adjoint to these equations. The set of equations can conveniently be represented in matrix notation

$$\mathbf{L} \cdot \mathrm{W} = aR^{1/2} \mathbf{M} \cdot \mathrm{W}, \tag{6.150}$$

where

$$\mathrm{W} = \{W, T\}, \qquad \mathbf{L} = \begin{Bmatrix} (D^2 - a^2)^2 & 0 \\ 0 & -(D^2 - a^2) \end{Bmatrix}$$

$$\mathbf{M} = \begin{Bmatrix} 0 & g(z) \\ f(z) & 0 \end{Bmatrix}. \tag{6.151}$$

The adjoint operator and boundary conditions are found by requiring

$$\langle \mathbf{V} \cdot (\mathbf{L} - aR^{1/2}\mathbf{M}) \cdot \mathbf{W} - \mathbf{W} \cdot (\mathbf{L}^* - aR^{1/2}\mathbf{M}^*) \cdot \mathbf{V} \rangle = \text{boundary terms} = 0. \tag{6.152}$$

and the result is

$$\mathbf{L}^* \cdot \mathbf{V} = aR^{1/2}\mathbf{M}^* \cdot \mathbf{V},$$

$$\mathbf{L}^* = \mathbf{L}, \qquad \mathbf{M}^* = \begin{Bmatrix} 0 & f(z) \\ g(z) & 0 \end{Bmatrix}. \tag{6.153}$$

The boundary conditions are (for $\mathbf{V} = \langle U, V \rangle$)

$$U = V = 0 \quad \text{at} \quad z = 0, 1, \qquad DU(0) = D^2U(1) = 0. \tag{6.154}$$

The boundary conditions are the same as in the original problem. Thus the same trial functions can be used for the adjoint function. This means that the Galerkin method corresponds to an adjoint variational principle, and the eigenvalue is stationary. In this case the variational principle is to make stationary

$$aR^{1/2} = \frac{\langle \mathbf{V} \cdot \mathbf{L} \cdot \mathbf{W} \rangle}{\langle \mathbf{V} \cdot \mathbf{M} \cdot \mathbf{W} \rangle} \tag{6.155}$$

among functions $\mathbf{W}$ and $\mathbf{V}$, at least one of which satisfies its boundary conditions.

Roberts (1960), Chandrasekhar (1961), and others have presented variational principles based on the adjoint for systems of equations. Some of the equations are solved exactly whereas the remaining equations are solved in a variational principle. For the problem (6.149), (6.153), and (6.154), Chandrasekhar would use the variational principle: make stationary

$$aR^{1/2} = \frac{\int_0^1 (DTDT^* + a^2TT^*)\, dz}{\int_0^1 [(D^2 - a^2)W][(D^2 - a^2)W^*]\, dz} \tag{6.156}$$

with respect to arbitrary variations in T and T^*, except that they are zero on the boundaries, and with variations in W and W^* determined as the unique solution to

$$\begin{gathered} (D^2 - a^2)^2\, \delta W = \delta T\, g(z), \\ (D^2 - a^2)^2\, \delta W^* = \delta T^* f(z), \\ W = W^* = 0 \qquad \text{at } z = 0, 1, \\ DW(0) = DW^*(0) = 0, \qquad D^2W(1) = D^2W^*(1) = 0. \end{gathered} \tag{6.157}$$

After taking the variation with respect to T, T^*, W, and W^* and requiring Eq. (6.157) we obtain as the first variation

$$-\langle \delta T^*, (D^2 - a^2)T + aR^{1/2}Wf\rangle - \langle \delta T, (D^2 - a^2)T^* + aR^{1/2}W^*g\rangle = 0. \tag{6.158}$$

The Euler equations are given in the angle brackets. When the variational principle is applied the equations are

$$\begin{aligned}\sum_j A_j\langle T_k^*, (D^2 - a^2)T_j + aR^{1/2}W_j f\rangle &= 0,\\ \sum_j B_j\langle T_k, (D^2 - a^2)T_j^* + aR^{1/2}W_j^* g\rangle &= 0.\end{aligned} \tag{6.159}$$

These are the same equations obtained by applying the Galerkin method and requiring that two of the equations (6.157) be satisfied exactly:

$$(D^2 - a^2)^2 W_j = g(z)T_j, \qquad (D^2 - a^2)^2 W_j^* = f(z)T_j^*. \tag{6.160}$$

Thus this adjoint variational method is also equivalent to a Galerkin method. Chandrasekhar (1961) also treats cases for which a variational principle of the form (6.156) is not written down. Instead, a trial function is assumed for one of the variables, it is substituted into one of the equations which is then solved exactly for the second variable, for example Eq. (6.160a). These trial functions are then substituted into the other equation which is made orthogonal to the trial function:

$$\sum_j A_j\langle T_k, (D^2 - a^2)T_j + aR^{1/2}W_j f\rangle = 0. \tag{6.161}$$

This method of solution is equivalent to a variational principle when $T_j = T_j^*$ and $f(z) = g(z)$, and it is equivalent in all cases to a Galerkin method in which one of the equations is satisfied exactly. It has been applied to many cases including the effects of magnetic fields (Chandrasekhar, 1961), convection in fluid spheres (Bisshopp, 1958) and Couette flow (Chow and Uberoi, 1965; Debler, 1966). DiPrima and Pan (1964) compare calculations made with this "modified Galerkin method," in which some equations are solved exactly, with the usual Galerkin method for a system of equations. They find that both give comparable results in higher approximations.

We next illustrate some computations using the variational principles. We give a summary of the computations for the buoyancy problem (6.149) with $f = g = 1$ and boundary conditions

$$W = DW = T = 0. \tag{6.162}$$

Pellew and Southwell (1940) give the following variational principle: make stationary

$$\lambda = \frac{\int_{-1/2}^{1/2} [(DT)^2 + a^2T^2]\, dz}{a^2 \int_{-1/2}^{1/2} [(D^2 - a^2)W]^2\, dz} \tag{6.163}$$

among functions T which satisfy the thermal boundary conditions and W satisfies

$$(D^2 - a^2)^2 W = T. \tag{6.164}$$

For the case of two rigid, good conducting boundaries they try

$$T = 1 + \cos 2\pi z + A \cos \pi z,$$

$$W = \frac{1}{a^4} + \frac{\cos 2\pi z}{(4\pi^2 + a^2)^2} + Pa \cosh az + Qa^2 z \sinh az + A \frac{\cos \pi z}{(\pi^2 + a^2)^2}. \tag{6.165}$$

After two pages of additional calculations Pellew and Southwell obtain the results listed in Table 6.10. Clearly this is a very accurate method of solution,

TABLE 6.10

RAYLEIGH NUMBERS DERIVED FROM VARIATIONAL PRINCIPLES

	Pellew and Southwell Eq. (6.165) $a = 3.117$	Reid and Harris $a = 3.117$	Chandrasekhar Eqs. (6.166), (6.167) $a = 3.117$	Galerkin for system Eqs. (6.16), (6.17) $a = 3.117$
First approximation	1707.87	1719.2	1715.1	1750.0
Second approximation	—	1708.8	1707.94	1708.5
Third approximation	—	1707.970	1707.775	1707.762
Exact	1707.762	1707.762	1707.762	1707.762

but requires considerable algebra, which must be repeated for each new problem.

Chandrasekhar (1961) used the same variational principle with the trial function

$$T = \sum_j A_j \cos[(2j + 1)\pi z], \tag{6.166}$$

where W_j is the solution to

$$(D^2 - a^2)^2 W_j = \cos(2j+1)\pi z, \qquad W_j = DW_j = 0 \quad \text{at} \quad z = 0, 1. \tag{6.167}$$

Reid and Harris (1958) expanded the velocity in trial functions, (6.8), and solved the temperature equation, $(D^2 - a^2)T = W$ approximately. Results are listed in Table 6.10. The first approximation is very good, although it requires a great deal of algebra. By contrast the first three approximations using the Galerkin method are listed for the polynomial trial functions as applied in Section 6.2. In these computations the *system* of equations is satisfied approximately with no single equation solved exactly. It is thus simpler to apply. While the first approximation is not quite as good, the third approximation is better. The polynomial trial functions are easily programmed for the computer and can be used in other problems as well. Consequently this author prefers to use polynomial trial functions. When the thermal boundary conditions are changed a variational principle such as (6.163) becomes very tedious to apply since the functions for T_m and W_m become very complicated.

Chandrasekhar (1961) applies the same method to the case of the rotating fluid layer including oscillatory instability. In that case the temperature is expanded in the series (6.166). The velocity and vorticity equations are combined and the velocity function is taken as the solution to

$$\begin{gathered}[(D^2 - a^2 - i\sigma)^2(D^2 - a^2) + TD^2]W_j = -C_j R^{1/2} a \cos(2j+1)\pi z, \\ C_j = (2j+1)^2\pi^2 + a^2 + i\sigma.\end{gathered} \tag{6.168}$$

The velocity and temperature functions are then substituted into the temperature equation, which is made orthogonal to the temperature trial function. The result is a complicated expression involving complex numbers. Roots of the equation give the critical Rayleigh numbers. This method is also clearly a Galerkin method in which some of the equations are solved exactly. In the Galerkin method applied in Section 6.3, we obtained considerably more than a numerical solution. We obtained relations (6.103)–(6.109), indicating how various parameters influenced the solution. None of these results are obtained for rigid boundaries using Chandrasekhar's method. They are obtained, however, for the exact solution for two free boundaries.

In conclusion, the Galerkin method applied to systems of equations is useful for delineating the role of various parameters, gives a first approximation with a minimum of algebra, and, is easily extended on the computer to calculate higher approximations. The Galerkin method is equivalent to adjoint variational principles, so that the eigenvalue is stationary. As such, the Galerkin method is a very powerful method.

6.5 Nonlinear Convective Instability

The problems above are for the linearized perturbation equations and therefore can only describe the onset of motion. When a system is unstable with respect to small perturbations it is unstable with respect to large disturbances, too. The development of a finite amplitude steady state is governed by the nonlinear equations. Expansion methods and the Galerkin method are applicable in these cases, too. The ideas are simple in principle but applications are limited primarily by computer storage and running times. We discuss some typical applications here.

For the nonlinear problem, the first possibility is to expand the unknown solution in terms of the eigenfunctions of the linear problem with coefficients which are unknown functions of time. The nonlinear residuals are then made orthogonal to the eigenfunctions, in the case of self-adjoint linear problems, or orthogonal to the adjoint eigenfunctions, in the case of non-self-adjoint problems. This leads to a set of nonlinear ordinary differential equations in time. These are integrated as an initial-value problem to obtain the steady-state solutions. This approach was taken by Platzman (1965) and Eckhaus (1965). More recently it has been extended to vector eigenfunctions by Kogelman and DiPrima (1970). An excellent description of this technique and its comparison to other techniques is given by DiPrima and Rogers (1969). For example, in the nonlinear Couette flow problem, the extension to Eqs. (6.92), the velocity component might be expanded in the form

$$u_\theta(r, z, t) = V(r) + \sum_{q=0}^{\infty} \sum_{p=1}^{\infty} A_{pq}(t) v_p(r) \cos q\lambda z, \tag{6.169}$$

where the $v_p(r)$ are the spatial eigenfunctions of the linear stability problem. Ordinary differential equations are then derived for the $A_{pq}(t)$.

Another possibility is to expand the velocity functions in a Fourier series with time-dependent coefficients (see Veronis 1966a,b, 1968a,b; Catton 1966). Foster (1969a) studied two-dimensional, finite amplitude thermal convection in a fluid with infinite Prandtl number. He chose velocity functions to satisfy

$$w = D^2 w = D^4 w = 0 \qquad \text{at} \quad z = 0, 1 \tag{6.170}$$

at the bottom and top free surface, where the temperature is specified. On lateral boundaries the boundary conditions were taken as free and perfectly insulating:

$$\begin{aligned} u = \partial^2 u/\partial x^2 &= 0, \\ \partial w/\partial x = \partial^3 w/\partial x^3 &= 0, \qquad x = 0,\ L, \\ \partial\theta/\partial x &= 0. \end{aligned} \tag{6.171}$$

The vertical component of velocity is then written as

$$w = \sum_{m,n} A_{mn}(t)\cos(m\pi x/L)\sin(n\pi z). \tag{6.172}$$

The horizontal velocity was determined from the continuity equation and (6.172). The temperature function was obtained from the equation

$$\left(\frac{\partial^2}{\partial x^2} + \frac{\partial^2}{\partial z^2}\right)^2 w = R\,\frac{\partial^2 \theta}{\partial x^2}, \tag{6.173}$$

which is obtained from the vorticity equation for infinite Prandtl number

$$\left(\frac{\partial^2}{\partial x^2} + \frac{\partial^2}{\partial y^2}\right)\left(\frac{\partial u}{\partial z} - \frac{\partial w}{\partial x}\right) = -R\,\frac{\partial \theta}{\partial x} \tag{6.174}$$

by differentiation with respect to x. The trial functions are then substituted into the temperature equation, which is made orthogonal to $\cos(k\pi x/L)\sin(r\pi z)$. This gives a set of ordinary differential equations for the $A_{mn}(t)$. This Fourier series method as well as the vector eigenfunction expansion, are special cases of Galerkin's method. For more general boundary conditions, when sines and cosines do not satisfy the appropriate boundary conditions, other expansion functions must be used and the method is a Galerkin method.

The next level of sophistication is to solve for the steady state, which develops when $R \geq R_c$, and study its stability. Busse (1967a) applied the Galerkin method of roll cells to three-dimensional disturbances in a layer of fluid heated from below. Busse treated fluids with infinite Prandtl number and deduced the band of wave numbers which give rise to stable steady states.

Using these types of expansions it is also possible to predict the variation of Nusselt number with Rayleigh and Prandtl numbers. This has been done for a single convective mode by Platzman (1965) and Edwards and Catton (1969) and with more than 12 temperature terms and 16 velocity terms by Poots (1958) for a square box, infinitely long. Davis (1968) considers the problem of wave number selection for the nonlinear problem in a three-dimensional box (see Section 6.2).

The choice of trial functions is somewhat specialized, depending on whether there are lateral boundaries, rigid or free boundaries at the top and bottom. Either two or three dimensional disturbances can be included. For problems with free boundaries, sines and cosines are useful, and combinations of sines are useful even for rigid boundaries [see Eqs. (6.28) and (6.122)]. Calculation of the integrals is tedious but not impossible, as the number of different choices demonstrates. The chief limitation in nonlinear problems is the number of modes that can be included, which is limited by computer storage and computation time. This limitation is more severe in the case of finite difference computations, due to the small grid that is necessary to resolve the higher Fourier modes (see DiPrima and Rogers, 1969).

6.6 Hydrodynamic Stability

Next we consider the stability of an established, laminar flow. Laminar, rectilinear flow between two flat plates, in a direction parallel to the plates, is called plane Poiseuille flow when it is established by an applied pressure gradient. Plane Couette flow is established by moving one of the plates with a fixed velocity. We study plane Poiseuille flow first, for both linear and nonlinear equations, and then treat linear and nonlinear plane Couette flow. Then we examine plane Couette flow with an established temperature gradient as well as the flow. The chief difference between these instabilities and convective instabilities is that here oscillatory instabilities are the rule, rather than the exception, and many more terms are needed in the Galerkin method for good results. A few of the methods described below are not identified by the authors as Galerkin methods, although it is clear they can be so interpreted.

Take a coordinate system with the flow in the x direction and the velocity gradient in the y direction. There is no flow in the z direction or dependence on the z coordinate. The pertinent equations are derived from the Navier–Stokes equations. For the basic, steady flow we have

$$u = (U(y), 0, 0), \qquad 0 = -\frac{\partial p}{\partial x} + \nu \frac{d^2 U}{dy^2}, \tag{6.175}$$

subject to either

$$\frac{\partial p}{\partial x} = -\gamma, \qquad U = 0 \quad \text{at } y = \pm d/2, \qquad \text{whence} \quad U = U_0(1 - (2y/d)^2) \tag{6.176a}$$

or

$$\frac{\partial p}{\partial x} = 0, \qquad \begin{array}{ll} U = U_0 & \text{at } y = +d, \\ U = 0 & \text{at } y = 0, \end{array} \qquad U = \frac{U_0 y}{d} \tag{6.176b}$$

Write the velocity as the sum of this basic flow and a perturbed flow, which is to be determined. We restrict attention to perturbations in the xy plane, and write the equations only for the linear case here:

$$\frac{\partial \mathbf{u}'}{\partial t} + \mathbf{u}' \cdot \nabla \mathbf{U} + \mathbf{U} \cdot \nabla \mathbf{u}' = -\frac{1}{\rho} \nabla p' + \nu \nabla^2 \mathbf{u}'. \tag{6.177}$$

Take the curl of this equation and write the stream function in the form

$$\begin{gathered} u_x' = \frac{\partial \psi}{\partial y}, \qquad u_y' = -\frac{\partial \psi}{\partial x}, \\ \psi(x, y, t) = \phi(y, t) e^{i\alpha x}, \\ \nabla \times \mathbf{u}' = -\mathbf{k}(\phi_{yy} - \alpha^2 \phi) e^{i\alpha x}. \end{gathered} \tag{6.178}$$

The vertical component of the dimensionless vorticity equation is then

$$\frac{\partial}{\partial t} L\phi = i\alpha(U'' - UL)\phi + \frac{1}{\text{Re}} L^2\phi, \qquad L = D^2 - \alpha^2 = \frac{\partial^2}{\partial y^2} - \alpha^2, \quad (6.179)$$

where $\text{Re} = U_0 d/\nu$, U_0 is the maximum speed, d is the separation between the two plates, and ν is the kinematic viscosity. This is the time-dependent Orr–Sommerfeld equation. If an exponential time-dependence is assumed, $\exp[-i\alpha\xi t]$, the Orr–Sommerfeld equation is obtained

$$L^2\phi - i\alpha \,\text{Re}[(U - \xi)L - U'']\phi = 0. \quad (6.180)$$

The flow is unstable if for some α, $\text{Im}(\xi) > 0$. The boundary conditions on the perturbation at the boundary are

$$\phi = \partial\phi/\partial y = 0 \quad (6.181)$$

for both the plane Poiseuille and plane Couette problems.

Some general features of the solution are summarized. A plot of wave number α versus Reynolds number gives a critical value for instability. For Reynolds numbers below the critical, all perturbations die out, whereas for Reynolds numbers above the critical, the disturbances grow until the nonlinear terms in the equations become important. An early application of the Galerkin method to plane Poiseuille flow by Dolph and Lewis (1958) was somewhat discouraging. They used eight expansion functions and obtained poor agreement with previous work. When using twenty expansion functions the agreement was more satisfactory, but required a large computational effort. More recently the Galerkin method has been successfully applied, but the early computations give two warnings: a great many expansion coefficients must be used and this means that the expansion functions must be orthonormal, to retain computational accuracy, and the integrals must be easy to compute. The same requirements in Chapter 5 led to the use of orthogonal polynomials, and similarly for this problem the Galerkin applications had led to the generation of special eigenfunction expansions. It is mainly these different trial functions which we examine below.

Dolph and Lewis (1958) use as expansion functions the eigenfunctions of the problem

$$L^2\phi_n = \lambda_n(\alpha^2 - D^2)\phi_n, \quad (6.182)$$

which satisfy the orthogonality requirement

$$\int_{-1}^{1} [\alpha^2\phi_n\phi_m + \phi_n'\phi_m']\, dy = \delta_{nm}. \quad (6.183)$$

The actual eigenfunctions are determined in the same manner used to solve Eqs. (6.119). Dolph and Lewis divide the eigenfunctions into even and odd functions, and use only the even eigenfunctions in the calculations. These eigenfunctions, $\phi_n(y)$, are used to expand the solution

$$\phi(y) = \sum_{i=1}^{N} a_i \phi_i(y). \tag{6.184}$$

The series is substituted into the differential equation (6.180) to form the residual, which is made orthogonal to the eigenfunctions, ϕ_n. This leads to a system of homogeneous linear algebraic equations, which have a solution if and only if the determinant of the coefficients vanishes. They report that a 2×2 matrix gives no instability at all, whereas a 6×6 matrix (using six terms in the stream function expansion) gave instability. The results were poor, however, and a 20×20 matrix was needed to obtain reasonable agreement with prior work.

Later Grosch and Salwen (1968) studied the stability of steady and time-dependent plane Poiseuille flow. They expanded the stream function in the series

$$\phi(y, t) = \sum_{i=1}^{N} A_i(t)\phi_i(y) \tag{6.185}$$

with time-dependent coefficients. The trial functions were taken as the eigenfunctions to the problem

$$L^2\phi_n = \lambda^4\phi_n, \qquad \phi_n = \phi_n{}' = 0 \qquad \text{at } y \pm \tfrac{1}{2}. \tag{6.186}$$

For example, the even functions are

$$\begin{gathered}\phi_n(y) = c_n\left[\frac{\cosh(\beta_n y)}{\cosh(\beta_n/2)} - \frac{\cos(\gamma_n y)}{\cos(\gamma_n/2)}\right],\\ \langle \phi_n, \phi_m \rangle = \delta_{nm},\\ \beta_n = (\lambda_n{}^2 + \alpha^2)^{1/2}, \qquad \gamma_n = (\lambda_n{}^2 - \alpha^2)^{1/2},\\ \beta_n \tanh(\beta_n/2) + \gamma_n \tan(\gamma_n/2) = 0.\end{gathered} \tag{6.187}$$

For the steady flow, the calculations are done with 20 to 50 terms in the expansion. The number of terms needed increases as the product αRe increases. For $\alpha = 2$ and Re $= 20{,}000$ the results for $N = 30$ gave the first ten eigenvalues within 1% and for $N = 40$ the accuracy was about 0.1%. (The accuracy was estimated by studying the convergence of the results as N increased.) For $\alpha = 2.2$ and Re $= 70{,}000$ the first eleven eigenvalues were within 1% using $N = 40$. Values obtained by Thomas (1953) using a finite difference method (100 grid points) are available for comparison for the first eigenvalue and a limited range of α and Re. The comparison is within four significant figures for the eigenvalues. The first eigenfunction compares within 3%, and the maximum deviation occurs near the wall where the function itself is small. Grosch and Salwen were able to compute higher eigenvalues and examine the behavior of the eigenvalues as a function of Reynolds number. This gives suggestive information on the relative importance of higher modes after the instability sets in. Grosch and Salwen then investigated the problem resulting from a pressure gradient which is the sum of a constant and a part varying sinusoidally in time. The basic velocity is then more complicated, and the problem is solved using an expansion (6.185) with time-dependent coefficients. The ordinary differential equations generated by the Galerkin method (not named) were integrated numerically using the Runge–Kutta method. The set of equations is in the form

$$\frac{dA_i}{dt} = \sum_{j=1}^{N} F_{ij} A_j, \tag{6.188}$$

and when the matrix F is periodic there exists a fundamental solution matrix $W(t)$ which satisfies

$$\frac{dW_{ik}}{dt} = \sum_{j=1}^{N} F_{ij} W_{jk}, \qquad W_{ik}(0) = \delta_{ik}. \tag{6.189}$$

Thus the solution matrix is found by integrating (6.188) N times, with $A_j(0) = 1$, all other $A_i(0) = 0$, and taking $j = 1, \ldots, N$. The eigenvalues of the solution matrix W thus generated determine the stability of the flow. The effect of modulation is to stabilize the flow, that is increase the critical Reynolds number.

Dowell (1969) studied plane Poiseuille flow including the full nonlinear equations (in two dimensions). His expansion functions are simpler than those of Grosch and Salwen, which is necessitated by the inclusion of the nonlinear terms and the necessity to calculate them efficiently on the computer:

$$\begin{aligned}
\psi(x, y, t) &= \sum_{m=1}^{M} \sum_{v=0}^{V} [A_{mv} \cos v\alpha x + B_{mv} \sin v\alpha x]\psi_m(y), \\
\psi_m(y) &= \cos(m-1)\pi y - \cos(m+1)\pi y, \\
\partial\psi/\partial x &= \partial\psi/\partial y = 0 \qquad \text{at} \quad y = 0, 1.
\end{aligned} \tag{6.190}$$

For the linear problem, comparison to Thomas's (1953) results showed good agreement provided 40 to 50 terms are retained in the expansion. In the nonlinear problem, it proved unfeasible to include this many terms in the numerical integration, which were limited to $V = 2$ and $M = 16$. Dowell found that the even modes (for velocity), which are stable in the linear theory, are excited by the odd modes, which are the unstable modes in the linear theory. The even modes exhibit a steady time behavior, whereas the odd modes exhibit an oscillatory behavior. The inclusion of the nonlinear terms caused the disturbance to exhibit a limit cycle.

Pekeris and Shkoller (1959) study the same problem using a method which can be interpreted as a method of weighted residuals. They expand the stream function

$$\psi = \sum_{n=0}^{V} f_n(y, t) e^{-i\alpha n x} \tag{6.191}$$

and take $V = 3$ in the calculations. Each f_n is then expanded in terms of eigenfunctions to the linear Orr–Sommerfeld equation:

$$\begin{gathered} D^4\phi_n{}^\nu - 2\alpha^2 n^2 D^2 \phi_n{}^\nu + \alpha^4 n^4 \phi_n{}^\nu \\ + i\alpha \operatorname{Re} n[(1 - y^2 - c_n{}^\nu)(D^2\phi_n{}^\nu - \alpha^2 n^2 \phi_n{}^\nu) + 2\phi_n{}^\nu] = 0, \\ \phi_n{}^\nu = D\phi_n{}^\nu = 0 \qquad \text{at} \quad y = \pm 1, \end{gathered} \tag{6.192}$$

$$f_n(y, t) = \sum_{\nu=1}^{\kappa} B_\nu^{(n)}(t)\phi_n{}^\nu(y), \qquad f_0{}'(y, t) = \sum_{\sigma=1}^{s} A_\sigma(t)\cos(\sigma - \tfrac{1}{2})\pi y.$$

These equations are substituted into the differential equation to form the residual, which is then made orthogonal to the adjoint to $\phi_n{}^\nu$, and $\cos(\sigma - 1)\pi y$ for $n = 0$. For the linear problem this would correspond to a variational method, using the adjoint equation, but for the nonlinear case it is a weighted residual method with the weighting function the adjoint, $\tilde{\phi}_n{}^\nu$. Calculations are done for $\kappa = 21$ and $s = 30$ or 60. Only even eigenfunctions (for the stream function) are used. They study several different types of interactions giving rise to instability which leads to turbulence, and show that for Re less than the critical value, disturbances can grow provided the amplitude is large enough. If the disturbance is initially only the first mode, $B_1^{(1)}$, the nonlinear terms in the equations cause other modes to be excited, such as $B_9^{(1)}$, thus leading to instability.

Consider next plane Couette flow. Gallagher and Mercer (1962) applied the Galerkin method, without naming it as such. They studied the linearized equations using the expansion function

$$\phi(y) = \sum_{i=1}^{N} a_i Y_i(y), \tag{6.193}$$

where a_i are complex numbers and the $Y_i(y)$ are orthogonal functions defined by (6.5) with the boundaries at $y = 0, \pi$. They made calculations with increasing values of N until the first latent root remained unchanged to the desired accuracy. Up to 20 terms could be included but this limited consideration to values of $\alpha\text{Re} \leq 1000$. Prior to this time only limited information was available: it was known that the problem was stable at small and large values of αRe and also for $\alpha = 1$. While Gallagher and Mercer's work could not be extended to large enough αRe to connect with the asymptotic results, the results do suggest that the problem is stable for all values of αRe.

The nonlinear problem was tackled by Kuwabara (1967) using the Galerkin method. He expanded the stream function in terms of the associate Legendre polynomials, $P^4_{n+4}(y)$, each of which satisfies the boundary conditions, although he used only two terms in the nonlinear analysis.

Ingersoll (1966) performed a linearized analysis of the plane Couette problem with a superposed temperature gradient established between the two plates. The equations are then

$$[L - i\alpha\varepsilon(y - c)]LW = \alpha^2 RT,$$
$$[L - i\alpha\varepsilon \,\mathrm{Pr}(y - c)]T = -W, \tag{6.194}$$
$$L = D^2 - \alpha^2, \qquad \varepsilon = \mathrm{Re}\, \alpha_x/\alpha.$$

He expands the trial solution in the series

$$W = \sum_{n=0}^{N} a_n f_n(y), \qquad T = \sum_{n=0}^{N} b_n g_n(y), \tag{6.195}$$

where the trial functions are defined by

$$L^2 f_n = \lambda_n g_n, \qquad Lg_n = -f_n, \tag{6.196a}$$

$$f_n = Df_n = g_n = 0 \qquad \text{at} \quad y = \pm\tfrac{1}{2}. \tag{6.196b}$$

These functions have convenient properties which simplify the solution. First we derive an orthogonality property. Multiply (6.196a) by f_m and (6.196b) by g_m and integrate over the region:

$$\langle f_m L^2 f_n \rangle = \lambda_n \langle f_m g_n \rangle, \tag{6.197a}$$

$$-\langle g_m L g_n \rangle = \langle g_m f_n \rangle. \tag{6.197b}$$

The second equation can be integrated by parts to deduce the fact that $\langle g_m f_n \rangle$ is symmetric in m and n. Interchange n and m in (6.197a) and subtract from (6.197a). Integration by parts gives

$$(\lambda_n - \lambda_m)\langle f_m g_n \rangle = 0. \tag{6.198}$$

The Galerkin method applied to (6.194) then gives

$$\lambda_{(n)} a_n - i\alpha\varepsilon B_{nm} a_m = \alpha^2 R b_n, \qquad -b_m - i\alpha\varepsilon \,\mathrm{Pr}\, D_{mn} b_n = -a_m, \tag{6.199}$$

where repeated indices are summed, except for $\lambda_{(n)}$. The first equation can be multiplied by D_{kn} and summed over n and this result for $D_{kn} b_n$ substituted into the second equation. This can be solved for b_m, which is substituted back into the first equation. The result is

$$[(\lambda_{(n)} - \alpha^2 R)\delta_{nm} + i\alpha\varepsilon(\mathrm{Pr}\, D_{nm}\lambda_{(m)} - B_{nm}) + \alpha^2\varepsilon^2 \,\mathrm{Pr}\, D_{nl} B_{lm}]a_m = 0. \tag{6.200}$$

This reduces by half the size of the matrix and shortens the computations by a factor of 4. Ingersoll expands the matrix explicitly for the case of small Reynolds number and obtains

$$R = 1707.76 + \varepsilon^2[0.5598\ \mathrm{Pr}^2 + 0.1270\ \mathrm{Pr} + 0.06451], \qquad (6.201)$$

which shows that the steady flow increases the critical Rayleigh number. The same trends are confirmed in the detailed calculations (using 20 terms): the Rayleigh number is an increasing function of Pr and ε. Results are also available using (6.193) (Gallagher and Mercer, 1965).

Rudakov (1967) and Birikh *et al.* (1968) treat similar problems using the Galerkin method. The trial functions are the solutions to

$$\begin{aligned} L^2 f_n &= -\mu_n L f_n, & L g_n &= -\mathrm{Pr}\, \nu_n g_n, \\ f_n &= f_n' = 0, & g_n &= 0 \quad \text{on boundary,} \end{aligned} \qquad (6.202)$$

and the flow is established on a plane inclined to gravity. From 8 to 14 terms were retained in the expansion and they found that with a small number of terms (such as 4) they obtained oscillatory instability which disappeared in higher approximations. This result, coupled with that of Dolph and Lewis mentioned above, points out that it is particularly important to examine the convergence of the results with successive approximations.

Salwen and Grosch (1968) applied an expansion method such as (6.185) to pipe Poiseuille flow and found the first five azimuthally varying modes are stable for $\alpha \le 10$, $\mathrm{Re} \le 10{,}000$.

EXERCISES

6.1. Apply the boundary conditions (6.2) to the polynomial (6.14) to derive the trial function (6.15). Apply the boundary conditions (6.58) to a third-order polynomial to obtain (6.66a).

6.2. Solve for the first approximation to the Rayleigh number for a fluid layer heated from below subject to the thermal boundary conditions

$$T(-\tfrac{1}{2}) = 0, \qquad -DT(\tfrac{1}{2}) = \mathrm{Nu}\ T(\tfrac{1}{2}).$$

For a first approximation best results are achieved if the trial function for T satisfies the boundary conditions. Choose a second-order polynomial to do this and derive the Rayleigh number as in (6.50). Compare the approximate R_c to the exact R_c for $\mathrm{Nu} = \infty$ ($R_c = 1708$), $\mathrm{Nu} = 3$ ($R_c = 1498$), $\mathrm{Nu} = 0$ ($R_c = 1296$). Then use the technique in Eq. (6.77) to predict an improved value of R_c for $\mathrm{Nu} = 3$.

Answer $R_c = 1750$, 1592, 1446, respectively. Using Eq. (6.77) gives $R_c = 1494$ for $\mathrm{Nu} = 3$.

6.3. Calculate the growth factor as a function of Nusselt number for Exercise 6.2 for a fluid such as water (take Pr = 7). See Eq. (6.51). Determine λ for Nu = 0, 0.3, 3, 30, ∞, and various $R - R_c$.

6.4. Consider the fluid layer heated from below and subject to both buoyancy and surface-tension driven convection [see Eq. (6.62b)]. For the boundary conditions (6.58), $T(0) = DT(1) = 0$, and $R = 350$, estimate M_c.
Answer: First approximation, $M_c = 43.5$.

6.5. Consider the eigenvalue problem

$$(D^2 - a^2)^3 W + Ra^2 W = 0,$$

subject to one of the following sets of boundary conditions

(a) $W = DW = (D^2 - a^2)^2 W = 0$ at $z = 0, 1,$
(b) $W = D^2 W = (D^2 - a^2)^2 W = 0$ at $z = 0, 1.$

For each set, is the problem self-adjoint? If not, what is the adjoint problem?
Answer: (a) No; (b) Yes.

6.6. Show that Eq. (6.82) is a self-adjoint problem for the boundary conditions $\theta = \mathbf{u} = 0$.

REFERENCES

Ames, W. F. (1965). "Nonlinear Partial Differential Equations in Engineering." Academic Press, New York.

Amundson, N. R. (1966). "Mathematical Methods in Chemical Engineering." Prentice-Hall, Englewood Cliffs, New Jersey.

Aris, R. (1962). "Vectors, Tensors, and the Basic Equations of Fluid Mechanics." Prentice-Hall, Englewood Cliffs, New Jersey.

Beal, T. R. (1965). Dynamic Stability of a Flexible Missile under Constant and Pulsating Thrusts, *AIAA J.* **3**, 486–494.

Birikh, R. V., Gershuni, G. Z., Zhukhovitskii, E. M., and Rudakov, R. N. (1968). Hydrodynamic and Thermal Instability of a Steady Convective Flow, *J. Appl. Math. Mech.* **32**, 246–252.

Bisshopp, F. E. (1958). On the Thermal Instability of a Rotating Fluid Sphere, *Phil. Mag.* [8] **3**, 1342–1360.

Bolotin, V. V. (1963). "Nonconservative Problems of the Theory of Elastic Stability." Macmillan, New York.

Busse, F. H. (1967a). On the Stability of Two-Dimensional Convection in a Layer Heated from Below, *J. Math. Phys.* **46**, 140–150.

Busse, F. H. (1967b). The Stability of Finite Amplitude Cellular Convection and its Relation to an Extremum Principle, *J. Fluid Mech.* **30**, 625–649.

Catton, I. (1966). Natural Convection in Horizontal Liquid Layers, *Phys. Fluids* **9**, 2521–2522.

Catton, I. (1970). Convection in a Closed Rectangular Region: The Onset of Motion, *J. Heat Transfer, Trans. ASME, Ser. C* **92**, 186–188.

Chandrasekhar, S. (1953). The Instability of a Layer of Fluid Heated Below and Subject to Coriolis Forces, *Proc. Roy. Soc. (London)* **A217**, 306–327.

Chandrasekhar, S. (1960). The Hydrodynamic Stability of Viscid Flow between Coaxial Cylinders, *Proc. Nat. Acad. Sci. (U.S.)* **46**, 141–143.

Chandrasekhar, S. (1961). "Hydrodynamic and Hydromagnetic Stability." Oxford Univ. Press (Clarendon), London and New York.

Chandrasekhar, S., and Elbert, D. D. (1955). The Instability of a Layer of Fluid Heated Below and Subject to Coriolis Forces, II. *Proc. Roy. Soc. (London)* **A231**, 198–210.

Chandrasekhar, S., and Elbert, D. D. (1958). On Orthogonal Functions Which Satisfy Four Boundary Conditions. III, Tables for Use in Fourier-Bessel-Type Expansions, *Astrophys. J. Suppl. Ser.* **3**, 453–458.

Chandrasekhar, S., and Reid, W. H. (1957). On the Expansion Functions Which Satisfy Four Boundary Conditions, *Proc. Nat. Acad. Sci. (U.S.)* **43**, 521–527.

Chow, C. Y., and Uberoi, M. S. (1965). Hydromagnetic Instability of Fluid between Two Coaxial Cylinders, *Phys. Fluids* **8**, 413–417.

Crouch, J. G., Anderson, W. J., and Greenwood, D. T. (1970). Eigenvalue Errors in the Method of Weighted Residuals, *AIAA J.* **8**, 2048–2054.

Davis, S. H. (1967). Convection in a Box: Linear Theory, *J. Fluid Mech.* **30**, 465–478.

Davis, S. H. (1968). Convection in a Box: On the Dependence of Preferred Wave-Number upon the Rayleigh Number at Finite Amplitude, *J. Fluid Mech.* **32**, 619–624.

Davis, S. H. (1969). On the Principle of Exchange of Stabilities, *Proc. Roy. Soc. (London)* **A310**, 341–358.

Debler, W. R. (1966). On the Analogy between Thermal and Rotational Hydrodynamic Stability, *J. Fluid Mech.* **24**, 165–176.

DiPrima, R. C. (1960). The Stability of a Viscous Fluid between Rotating Cylinders with an Axial Flow, *J. Fluid Mech.* **9**, 621–631.

DiPrima, R. C. (1961). Stability of Nonrotationally Symmetric Disturbances for Viscous Flow between Rotating Cylinders, *Phys. Fluids* **4**, 751–755.

DiPrima, R. C., and Pan, C. H. T. (1964). The Stability of Flow between Concentric Cylindrical Surfaces with a Circular Magnetic Field, *Z. Angew. Math. Phys.* **15**, 560–567.

DiPrima, R. C., and Rogers, E. H. (1969). Computing Problems in Nonlinear Hydrodynamic Stability, *Phys. Fluids Suppl.* **12**, II 155–165.

Dolph, C. L., and Lewis, D. C. (1958). On the Application of Infinite Systems of Ordinary Differential Equations to Perturbations of Plane Poiseuille Flow, *Quart. Appl. Math.* **16**, 97–110.

Dowell, E. H. (1969). Non-Linear Theory of Unstable Plane Poiseuille Flow, *J. Fluid Mech.* **38**, 401–414.

Duncan, W. J. (1952). "The Principles of the Control and Stability of Aircraft." Cambridge Univ. Press, London and New York.

Eckhaus, W. (1965). "Studies in Non-Linear Stability Theory." Springer-Verlag, Berlin.

Edwards, D. K., and Catton, I. (1969). Prediction of Heat Transfer by Natural Convection in Closed Cylinders Heated from Below, *Int. J. Heat Mass Transfer* **12**, 23–30.

Finlayson, B. A. (1968). The Galerkin Method Applied to Convective Instability Problems, *J. Fluid Mech.* **33**, 201–208.

Finlayson, B. A. (1970). Convective Instability of Ferromagnetic Fluids, *J. Fluid Mech.* **40**, 753–767.

Finlayson, B. A. and Scriven, L. E. (1969), Convective Instability by Active Stress, *Proc. Roy. Soc. (London)* **A310**, 183–219.

Foster, T. D. (1968). Effect of Boundary Conditions on the Onset of Convection, *Phys. Fluids* **11**, 1257–1262.

Foster, T. D. (1969a). The Effect of Initial Conditions and Lateral Boundaries on Convection, *J. Fluid Mech.* **37**, 81–94.

Foster, T. D. (1969b). Onset of Manifest Convection in a Layer of Fluid with a Time-Dependent Surface Temperature, *Phys. Fluids* **12**, 2482–2487.

Gallagher, A. P., and Mercer, A. McD. (1962). On the Behavior of Small Disturbances in Plane Couette Flow, *J. Fluid Mech.* **13**, 91–100.

Gallagher, A. P., and Mercer, A. McD. (1965). On the Behavior of Small Disturbances in Plane Couette Flow with a Temperature Gradient, *Proc. Roy. Soc.* (*London*) **A286**, 117–128.

Gantmacher, F. R. (1959). "Applications of the Theory of Matrices." Wiley (Interscience), New York.

Gershuni, G. Z., and Zhukhovitskii, E. M. (1968). Convective Instability of Horizontal Fluid Layers Bound by Thermal Interaction, *J. Appl. Math. Mech.* **32**, 484–488.

Gresho, P. M., and Sani, R. L. (1970). The Effects of Gravity Modulation on the Stability of a Heated Fluid Layer, *J. Fluid Mech.* **40**, 783–806.

Grosch, C. E., and Salwen, H. (1968). The Stability of Steady and Time-Dependent Plane Poiseuille Flow, *J. Fluid Mech.* **34**, 177–205.

Harris, D. L., and Reid, W. H. (1958). On Orthogonal Functions Which Satisfy Four Boundary Conditions. I, Tables for Use in Fourier-Type Expansions, *Astrophys. J. Suppl. Ser.* **3**, 429–447.

Hurle, D. T. J., Jakeman, E., and Pike, E. R. (1967). On the Solution of the Benard Problem with Boundaries of Finite Conductivity, *Proc. Roy. Soc.* (*London*) **A296**, 469–475.

Ingersoll, A. P. (1966). Convective Instabilities in Plane Couette Flow, *Phys. Fluids* **9**, 682–689.

Jakeman, E. (1968). Convective Instability in Fluids of High Thermal Diffusivity, *Phys. Fluids* **11**, 10–14.

Kaplan, S. (1963). On the Best Method for Choosing the Weighting Functions in the Method of Weighted Residuals, *Trans. Am. Nucl. Soc.* **6**, 3–4.

Kogelman, S., and DiPrima, R. C. (1970). Stability of Spatially Periodic Supercritical Flows in Hydrodynamics, *Phys. Fluids* **13**, 1–11.

Krueger, E. R., and DiPrima, R. C. (1964). The Stability of a Viscous Fluid between Rotating Cylinders with an Axial Flow, *J. Fluid Mech.* **19**, 528–538.

Krueger, E. R., Gross, A., and DiPrima, R. C. (1966). On the Relative Importance of Taylor–Vortex and Nonaxisymmetric Modes in Flow between Rotating Cylinders, *J. Fluid Mech.* **24**, 521–538.

Kurzweg. U. H. (1964). The Stability of Dissipative Couette Flow between Rotating Cylinders in the Presence of an Axial Magnetic Field, *Z. Angew. Math. Phys.* **15**, 39–45.

Kurzweg, U. H. (1965). Convective Instability of a Hydromagnetic Fluid within an Enclosed Rectangular Cavity, *Int. J. Heat Mass Transfer* **8**, 35–41.

Kuwabara, S. (1967). Nonlinear Instability of Plane Couette Flow, *Phys. Fluids* **10**, S115–6.

Mahler, E. G., and Schechter, R. S. (1970). The Stability of a Fluid Layer with Gas Absorption, *Chem. Eng. Sci.* **25**, 955–968.

Mahler, E. G., Schechter, R. S., and Wissler, E. H. (1968). Stability of a Fluid Layer with Time-Dependent Density Gradients, *Phys. Fluids* **11**, 1901–1912.

McConaghy, G. A., and Finlayson, B. A. (1969). Surface Tension Driven Oscillatory Instability in a Rotating Fluid Layer, *J. Fluid Mech.* **39**, 49–55.

Meister, B. (1963). Die Anfangswertaufgabe für die Störungsdifferentialgleichungen des Taylorschen Stabilitätsproblem, *Arch. Rat. Mech. Anal.* **14**, 81–107.

Mikhlin, S. G. (1964). "Variational Methods in Mathematical Physics." Macmillan, New York.

Nakagawa, Y., and Frenzen, P. (1955). A Theoretical and Experimental Study of Cellular Convection in Rotating Fluids, *Tellus* **7**, 1–21.

Pearson, J. R. A. (1958). On Convection Cells Induced by Surface Tension, *J. Fluid Mech.* **4**, 489–500.

Pekeris, C. L., and Shkoller, B. (1969). Stability of Plane Poiseuille Flow to Periodic Disturbances of Finite Amplitude, *J. Fluid Mech.* **39**, 611–627.

Pellew, A., and Southwell, R. V. (1940). On Maintaining Convective Motion in a Fluid Heated from Below, *Proc. Roy. Soc.* (*London*) **A176**, 312–343.

Platzman, G. W. (1965). The Spectral Dynamics of Laminar Convection, *J. Fluid Mech.* **23**, 481–510.

Poots, G. (1958). Heat Transfer by Laminar Free Convection in Enclosed Plane Gas Layers, *Quart. J. Mech. Appl. Math.* **11**, 257–273.

Powell, E. A., and Zinn, B. T. (1969). Solution of Linear Combustion Instability Problems Using the Galerkin Method, *Israel J. Tech.* **7**, 79–89.

Reid, W. H., and Harris, D. L. (1958). On Orthogonal Functions Which Satisfy Four Boundary Conditions. II, Integrals for Use with Fourier-Type Expansions, *Astrophys. J. Suppl. Ser.* **3**, 448–452.

Reid, W. H., and Harris, D. L. (1958). Some Further Results on the Benard Problem, *Phys. Fluids* **1**, 102–110; (1959). **2**, 716–717. (Correction).

Rintel, L. (1967). Penetrative Convective Instabilities, *Phys. Fluids* **10**, 848–854.

Ritchie, G. S. (1968). On the Stability of Viscous Flow between Eccentric Rotating Cylinders *J. Fluid Mech.* **32**, 131–144.

Roberts, P. H. (1960). Characteristic Value Problems Posed by Differential Equations Arising in Hydrodynamics and Hydromagnetics, *J. Math. Anal. Appl.* **1**, 195–214.

Roberts, P. H. (1969). On the Thermal Instability of a Rotating-Fluid Sphere Containing Heat Sources, *Phil. Trans. Roy. Soc.* (*London*) **A263**, 93–117.

Roja, D. S. C. (1969). The Effect of Property Variations on the Convective Instability of Gases, MS Thesis, Univ. of Washington, Seattle, Washington.

Roja, D. S., and Finlayson B. A. (1970). The Effect of Property Variations on the Convective Instability of Gases, *AIChE J.* **16**, 876–877.

Rudakov, R. N. (1967). Spectrum of Perturbations and Stability of Convective Motion between Vertical Plates, *J. Appl. Math. Mech* **31**, 376–383.

Salwen, H., and Grosch, C. (1968). Stability of Poiseuille Flow in a Circular Pipe, *Bull. Amer. Phys. Soc.* **13**, 814.

Sani, R. L. (1963). Convective Instability, Ph.D. Thesis, Univ. of Minnesota, Minneapolis, Minnesota.

Sani, R. (1965). On Finite-Amplitude Roll-Cell Disturbances in a Fluid Layer Subjected to Mass and Enthalpy Transfer, *AIChE J.* **11**, 971–980.

Sparrow, E. M., Goldstein, R. J., and Jonsson, V. K. (1964). Thermal Instability in a Horizontal Fluid Layer: Effect of Boundary Conditions and Nonlinear Temperature Profile, *J. Fluid Mech.* **18**, 513–528.

Thomas, L. H. (1953). The Stability of Plane Poiseuille Flow, *Phys. Rev.* **91**, 780–783.

Walowit, J. A. (1966). The Stability of Couette Flow between Rotating Cylinders in the Presence of a Radial Temperature Gradient, *AIChE J.* **12**, 104–109.

Walowit, J. A., Tsao, S., and DiPrima, R. C. (1964). Stability of Flow between Arbitrarily Spaced Concentric Cylindrical Surfaces Including the Effect of a Radial Temperature Gradient, *J. Appl. Mech. Trans. ASME, Ser. E* **86**, 585–593.

Veronis, G. (1966a) Motions at Subcritical Values of Rayleigh Number in a Rotating Fluid, *J. Fluid Mech.* **24**, 545–554.

Veronis, G. (1966b). Large-Amplitude Bénard Convection, *J. Fluid Mech.* **26**, 49–68.

Veronis, G. (1968a). Large-Amplitude Bénard Convection in a Rotating Fluid, *J. Fluid Mech.* **31**, 113–139.

Veronis, G. (1968b). Effect of a Stabilizing Gradient of Solute on Thermal Convection, *J. Fluid Mech.* **34**, 315–336.

Vidal, A., and Acrivos, A. (1966). Nature of the Neutral State in Surface-Tension Driven Convection, *Phys. Fluids* **9**, 615–616.

Zinn, B. T., and Powell, E. A. (1970). Application of the Galerkin Method in the Solution of Combustion-Instability Problems, *Int. Astronaut. Congr. 29th, Proc.* **3**, 59–73.

Part

II

VARIATIONAL PRINCIPLES

Chapter 7

Introduction to Variational Principles

Many problems in engineering can be characterized by variational principles. The variational principle may succinctly summarize the equations, allow insight into the effect of parameters, and provide a means for approximating the solution. The variational method is in many respects similar to the method of weighted residuals: the solution is expanded in terms of a trial function with undetermined constants or functions, which are found according to some prescription. A variational integral is made stationary, and possibly minimized or maximized with respect to the undetermined constants. The results are identical to those obtained by the Galerkin method. The functional in the variational principle may be the only part of the solution which is of interest, and the variational method approximates its value closely. For example, when certain non-Newtonian fluids flow past a sphere, upper and lower bounds can be derived for the drag using variational methods, even though the governing equations cannot be solved exactly. Many problems are not characterized by variational principles, others do not entail minimum or maximum principles, and the ones that are amenable to variational treatment are often classical problems or ones which have already been solved.

An important question is then: does a variational principle exist for a given problem? In Chapter 8 we answer the question for applications in fluid mechanics, and in Chapter 9 we treat the question systematically using Fréchet differentials. Variational principles are discussed for heat and mass transfer problems in Chapter 9 before discussing in Chapter 10 the various attempts to derive variational principles based on a principle of minimum rate of entropy production.

Variational principles are very useful for approximating the solution for the wave function governing the molecular behavior of atoms and molecules. The trial functions are restricted to have the symmetry of the molecule in question, and calculations are made using as many as 100 terms in the trial function (see Kauzmann, 1957; Slater, 1960; Löwden, 1966). Variational principles are also used to solve the Boltzman equation (Hirschfelder *et al.*, 1954). In elasticity, the equations governing the position of membranes and plates can be derived from a variational principle. A great body of literature on variational methods has developed for these applications, which are discussed elsewhere. See Weinstock (1952), Moiseiwitsch (1966), and Forray (1968). In the field of nuclear engineering, the equations governing the distribution of neutrons in a reactor can be derived from a variational principle and many approximate variational methods have been developed for this case (Lewins, 1965; Kaplan, 1969). The equations governing the equilibrium position of a curved interface can be derived from a variational principle (Erle *et al.*, 1970, 1971; Schechter, 1967, p. 38). Optimization problems often employ variational methods, as discussed by Leitman (1962), Petrov (1968), and Denn (1969). For optimal control and optimization problems governed by partial differential equations (distributed parameter systems) the MWR can be used to reduce the problem to sets of ordinary differential equations. These are solved using standard methods. The Galerkin method has been used to reduce these ordinary differential equations to algebraic equations (Lynn *et al.*, 1970). See Wexler (1969) for a discussion of finite element and variational calculations in electrical engineering, as well as Harrington (1968) for electrical problems without variational principles.

7.1 Calculus of Variations

The calculus of variations is introduced briefly since more detailed accounts are available (Courant and Hilbert, 1953; Gelfand and Fomin, 1963). The calculus of variations is concerned with changes in functionals. A functional is a correspondence between a function in some class and the set of real numbers. Examples are

$$\Phi(y) = \int_a^b f(x, y, y')\, dx, \qquad \Phi(y) = \max_{a \le x \le b} |y(x)|. \tag{7.1}$$

The function f is a known function of its arguments. Once a function $y(x)$ is specified the functional Φ, which is a real number, can be calculated. The domain of a functional is the space of admissible functions, which may be restricted to functions which satisfy certain continuity restrictions or boundary conditions. For the functional given above the space of admissible functions might be those functions which are continuous and have continuous first derivatives on the interval a to b. Call this class $c^1(a,b)$. In the calculus of functions a function is stationary when its first derivative vanishes. Similar ideas are applicable to functionals. We ask the following question.

Consider the class of functions (called $\mathscr{C}$) in c^1 which satisfy $y(a) = y_1$, $y(b) = y_2$. For what functions $y(x)$ in $\mathscr{C}$ is the functional $\Phi(x)$ defined in Eq. (7.1) stationary?

The stationary property is defined by analogy with functions. A function is stationary at the point $\bar{x}$ when

$$\left.\frac{dg(x)}{dx}\right|_{x=\bar{x}} = 0, \qquad \lim_{\varepsilon\to 0} \frac{g(\bar{x}+\varepsilon) - g(\bar{x})}{\varepsilon} = 0. \tag{7.2}$$

Consider the function $y + \varepsilon\eta(x)$ where y is in $\mathscr{C}$ and η is zero at a and b and is in c^1. Then the sum $y + \varepsilon\eta$, is in $\mathscr{C}$, too, for all values of ε. Define the derivative of the functional, called the variation, as

$$\varepsilon \lim_{\varepsilon\to 0} \frac{\Phi(y+\varepsilon\eta) - \Phi(y)}{\varepsilon}. \tag{7.3}$$

The functions which make this variation zero are called extremals. The conditions the extremals satisfy can be found from Eq. (7.3), but it is more instructive to obtain the result in a more direct way. Consider the functional

$$\Phi(\varepsilon) = \int_a^b f(x, \bar{y} + \varepsilon\eta, \bar{y}' + \varepsilon\eta')\, dx, \tag{7.4}$$

where $\bar{y}$ and η are specified. Then Φ is a function of ε. Differentiate the function with respect to ε and evaluate it at $\varepsilon = 0$:

$$\left.\frac{d\Phi}{d\varepsilon}\right|_{\varepsilon=0} = \int_a^b (f_y\eta + f_{y'}\eta')\, dx, \qquad f_y = \frac{\partial f}{\partial y}, \quad f_{y'} = \frac{\partial f}{\partial y'}. \tag{7.5}$$

Integrate the last term by parts:

$$\begin{aligned}\int_a^b f_{y'} \frac{d\eta}{dx}\, dx &= \int_a^b \frac{d}{dx}[f_{y'}\eta]\, dx - \int_a^b \eta \frac{d}{dx} f_{y'}\, dx \\ &= [\eta f_{y'}]_a^b - \int_a^b \eta \frac{d}{dx} f_{y'}\, dx. \end{aligned} \tag{7.6}$$

The first term vanishes, however, because of the conditions on η and the derivative of Eq. (7.5) is

$$\Phi'(0) = \int_a^b \eta(x) \left[f_y - \frac{d}{dx} f_{y'} \right] dx = 0. \tag{7.7}$$

Next we must use the Fundamental Lemma of the Calculus of Variations, which says:

If $M(x)$ is in c, $\eta(x)$ is in c^1 and vanishes at a,b and if

$$\int_a^b \eta(x) M(x)\, dx = 0 \tag{7.8}$$

for all possible functions η, then

$$M(x) = 0, \qquad a \le x \le b. \tag{7.9}$$

We only sketch the proof. If M differed from zero in some small region of x, we could construct an η which was zero everywhere except near where $M \neq 0$ and there had the same sign as M. The value of the integral would necessarily be positive, leading to a contradiction, so the original assumption, $M \neq 0$, is false. If Φ is stationary to all possible variations η, then

$$[f]_y \equiv f_y - \frac{d}{dx} f_{y'} = 0. \tag{7.10}$$

When written in full this is

$$\frac{\partial f}{\partial y} - \left(\frac{\partial^2 f}{\partial x\, \partial y'} + \frac{\partial^2 f}{\partial y\, \partial y'} \frac{d\bar{y}}{dx} + \frac{\partial^2 f}{\partial (y')^2} \frac{d^2 \bar{y}}{dx^2} \right) = 0, \tag{7.11}$$

which is a second-order differential equation for $\bar{y}(x)$. This is called the Euler–Lagrange equation, or the Euler equation, after the mathematicians who pioneered the calculus of variations in the eighteenth century.† To relate this property of the function $\Phi(\varepsilon)$ to a property of the functional $\Phi(y)$, write a Taylor series about $\varepsilon = 0$:

$$\Phi(\varepsilon) = \Phi(0) + \left.\frac{d\Phi(\varepsilon)}{d\varepsilon}\right|_{\varepsilon=0} \varepsilon + \left.\frac{d^2\Phi(\varepsilon)}{d\varepsilon^2}\right|_{\varepsilon=0} \frac{\varepsilon^2}{2} + \cdots. \tag{7.12}$$

† Lagrange solved the isoperimetric problem (7.26) at the age of 19. He sent his work to Euler, who was so impressed with the generality that he withheld his own publication until Lagrange could publish the work (Williams, 1969).

The functional can also be expanded:

$$\Phi(\bar{y} + \varepsilon\eta) = \int_0^1 f(x, \bar{y}, \bar{y}')\, dx + \varepsilon \int_0^1 \left[\frac{\partial f}{\partial y}\eta + \frac{\partial f}{\partial y'}\eta'\right]\Bigg|_{\bar{y}} dx$$

$$+ \frac{1}{2}\varepsilon^2 \int_0^1 \left[\frac{\partial^2 f}{\partial y^2}\eta^2 + \frac{\partial^2 f}{\partial y\, \partial y'} 2\eta\eta' + \frac{\partial^2 f}{\partial y'^2}(\eta')^2\right]\Bigg|_{\tilde{y}} dx, \quad (7.13)$$

$$\Phi(\bar{y} + \varepsilon\eta) \equiv \Phi(\bar{y}) + \varepsilon\Phi_1(\bar{y}, \eta) + \tfrac{1}{2}\varepsilon^2\Phi_2(\tilde{y}, \eta),$$

where $\tilde{y} = \bar{y} + \rho\eta$ is evaluated for some $0 \le \rho \le \varepsilon$. By the Mean Value Theorem this accounts for all higher terms in the Taylor expansion. We equate equivalent terms in Eqs. (7.12) and (7.13):

$$\frac{d\Phi(\varepsilon)}{d\varepsilon}\bigg|_{\varepsilon=0} \equiv \Phi_1(\bar{y},\eta), \quad (7.14a)$$

and define

$$\delta\Phi \equiv \varepsilon\Phi_1(\bar{y},\eta). \quad \textbf{(7.14b)}$$

$\delta\Phi$ is called the first variation of Φ. Comparing Eqs. (7.3) and (7.13) shows that a functional is stationary if its first variation vanishes. Thus Eq. (7.10) represents a necessary condition for Φ to be stationary.

Is the extremum a minimum or maximum, or just a stationary property? From Eq. (7.13) we can say that Φ is a minimum for an extremal $\bar{y}$ only if

$$\Phi_2(\bar{y},\eta) \ge 0 \qquad \text{(necessary condition for minimum).} \quad (7.15)$$

We define the second variation

$$\delta^2\Phi \equiv \tfrac{1}{2}\varepsilon^2\Phi_2(\bar{y},\eta), \quad (7.16)$$

and evaluate it for an extremal:

$$\frac{d^2\Phi}{d\varepsilon^2} = \int_a^b [f_{yy}\eta^2 + 2f_{yy'}\eta\eta' + f_{y'y'}(\eta')^2]\, dx,$$

$$\Phi_2(\tilde{y}, \eta) = \int_a^b [\tilde{f}_{yy}\eta^2 + 2\tilde{f}_{yy'}\eta\eta' + \tilde{f}_{y'y'}(\eta')^2]\, dx, \quad (7.17)$$

where $\tilde{f}$ means the function $f(x,y,y')$ evaluated for $\tilde{y} = \bar{y} + \rho\eta$. But for an extremal

$$\Phi_2(\bar{y},\eta) = \int_a^b (\bar{f}_{yy}\eta^2 + 2\bar{f}_{yy'}\eta\eta' + \bar{f}_{y'y'}(\eta')^2]\, dx. \quad (7.18)$$

This is integrated by parts to get

$$\Phi_2(\bar{y}, \eta) = \int_a^b \left[\eta^2\left(\bar{f}_{yy} - \frac{d}{dx}\bar{f}_{yy'}\right) + \bar{f}_{y'y'}(\eta')^2\right] dx. \quad (7.19)$$

The first two terms vanish for an extremal, which satisfies Eq. (7.10):

$$\Phi_2(\bar{y}, \eta) = \int_a^b (\eta')^2 \bar{f}_{y'y'}\, dx. \tag{7.20}$$

A necessary condition for a minimum is then

$$\bar{f}_{y'y'} \geq 0, \qquad a \leq x \leq b, \tag{7.21}$$

which is called the Legendre condition. A more detailed treatment shows (see Gelfand and Fomin, 1963, p. 100) that a *sufficient condition* for Φ to be a minimum is that there exists a positive number k such that

$$\Phi_2(\bar{y}, \eta) \geq k \|\eta\| \tag{7.22}$$

for all η in some normed space. The inequalities (7.15, 7.21) do not ensure a minimum principle, but if they are satisfied, a minimum principle may be possible. Another way to prove the minimum nature of the variational principle is to show that for the extremal,

$$\Phi(\bar{y} + \eta) = \Phi(\bar{y}) + k(\eta), \tag{7.23}$$

where $k(\eta) \geq 0$ and the equality holds only when $\eta = 0$.

This brief treatment of the calculus of variations illustrates the important features: given a functional, the extremals which make it stationary are solutions to the Euler equation. Whether the extremal makes the functional a minimum or a maximum is a separate question that is not easily answered definitively, although Eq. (7.15) or (7.21) are necessary conditions.

Boundary Conditions

Before the variation in a functional can be calculated it is necessary to specify the admissible class of functions. The boundary conditions need not always be satisfied, and it is convenient to distinguish between what are called essential and natural boundary conditions. For the functional defined by Eq. (7.1) suppose the admissible class of functions is defined as those functions in $c^1(a,b)$, which satisfy $y(a) = y_1$. Then in Eqs. (7.6) and (7.7) the boundary terms do not vanish since $\eta(b) \neq 0$:

$$\Phi'(0) = \int_a^b \eta \left[f_y - \frac{d}{dx} f_{y'} \right] dx + [\eta f_{y'}]_{x=b} = 0. \tag{7.24}$$

If this equation is to be satisfied for *arbitrary* variations in η then it is necessary that *both* terms vanish. In addition to the Euler equation, it is necessary that

$$f_{y'} = \partial f / \partial y' = 0 \qquad \text{at} \quad x = b. \tag{7.25}$$

This is a natural boundary condition. In applications trial functions need not satisfy the natural boundary conditions since the variational principle forces them to be satisfied. This is especially convenient when it is hard to find trial functions satisfying the boundary conditions. If the Euler equation is of order $2n$, the natural boundary conditions are generally those including derivatives of order n to $2n-1$, while those boundary conditions involving derivatives of order $n-1$ and less are called essential boundary conditions.

Given a variational integral, the same steps are always followed to obtain the Euler equation, essential and natural boundary conditions. The first variation is calculated and the term in the integral (7.24) is the Euler equation. The natural boundary conditions come from the boundary term and the remaining boundary conditions are essential. A few of the more common variational integrals are listed in Table 7.1 together with Euler equations and natural boundary conditions.

Lagrange Multipliers

Variations can be taken subject to constraints and these are easily handled using Lagrange multipliers.

Make the functional J stationary among all functions in $\mathscr{C}$ subject to the condition that the functional K has a prescribed value K_1:

$$J \equiv \int_a^b F(x, y, y')\, dx, \qquad K \equiv \int_a^b G(x, y, y')\, dx. \tag{7.26}$$

We solve this problem following the treatment of Courant and Hilbert (1953). Suppose $\bar{y} = \bar{y}(x)$ is the desired extremal. Consider the family of curves $y = \bar{y} + \varepsilon_1 \eta(x) + \varepsilon_2 \zeta(x)$, where η and ζ satisfy the homogeneous boundary conditions so that y is an admissible function. Then the function

$$\Phi(\varepsilon_1, \varepsilon_2) = \int_a^b F(x, \bar{y} + \varepsilon_1 \eta + \varepsilon_2 \zeta, \bar{y}' + \varepsilon_1 \eta' + \varepsilon_2 \zeta')\, dx \tag{7.27}$$

is to be made stationary at $\varepsilon_1 = \varepsilon_2 = 0$, subject to the constraint that

$$\psi(\varepsilon_1, \varepsilon_2) = \int_a^b G(x, \bar{y} + \varepsilon_1 \eta + \varepsilon_2 \zeta, \bar{y}' + \varepsilon_1 \eta' + \varepsilon_2 \zeta')\, dx = K_1 \tag{7.28}$$

for sufficiently small values of ε_1 and ε_2. As discussed by Courant and Hilbert (1953, p. 165) there exist two constants λ_0 and λ, not both zero, such that

$$\begin{aligned} \frac{\partial}{\partial \varepsilon_1} [\lambda_0 \Phi(\varepsilon_1, \varepsilon_2) + \lambda \Psi(\varepsilon_1, \varepsilon_2)]_{\varepsilon_1 = \varepsilon_2 = 0} &= 0, \\ \frac{\partial}{\partial \varepsilon_2} [\lambda_0 \Phi(\varepsilon_1, \varepsilon_2) + \lambda \Psi(\varepsilon_1, \varepsilon_2)]_{\varepsilon_1 = \varepsilon_2 = 0} &= 0. \end{aligned} \tag{7.29}$$

TABLE 7.1

SUMMARY OF FUNCTIONALS, EULER EQUATIONS, AND BOUNDARY CONDITIONS

Functional	Euler equation	Essential boundary conditions	Natural boundary conditions
$\Phi = \int_{x_1}^{x_2} f(x, y, y')\, dx + g(y)\|_{x_2}$	$f_y - \frac{d}{dx} f_y' = 0$	$y(x_1) = y_1$	$\left[f_{y'} + \frac{dg}{dy}\right]_{x=x_2} = 0$
$\Phi = \iint_V f(x, y, z, z_x, z_y)\, dx\, dy$	$f_z - \frac{\partial}{\partial x} f_{z_x} - \frac{\partial}{\partial y} f_{z_y} = 0$	z specified	
$\Phi = \int_{x_1}^{x_2} f(x, y, z, y', z')\, dx$	$f_y - \frac{d}{dx} f_{y'} = 0$	$y(x_1) = y_1$	$\left[f_{y'} + \frac{\partial g}{\partial y}\right]_{x=x_2} = 0$
$+ g(y, z)\|_{x_2}$	$f_z - \frac{d}{dx} f_{z'} = 0$	$z(x_1) = z_1$	$\left[f_{z'} + \frac{\partial g}{\partial z}\right]_{x=x_2} = 0$
$\Phi = \int_0^1 [p(x)(dy/dx)^2 - q(x)y^2]\, dx$	$\frac{d}{dx}\left(p \frac{dy}{dx}\right) + qy = 0$	$y(0) = y_0$	$p(1)y'(1) + h[y(1) - y_1] = 0$
$+ h(y(1) - y_1)^2$			
$\Phi = \int_V [\frac{1}{2}k(\mathbf{x})\nabla T \cdot \nabla T - Tf(\mathbf{x})]\, dV$	$\nabla \cdot (k \nabla T) = f$	$T = T_s$ on S_1	$k\mathbf{n} \cdot \nabla T + h(T - T_s) = 0$ on S_2
$+ \frac{1}{2}\int_{s_2} h(\mathbf{x})(T - T_s)^2\, ds$			

From these equations we obtain

$$\int_a^b [\lambda_0[F]_y + \lambda[G]_y]\eta \, dx = 0, \tag{7.30a}$$

$$\int_a^b [\lambda_0[F]_y + \lambda[G]_y]\zeta \, dx = 0, \tag{7.30b}$$

where $[F]_y$ and $[G]_y$ represent the Euler equation corresponding to the functionals F and G, respectively. Since the first equation does not contain the arbitrary function ζ, the ratio of λ_0 to λ does not depend on ζ. Since ζ is arbitrary the second equation gives $\lambda_0[F]_y + \lambda[G]_y = 0$. If $\lambda_0 \neq 0$, or

$$[G]_y = \frac{d}{dx} G_{y'} - G_y \neq 0, \tag{7.31}$$

we may set $\lambda_0 = 1$ and Eq. (7.30a) gives for arbitrary η

$$\frac{d}{dx} [F_{y'} + \lambda G_{y'}] - \frac{\partial}{\partial y} [F + \lambda G] = 0. \tag{7.32}$$

To derive the Euler equation for a constrained problem the Euler equation is derived for the integrand $F^* = F + \lambda G$, disregarding the subsidiary condition. The solution to Eq. (7.32), has two undetermined constants plus the unknown parameter λ. These are determined by the two boundary conditions and $K = K_1$.

If the integrand is of the type $F(x, y, y', z, z')$ and the constraint is

$$G(x, y, z) = 0, \tag{7.33}$$

the Euler equation can be found using Lagrange multipliers in the same way. In this case, however, the λ is a function of x rather than a constant. The type of constraint used in fluid mechanics problems is usually a differential equation,

$$G(x, y, y'z, z') = 0. \tag{7.34}$$

If this cannot be solved to express $z = f(y)$ then the use of Lagrange multipliers is essential. The integrand is again $F^* = F + \lambda G$, with $\lambda = \lambda(x)$.

Reciprocal Variational Principles

Reciprocal variational principles are useful for obtaining upper and lower bounds.

PROBLEM I. Make J stationary subject to the usual continuity restrictions, the boundary conditions

$$u(a) - u_1 = 0, \qquad u(b) - u_2 = 0 \tag{7.35}$$

and the subsidiary condition

$$\frac{du}{dx} - z = 0, \tag{7.36}$$

$$J = \int_a^b F(x, u, z)\, dx. \tag{7.37}$$

We can employ Lagrange multipliers to form the following problem.

PROBLEM II. Make stationary the functional

$$H[u, z, \lambda; \mu_1, \mu_2] = \int_a^b \left[F + \lambda\left(\frac{du}{dx} - z\right)\right] dx - \mu_1[u(a) - u_1] + \mu_2[u(b) - u_2]. \tag{7.38}$$

Now there are no subsidiary conditions. The Euler equations are Eq. (7.36) and

$$F_z - \lambda = 0, \qquad F_u - \frac{d\lambda}{dx} = 0, \tag{7.39}$$

and the natural boundary conditions are Eqs. (7.35) and

$$\lambda(a) + \mu_1 = 0, \qquad \lambda(\mathrm{b}) + \mu_2 = 0. \tag{7.40}$$

When this variational principle is applied subject to the constraints (7.35) and (7.36) we obtain Problem I. If, on the other hand, we use as constraints (7.39) and (7.40) we obtain (7.36) as Euler equation and (7.35) as natural boundary conditions. In this case eliminate du/dx from the integral H by integration by parts. Then introduce new functions p and p' and $\psi(x, p, p')$ defined by the Legendre transformation

$$F_z = p, \qquad F_u = p', \qquad pz + p'u - F = \psi. \tag{7.41}$$

The reader can verify by calculating ψ_z and ψ_u that ψ is a function of only x, p, and p'. Then the reciprocal problem is the following problem.

PROBLEM III. Make stationary

$$I(p, p') = -\int_a^b \psi(x, p, p')\, dx - p(a)u_1 + p(b)u_2 \tag{7.42}$$

subject to the conditions

$$\frac{dp}{dx} - p' = 0 \tag{7.43}$$

and no boundary conditions are imposed.

The Euler equation is

$$\frac{d}{dx}\psi_{p'} - \psi_p = 0 \tag{7.44}$$

and the natural boundary conditions are

$$\psi_{p'}|_a - u_1 = 0, \qquad \psi_{p'}|_b - u_2 = 0. \tag{7.45}$$

To show that these are (7.35) and (7.36), invert the Legendre transformation (7.41):

$$\psi_{p'} = u, \qquad \psi_p = z, \qquad pz + p'u - \psi = F. \tag{7.46}$$

The reciprocal variational problems are formed from a variational principle with constraints, which are incorporated into the variational principle using Lagrange multipliers. The Euler equations of the first principle are then used as constraints to obtain the variational principle. This is especially important when one of the principles is a minimum principle and the transformed problem is a maximum principle. In the calculations we then obtain upper and lower bounds for the functional H, as is illustrated below. The reciprocal variational principle is also useful when the trial functions are discontinuous (see Section 7.4). Quasilinearization can also be used to construct the complementary, or reciprocal, variational principle (Bellman, 1962).

7.2 Steady-State Heat Conduction

A variational principle is presented for steady-state heat conduction. Consider (2.48) with the following restrictions: no velocity, $\mathbf{u} = 0$, thermal conductivity and heat transfer coefficient functions of position but not of temperature. If the thermal conductivity depends on temperature the transformation (1.49) reduces the problem to the form of (7.47) provided there are no boundary conditions of the third kind (2.50). These are treated as a special case in (7.52):

$$\nabla \cdot (k\, \nabla T) = f(\mathbf{x}, T) \qquad \text{in } V, \tag{7.47a}$$

$$T = T_1 \qquad \text{on } S_1, \tag{7.47b}$$

$$-k\mathbf{n} \cdot \nabla T = q_2 \qquad \text{on } S_2, \tag{7.47c}$$

$$-k\mathbf{n} \cdot \nabla T = h(T - T_3) \qquad \text{on } S_3. \tag{7.47d}$$

To establish a variational principle with these equations as the Euler equations and natural boundary conditions, we must find a suitable functional. This can often be found by multiplying the differential equation by T, integrating over the volume, and integrating by parts. Various boundary terms can also be added. Fréchet differentials (Chapter 9) provide a systematic way of finding the functional, and here we merely present the result.

Minimize $\Phi(T)$ among all functions satisfying $T = T_1$ on S_1, which are continuous and have continuous second derivatives:

$$\Phi(T) = \int_v \left[\tfrac{1}{2} k \, \nabla T \cdot \nabla T + \int_{T_0}^{T} f(\mathbf{x}, T') \, dT' \right] dV$$
$$+ \int_{S_2} q_2 T \, dS + \tfrac{1}{2} \int_{S_3} h(T - T_3)^2 \, dS. \tag{7.48}$$

To verify that this is the proper variational principle we need to calculate the first and second variations. Substitute $\overline{T} + \varepsilon\eta$ into the variational integral and calculate $\Phi'(0)$. After applying the divergence theorem (see p. 29) we obtain

$$\delta\Phi = \int_v \varepsilon\eta[-\nabla \cdot k \, \nabla\overline{T} + f(\mathbf{x}, \overline{T})] \, dV + \int_{S_2} \varepsilon\eta[k\mathbf{n} \cdot \nabla\overline{T} + q_2] \, dS$$
$$+ \int_{S_3} \varepsilon\eta[k\mathbf{n} \cdot \nabla\overline{T} + h(\overline{T} - T_3)] \, dS. \tag{7.49}$$

This gives (7.47a) as the Euler equations and (7.47c,d) as the natural boundary conditions. The second variation is

$$\delta^2\Phi = \tfrac{1}{2}\varepsilon^2 \int \left[k \, \nabla\eta \cdot \nabla\eta + \eta^2 \frac{\partial f}{\partial \overline{T}} \right] dV. \tag{7.50}$$

This satisfies the necessary condition for a minimum principle, Eq. (7.15), if $k > 0$ (as assumed) and $\partial f/\partial T \geq 0$. For $k = 1$ and $f = 0$ Mikhlin (1964, Section 22) shows that the sufficient condition is satisfied as well. If $k(\mathbf{x})$ is bounded below by a positive constant, $k \geq c > 0$, and $\partial f/\partial \overline{T} \geq 0$, then (7.50) satisfies (7.22) and the variational integral is a minimum. Having verified that the variational principle (7.48) gives the equations (7.47), we can regard (7.48) as an alternate statement of the problem.

If $k = k(T)$ and there is a boundary condition of the third kind, the last boundary term in (7.48) must be modified. After applying the transformation (2.49) the condition becomes

$$-\mathbf{n} \cdot \nabla\phi = h(\phi)[f(\phi) - f(\phi_3)] \qquad \text{on } S_3. \tag{7.51}$$

The boundary term on S_3 in (7.48) is then

$$\int_{S_3} \int_{\phi_3}^{\phi} h(\phi')[f(\phi') - f(\phi_3)] \, d\phi' \, ds. \tag{7.52}$$

Similarly, if the boundary condition is

$$-k(\mathbf{x})\mathbf{n}\cdot\nabla T = h(\mathbf{x})(T^n - T_3{}^n) \tag{7.53}$$

the boundary term in (7.48) is

$$\int_{S_3}\int_{T_3}^{T} h(T'^n - T_3{}^n)\,dT'\,dS = \int_{S_3}\left[\frac{h}{n+1}(T^{n+1} - T_3^{n+1}) - hT_3{}^n(T - T_3)\right] dS. \tag{7.54}$$

Thus the completely general problem (7.47) can be treated using variational methods, even when $k = k(T)$ and $h = h(T)$ and the boundary conditions are nonlinear.

The variational method is applied by expanding the solution in a trial function [see (2.51)]:

$$T_N = T_1 + \sum_{i=1}^{N} c_i T_i(\mathbf{x}), \tag{7.55}$$

where $T_i = 0$ on S_1. This trial function is substituted into (7.48) which is minimized with respect to the constants c_i:

$$\frac{\partial\Phi}{\partial c_j} = \int_v [k\,\nabla T_N\cdot\nabla T_j + T_j f(\mathbf{x}, T_N)]\,dV + \int_{S_2} q_2 T_j\,dS$$

$$+ \int_{S_3} h(T_N - T_3)T_j\,dS = 0. \tag{7.56}$$

This is the equation which determines the approximate solution. Comparing it to (2.54) we see that they are the same when the weighting function $w_j = T_j$ in (2.54). That choice corresponds to the Galerkin method, so that the variational and Galerkin methods give identical results, even though the trial functions do not satisfy the natural boundary conditions. The equivalence is not restricted to this problem: *there is always a Galerkin method that corresponds to the variational method.* There may be several ways to apply the Galerkin method, only one of which corresponds to the variational method. For example, in the variational method if a more general trial function were used, $T_N = g(x, c_i)$ then in (7.56) we would have $\partial g/\partial c_j$ everywhere T_j appears. The corresponding Galerkin method would be equivalent only if the weighting function in (2.54) was $\partial g/\partial c_j$.

The variational method is known as the Rayleigh–Ritz method, although some authors call it the Ritz method. Close examination of Rayleigh's works indicates Rayleigh used the variational method to calculate successive approximations to both boundary and eigenvalue problems.†

In the variational method the approximate solution gives a value of Φ which is above the true value, since Φ obtains its minimum for the exact solution. The functional Φ decreases (or remains the same) as terms are added in the series, since the minimum with N parameters is going to be the same or lower than the minimum with $N-1$ parameters, the Nth one being restricted to be zero. For some problems error bounds can be determined in terms of the value of the functional $\Phi(T_N)$ [see (11.44)]. Thus $\Phi(T_N)$ and its change as N is increased provide a means for assessing the approximate solution.

In the variational method, as in MWR, the trial function need not depend on all the independent variables. Trial functions of the form (2.23) are possible, and the variational method leads to ordinary differential equations as in MWR. In the example treated in Section 2.2 the Galerkin calculations are equivalent to the variational calculations. Shown in Table 2.2 are values of the functional. For the Galerkin and variational method, as N is increased the functional Φ decreases. Furthermore, the variational method gives the smallest value of Φ, compared to other MWR, which should happen since the variational method minimizes Φ. Note also that the best values of Φ do not always correspond to the best values of flux at the wall. Thus the variational integral, while it gives a means for assessing the accuracy of the solution, does not always give the "best" solution in all respects.

In special cases the functional can be given a physical interpretation. Consider the simpler problem

$$\nabla \cdot k \nabla T = 0, \qquad T = 0 \quad \text{on } S_1, \qquad T = a = \text{constant on } S_2. \tag{7.57}$$

The functional in this case is the Dirichlet integral. Integration by parts and application of (7.57) gives

$$\Phi(\bar{T}) = (a/2) \int_{S_2} k\mathbf{n} \cdot \nabla \bar{T} \, ds. \tag{7.58}$$

† For example Rayleigh (1873) treats eigenvalues which are stationary, although he does not give any approximate method of determining them. He added the comment (Rayleigh, 1896, p. 110) that the eigenvalue is a minimum (the equilibrium is absolutely stable) and gives a method for approximating the eigenvalues. In 1899 Rayleigh clearly applies the variational method to calculate the frequency of vibration of a fluid partially filling a horizontal cylinder, including a second approximation (Rayleigh, 1899). Even earlier Rayleigh (1871) applied the variational method to a boundary value problem. He calculated the approximate solution for flow through a cylindrical hole in a flat plate, including a second approximation, using the variational principle (8.20). See also Rayleigh (1896, Section 307 and Appendix A). Rayleigh himself (1911) commented on Ritz's paper (1908), saying he was surprised that Ritz thought his method new. As Courant (1943) remarked, it was probably the tragic circumstances of Ritz's work that caught the general interest. Ritz wrote his papers (1908, 1909) while aware that he was soon to die of tuberculosis. The variational method is also more clearly presented in Ritz's papers.

Thus the functional is proportional to the average heat flux on S_2. For more general situations this interpretation is not valid.

Lower bounds on the functional can be found using a reciprocal variational principle. Yasinsky (1966) presented such a principle applicable to heat conduction. Stewart (1962) and Sani (1963) had used the same principle in a similar problem (Section 7.3). The reader can derive the reciprocal principle as outlined above. The result is:

Maximize $\psi(\mathbf{q}, T)$ among functions T and $\mathbf{q}$ which are continuous and have continuous first derivatives and which satisfy

$$f(T) = -\nabla \cdot \mathbf{q} \qquad \text{in } V,$$

$$\mathbf{n} \cdot \mathbf{q} = q_2 \qquad \text{on } S_2,$$

$$\mathbf{n} \cdot \mathbf{q} = h(T - T_3) \qquad \text{on } S_3, \tag{7.59}$$

$$\Psi(\mathbf{q}, T) = \int_v \left[-\frac{1}{2k}\mathbf{q} \cdot \mathbf{q} - Tf(T) + \int_{T_0}^{T} f(T')\,dT' \right] dV$$

$$-\frac{1}{2}\int_{S_2} h(T^2 - T_3{}^2)\,dS - \int_{S_1} \mathbf{n} \cdot \mathbf{q}\, T_1\,dS.$$

The Euler equation and natural boundary condition are

$$\mathbf{q}/k + \nabla T = 0 \qquad \text{in } V, \tag{7.60a}$$

$$T - T_1 = 0 \qquad \text{in } S_1, \tag{7.60b}$$

and the necessary condition for a maximum is satisfied if $\partial f/\partial T \geq 0$, $k, h > 0$. In that case we have the bounds

$$\Psi(\mathbf{q}, T) \leq \Psi(\bar{\mathbf{q}}, \bar{T}) = \Phi(\bar{T}) \leq \Phi(T). \tag{7.61}$$

Similar principles are applicable to heat conduction in anisotropic media when $\mathbf{q} = -\mathbf{K} \cdot \nabla T$, provided the matrix $\mathbf{K}$ is symmetric. Herrmann (1963) has presented such a principle and the variational integral (7.48) is easily generalized to this case by replacing $\frac{1}{2}k\,\nabla T \cdot \nabla T$ by $\frac{1}{2}\,\nabla T \cdot \mathbf{K} \cdot \nabla T$.

7.3 Laminar Flow through Ducts

Laminar flow through a duct is also amenable to variational treatment. Write Eq. (4.1) in dimensionless variables:

$$\nabla^2 u = -1 \qquad \text{in } A, \tag{7.62a}$$

$$u = 0 \qquad \text{on C.} \tag{7.62b}$$

The variational principle for this equation can be derived as a special case of Eq. (7.48). For reasons which become apparent below we define the functional to be the negative of (7.48).

Maximize $\Phi(u)$ among all functions satisfying the boundary condition $u = 0$ on C, which are continuous and have continuous first derivatives:

$$\Phi(u) = \int_A [u - \tfrac{1}{2} \nabla u \cdot \nabla u] \, dA. \tag{7.63}$$

The functional can be related to the flow rate. Integrate by parts and apply the divergence theorem:

$$\Phi(u) = \tfrac{1}{2} \int_A u \, dA + \tfrac{1}{2} \int_A u[1 + \nabla^2 u] \, dA - \tfrac{1}{2} \int_C u\mathbf{n} \cdot \nabla u \, ds. \tag{7.64}$$

Evaluate the result for the exact solution. The second and third terms vanish. Since the flow rate is defined

$$Q = \int_A u \, dA/A, \qquad A \equiv \int_A dA, \tag{7.65}$$

the functional is proportional to the flow rate. Any trial function substituted into (7.63) gives a lower bound on the true maximum, and hence a lower bound on the flow rate:

$$\Phi(u) \leq \Phi(\bar{u}) = \tfrac{1}{2} QA. \tag{7.66}$$

The variational method is applied as before. For the problem in Section 4.1 the Galerkin results are identical to the Rayleigh–Ritz results. Thus the orthogonal collocation method, which is equivalent to the Galerkin method for this problem, provides lower bounds on the flow rate. The results shown in Table 4.1 indicate the lower bound is very close (0.05%) to the exact answer.

Upper bounds can be derived using the reciprocal variational principle. Sani (1963) has presented a principle similar to (7.59) for the flow problem, but we simplify it and present the treatment due to Stewart (1962). Suppose we require (7.60a) to be satisfied in the variational principle (7.59).

Minimize $\Psi(u)$ among all continuous functions u which have continuous first derivatives and satisfy (7.62a):

$$\Psi(u) \equiv \tfrac{1}{2} \int_A \nabla u \cdot \nabla u \, dA. \tag{7.67}$$

The only equation left to be satisfied is the natural boundary condition $u = 0$, (7.60b). The fact that this is an upper bound follows from (7.61) (remember we have multiplied the functionals by -1 which changes the direction of the inequalities). It is instructive to also prove the inequality using the Cauchy inequality.

Define the inner product $D(u, v)$ for any functions u, v which are continuous along with their first derivatives

$$D(u, v) \equiv \int_A \nabla u \cdot \nabla v \, dA. \tag{7.68}$$

Evaluate the inner product for the solution to (7.62). Thus

$$D(\bar{u}, \bar{u}) = \int_A (-\bar{u} \, \nabla^2 \bar{u}) \, dA + \int_C \bar{u}\mathbf{n} \cdot \nabla \bar{u} \, ds = \int_A \bar{u} \, dA = QA. \tag{7.69}$$

Suppose we have a function u_1 which satisfies $\nabla^2 u_1 = -1$. The inner product

$$D(\bar{u} - u_1, \bar{u}) = \int_A [-\bar{u} \, \nabla^2(\bar{u} - u_1)] \, dA + \int_C \bar{u}\mathbf{n} \cdot \nabla(\bar{u} - u_1) \, ds \tag{7.70}$$

is then zero. The first term vanishes because both $\bar{u}$ and u_1 satisfy the differential equation, and the second term vanishes because $\bar{u} = 0$ on the boundary. Using the properties of scalar products we have

$$D(\bar{u}, \bar{u}) = D(u_1, \bar{u}). \tag{7.71}$$

The Schwarz inequality,† combined with (7.67) gives the bounds:

$$D(\bar{u}, \bar{u}) \leq D(u_1, u_1). \tag{7.72}$$

The upper and lower bounds can be combined:

$$\Phi(u) \leq \Phi(\bar{u}) = \tfrac{1}{2}QA = \Psi(\bar{u}) \leq \Psi(u_1). \tag{7.73}$$

Use of both variational principles allows a direct measure of the error in the functional, which in this case is directly related to the error in the solution (11.44).

The application of the principle is called the Trefftz method, after Trefftz (1927) who applied variational boundary methods. For the example in

† It is easily verified by direct substitution that

$$D(u + w, v) = D(u, v) + D(w, v)$$

and the inequality

$$D^2(u, v) \leq D(u, u) \, D(v, v)$$

is proved by considering $D(u + kv, u + kv)$ and using the fact that the quadratic in k must be positive. The quantity $a + 2bk + ck^2$ is positive for any real k if $a > 0$ and $(2b)^2 < 4ac$.

Section 4.1, the same trial function (4.7) is used except that the c_1 is dropped because it makes no contribution to the integral (7.67):

$$u_N = u_0 + \sum_{i=1}^{N} c_i u_i = -\tfrac{1}{4}(x^2 + y^2) + \sum_{i=l}^{N} c_i u_i. \tag{7.74}$$

Minimization of $D(u_N, u_N)$ with respect to c_i gives

$$\frac{\partial D}{\partial c_j} = 2\int\left(\sum_{i=1}^{N} c_i u_i + u_0\right)\mathbf{n}\cdot\nabla u_j\, dA = 0, \qquad j = 1, \ldots, N. \tag{7.75}$$

The result for $N = 1$ happens to be the same as that achieved using boundary collocation with $N = 2$ (Villadsen and Stewart, 1967), and is listed in Table 4.1. Combination of the upper and lower bounds on flow rate gives

$$0.5622 \leq QL/(\Delta p\, a^4) \leq 0.5630. \tag{7.76}$$

Note that the same result (7.75) would have been achieved using a boundary method of MWR with weighting function $\mathbf{n}\cdot\nabla u_j$, but there is no apparent reason to pick such a weighting function without knowledge of the variational principle. The reason the orthogonal collocation method gave the same results as (7.75) is because integrals are evaluated as follows:

$$\int_0^1 f(x)\, dx = \sum_{i=1}^{N} W_i f(x_i), \tag{7.77}$$

where $0 < x_i < 1$ are the interior roots to the Legendre polynomial $P_N(x^2) = 0$. The integral is evaluated exactly if $f(x)$ is a polynomial of degree $4N - 1$. Since the trial function is of degree $2N$ and the derivative is of degree $2N - 1$, (7.75) can be rewritten as

$$\sum_{k=1}^{N} W_k\left[\left(\sum_{i=1}^{N} c_i u_i + u_0\right)(\mathbf{n}\cdot\nabla u_j)\right]\bigg|_{x_k} = 0. \tag{7.78}$$

Provided the matrix $A_{kj} = W_k \mathbf{n}\cdot\nabla u_j|_{x_k}$ has an inverse, this gives just

$$\sum_{i=1}^{N} c_i u_i + u_0 = 0 \qquad \text{at } x_k, \tag{7.79}$$

which are the equations used to solve (4.7) and (4.8).

The same trial functions used in MWR (Section 4.1) are suitable for the variational method as well. Delleur and Sooky (1961) suggest solving (7.62) for a convex polygon using the trial function

$$u = \prod_{i=1}^{n}(A_i x + B_i y + C_i) \sum_{j,k=1}^{N} A_{jk} x^j y^k, \tag{7.80}$$

where $A_i x + B_i y + C_i = 0$ defines boundary of the ith side. Then (7.80) satisfies the boundary conditions and is a suitable trial function for (7.63).

The integrals needed are tedious to calculate, however. Probably a better strategy is to do the computations using either orthogonal collocation in the interior or on the boundary, or by point matching on the boundary. Since the interior collocation uses trial functions which satisfy the boundary conditions, it provides an admissible function for (7.63). Similarly, the boundary collocation trial function satisfies the differential equation, and hence is an admissible function for (7.67). Thus the inequality (7.73) is valid even though the bounds are not the best possible using those trial functions, since the best bounds are obtained only by the variational principle. If the integrals are tedious to evaluate in the variational method, the value of (7.63) and (7.67) for the collocation solutions can be obtained by numerical integration once the approximate solution is determined.

7.4 Relation to Galerkin and Finite Element Methods

There is always a Galerkin method which is equivalent to the variational method. This was first noted by Galerkin himself (1915) and has been proved several times since. To show the equivalence consider a nonlinear equation $N(u) = 0$. The variational method always reduces to the form

$$\int \delta u \, N(u) \, dV = 0 \tag{7.81}$$

for the Euler equation. Application of the variational method gives

$$\int \frac{\partial u}{\partial c_k} N(u) \, dV = 0 \tag{7.82}$$

to determine the approximate solution. This is the same as the Galerkin method provided the weighting function is $\partial u/\partial c_k$. There may be other ways to apply the Galerkin method, and these would not necessarily be the same as a variational method. This is particularly true in the handling of natural boundary conditions, as illustrated in (7.56), (7.75), and (2.54). The variational method cannot always be applied because there may be no variational principle for the problem, but the Galerkin method is always applicable because it does not depend on the existence of a variational principle.

The finite element method is easily applied to variational principles. Consider the same problem (5.133) with different boundary conditions:

$$\nabla \cdot k \, \nabla T - Q = 0, \tag{7.83a}$$

$$k\mathbf{n} \cdot \nabla T + q + hT = 0, \tag{7.83b}$$

where k, Q, q, and h are constants. The variational principle is a special case of (7.48).

Minimize $\Phi(T)$ among all functions which are continuous and have piecewise continuous first derivatives:

$$\Phi(T) = \int_A [\tfrac{1}{2}k\,\nabla T \cdot \nabla T + QT]\,dA + \int_C qT\,dS + \tfrac{1}{2}\int_C hT^2\,dS. \quad (7.84)$$

In the finite element method this variational integral is minimized with respect to the nodal temperature; consider the contribution of an interior element:

$$\frac{\partial \Phi^e}{\partial T_i} = \int_\Delta \left[k\frac{\partial T}{\partial x}\frac{\partial}{\partial T_i}\left(\frac{\partial T}{\partial x}\right) + k\frac{\partial T}{\partial y}\frac{\partial}{\partial T_i}\left(\frac{\partial T}{\partial y}\right) + Q\frac{\partial T}{\partial T_i}\right] dx\,dy$$

$$\frac{\partial T}{\partial x} = \sum \frac{\partial N_i}{\partial x} T_i, \qquad \frac{\partial N_i}{\partial x} = \frac{b_i}{2\Delta}, \qquad \frac{\partial T}{\partial T_i} = N_i, \quad (7.85)$$

$$\frac{\partial T}{\partial x} = \frac{1}{2\Delta}\sum b_i T_i, \qquad \frac{\partial T}{\partial y} = \frac{\sum c_i T_i}{2\Delta},$$

$$\frac{\partial \Phi^e}{\partial T_i} = \frac{1}{(2\Delta)^2}\int_\Delta k[(\sum b_j T_j)b_i + (\sum c_j T_j)c_i]\,dx\,dy + \int_\Delta QN_i\,dx\,dy = 0.$$

The summation is taken over all nodes of the element Φ^e. The last equation is the same as that derived by the Galerkin method (5.140) and (5.141) so that the result can be used directly:

$$\sum_{\substack{\text{all interior}\\ \text{nodes}}} [\sum h_{mq} T_q + F_m] = 0, \quad (7.86)$$

$$h_{mq} = (k/4\Delta)(b_m b_q + c_m c_q), \qquad F_m = Q\,\Delta/3.$$

For an element on the boundary there is an additional contribution to (7.85). Along the boundary l_j write

$$T = d + es, \qquad T_j = d + es_j, \qquad T_l = d + es_l. \quad (7.87)$$

Integration of the boundary terms gives

$$\int_{s_l}^{s_j} qT\,ds = \frac{Lq}{2}(T_j + T_l); \int_{s_l}^{s_j} \frac{1}{2}hT^2\,ds = \frac{Lh}{6}(T_j^2 + T_j T_l + T_l^2), \quad (7.88)$$

where $L = s_j - s_l$ is the length of the boundary for that element. Differentiation with respect to the boundary nodal temperature T_j gives

$$\frac{\partial \Phi_B{}^e}{\partial T_j} = q\frac{L}{2} + \frac{hL}{3}(T_j + \frac{1}{2}T_l) \quad (7.89)$$

to add to (7.86) whenever the nodal point is on the boundary.

Another approach which is similar to the finite difference method is to approximate the derivatives in the functional (7.84) with difference expressions, and then minimize the functional with respect to nodal temperatures. For (7.83) with boundary conditions of the first kind this gives

$$\Phi = \sum_{j=1}^{N} \sum_{i=1}^{N} \left\{ \frac{k}{2} \left(\frac{T_{i+1,j} - T_{i,j}}{\Delta x} \right)^2 + \frac{k}{2} \left(\frac{T_{i,j+1} - T_{i,j}}{\Delta y} \right)^2 + QT_{i,j} \right\} \Delta x \, \Delta y,$$

$$\frac{1}{\Delta x \, \Delta y} \frac{\partial \Phi}{\partial T_{i,j}} = 0 = k \frac{-T_{i+1,j} + 2T_{i,j} - T_{i-1,j}}{\Delta x^2} + k \frac{-T_{i,j+1} + 2T_{i,j} - T_{i,j-1}}{\Delta y^2} + Q = 0, \tag{7.90}$$

which is the second-order difference expression for (7.83a). This approach has been used by Greenspan and Jain (1966; see also Greenspan, 1967) to study the subsonic flow of a compressible gas past a cylinder. The variational principle for this problem is given below (Section 8.2). Atkinson *et al.* (1969) apply it to developing flow entry-length problems.

Another variation is to use reciprocal variational principles to allow discontinuous trial functions. We present the treatment of Kaplan (1969). Take the principle (7.38) with $F = \frac{1}{2}kz^2 + \frac{1}{2}au^2 + Qu$ and $\lambda = F_z = kz$. Remember that z is a substitute for u'. Expand both z and u in terms of piecewise continuous trial functions, so that the variational integral must be modified for discontinuities at x_i:

$$H(u, z) = \int_a^b [-(\tfrac{1}{2}kz^2) + \tfrac{1}{2}au^2 + Qu + kzu'] \, dx + \Sigma k \tfrac{1}{2}[z(x_i+) + z(x_i-)][u(x_i+) - u(x_i-)] \tag{7.91}$$

The trial solution is taken in the form

$$u(x) = \sum_{n=1}^{N} U_n(x) u_n, \qquad z(x) = \sum_{n=0}^{N} Z_n(z) z_n. \tag{7.92}$$

The functions are pictured in Fig. 7.1. The functional is then

$$H = \sum_{n=0}^{N} [-\tfrac{1}{2}k \, h_n z_n{}^2 + \tfrac{1}{2}(h_n + h_{n-1})(\tfrac{1}{2}au_n{}^2 + Qu_n) + kz_n(u_{n+1} - u_n)],$$
$$u_0 = u_{N+1} = 0, \tag{7.93}$$

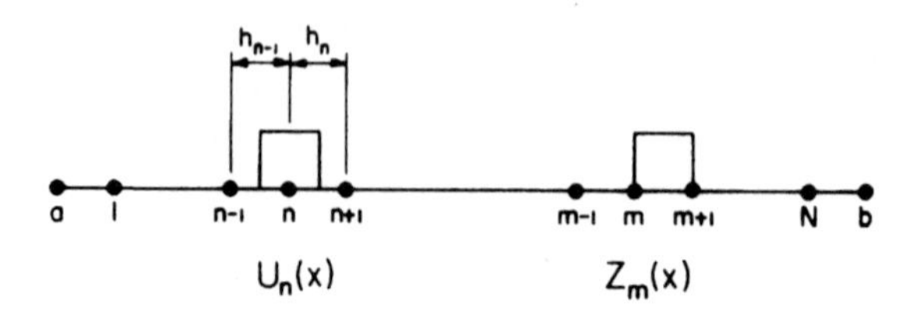

Fig. 7.1. Discontinuous trial functions.

and the stationary conditions yield

$$\begin{aligned} \partial H/\partial z_n: &\quad -kh_n z_n + k(u_{n+1} - u_n) = 0, \\ \partial H/\partial u_n: &\quad \tfrac{1}{2}(h_n + h_{n-1})(au_n + Q) + k(z_{n-1} - z_n) = 0. \end{aligned} \tag{7.94}$$

These are recognized as difference expressions for the Euler equations for the variational principle:

$$-kz + ku' = 0, \qquad au + Q - kz' = 0. \tag{7.95}$$

Variational principles using discontinuous trial functions have proved very useful in nuclear engineering applications (Yasinsky and Kaplan, 1967; Yasinsky, 1968; Becker, 1969; Kaplan, 1969).

7.5 Variational Principles for Eigenvalue Problems

Variational methods are highly developed for eigenvalue problems, principally because many problems of mathematical physics yield minimum principles, and thus the variational methods provide upper or lower bounds (or both) on the eigenvalue. Eigenvalue problems arise in heat and mass transfer problems after applying separation of variables to an initial-value or entry-length problem (Section 3.3) or from some of the stability problems treated in Chapter 6. We present a general treatment first and then discuss several applications.

Consider the general linear self-adjoint partial differential equation

$$(pu_x)_x + (pu_y)_y - qu + \lambda \rho u = 0, \qquad p, \rho > 0, \tag{7.96}$$

where p, q, and ρ are known functions of position. Obviously the simpler case of a second-order ordinary differential equation is a special case. The boundary conditions are

$$\frac{\partial u}{\partial n} + \sigma u = 0, \tag{7.97}$$

where σ is a piecewise continuous function which is positive on the boundary. In order to simplify the notation, the following bilinear functionals are defined for continuous functions ϕ and ψ with piecewise continuous first derivatives. A bilinear functional is just a functional depending upon two functions:

$$\mathscr{D}[\phi, \psi] = D[\phi, \psi] + \int p\sigma\phi\psi \, ds, \qquad \mathscr{D}[\phi] \equiv \mathscr{D}[\phi, \phi],$$

$$D[\phi, \psi] = \iint p(\phi_x \psi_x + \phi_y \psi_y) \, dx \, dy + \iint q\phi\psi \, dx \, dy, \tag{7.98}$$

$$H[\phi, \psi] = \iint \rho\phi\psi \, dx \, dy, \qquad H[\phi] \equiv H[\phi, \phi].$$

Note that the bilinear functionals are symmetric, which results from the self-adjointness of the problem (see Section 9.2) and they satisfy

$$\mathscr{D}[\phi + \psi] = \mathscr{D}[\phi] + 2\mathscr{D}[\phi, \psi] + \mathscr{D}[\psi] \tag{7.99}$$

with a similar equation for H. Courant and Hilbert (1953, p. 399) give the following variational principle.

The admissible function, which minimizes the Rayleigh quotient†

$$\lambda = \mathscr{D}(u)/H(u), \tag{7.100}$$

is an eigenfunction u_1 for the differential equation (7.96) and satisfies the natural boundary condition (7.97). The minimum value of $\mathscr{D}/H$ is the corresponding eigenvalue λ. If the orthogonality conditions

$$H[\phi, u_i] = 0 \qquad i = 1, \ldots, n-1, \tag{7.101}$$

are imposed then the solution is an eigenfunction u_n of (7.96) satisfying the same boundary condition,

$$\mathscr{D}(u_n)/H(u_n) = \lambda_n . \tag{7.102}$$

An alternative statement is to minimize $\mathscr{D}(u)$ subject to $H(u) = 1$. First verify that the Euler equation is (7.96) with natural boundary conditions (7.97). Consider the functional for the variable $\bar{u} + \varepsilon\eta$:

$$R(\varepsilon) = \frac{\mathscr{D}(\bar{u} + \varepsilon\eta)}{H(\bar{u} + \varepsilon\eta)}, \qquad R'(0) = \frac{2}{H(\bar{u})}\{\mathscr{D}[\bar{u}, \eta] - R(0)\, H[\bar{u}, \eta]\}. \tag{7.103}$$

This can be integrated by parts to obtain

$$\iint \eta[-(p\bar{u}_x)_x - (p\bar{u}_y)_y + q\bar{u} - R(0)\rho\bar{u}] \, dx \, dy + \int \eta p \left(\frac{\partial \bar{u}}{\partial n} + \sigma\bar{u}\right) ds = 0, \tag{7.104}$$

† The Rayleigh quotient is named after Lord Rayleigh, who first realized that the eigenvalue is the ratio of kinetic to potential energy (Rayleigh, 1873, 1896).

thus yielding the correct Euler equation and natural boundary condition with $R(0) = \lambda$.

To show that the principle is a minimum principle, we show that the value of the functional for any admissible function v is above the value λ_1. The eigenfunctions $\{u_n\}$ form a complete system of functions; expand the v in terms of that set. Furthermore due to (7.101) the eigenfunctions are orthogonal. Then

$$v = \sum_{i=1}^{\infty} c_i u_i, \qquad c_i = (v, u_i),$$

$$\frac{\mathscr{D}(v)}{H(v)} = \frac{\sum_{i,j=1}^{\infty} c_i c_j \mathscr{D}(u_i, u_j)}{\sum_{i,j=1}^{\infty} c_i c_j H(u_i, u_j)} = \frac{\sum_{i=1}^{\infty} c_i^2 \lambda_i H(u_i)}{\sum_{i=1}^{\infty} c_i^2 H(u_i)} \geq \lambda_1 = \frac{\mathscr{D}(u_1)}{H(u_1)}, \tag{7.105}$$

since

$$\mathscr{D}(u_i, u_j) = \lambda_i H(u_i, u_j) = \lambda_i H(u_i)\, \delta_{ij}, \qquad \lambda_i \geq \lambda_1.$$

Thus the functional achieves its minimum value for the eigenfunction u_1. Similar proofs hold for the other eigenvalues due to (7.101). The fact that the minimum is achieved and that a minimizing eigenfunction exists is a difficult theoretical question (see Chapter 11).

When the boundary condition (7.97) is replaced by $u = 0$, the condition $u = 0$ is added to the admissibility condition for the variational principle.

Next consider the Rayleigh–Ritz method of calculating approximations to the eigenvalues. Consider the first eigenvalue. Using the trial function

$$v_N = \sum_{i=1}^{N} c_i v_i(x, y), \tag{7.106}$$

the expansion functions each satisfy the admissibility conditions for the variational principle, so that v_N is admissible for all values of the constants c_i. Substitute this into (7.100) and minimize with respect to the c_i:

$$\frac{\partial \lambda}{\partial c_k} = 0 = \frac{2}{\sum c_i c_j H(v_i, v_j)} \sum_{i=1}^{N} c_i \{\mathscr{D}(v_i, v_k) - \lambda H(v_i, v_k)\}. \tag{7.107}$$

If this set of linear homogeneous equations in c_i is to have a solution the determinant of coefficients must vanish

$$|\mathscr{D}(v_i, v_k) - \lambda H(v_i, v_k)| = 0, \tag{7.108}$$

which provides N values of λ, with $\lambda_{(1)}$ being an upper bound on λ_1. Call $v^{(1)}$ the eigenfunction corresponding to $\lambda_{(1)}$.

Next consider the higher eigenfunctions and eigenvalues. The variational principle (7.100) cannot be used since we do not know the first eigenfunction, only an approximation to it. Instead we use the maximum-minimum principle of Courant and Hilbert (1953, p. 406).

Given $n-1$ functions $v_1, v_2, \ldots, v_{n-1}$ which are piecewise continuous in the domain, let $d(v_1, v_2, \ldots, v_{n-1})$ be the greatest lower bound of the set of values assumed by the functional $\mathscr{D}(\phi)/H(\phi)$, where ϕ is any admissible function which satisfies

$$H(\phi, v_i) = 0, \qquad i = 1, \ldots, n-1. \tag{7.109}$$

Then λ_n is equal to the largest value which this lower bound d assumes if the functions $v_1, v_2, \ldots, v_{n-1}$ range over all sets of admissible functions. This maximum–minimum is attained for $u = u_n$ and $v_1 = u_1, v_2 = u_2, \ldots, v_{n-1} = u_{n-1}$.

We next show that the application of this principle gives the same results obtained from (7.108), following the treatment of Mikhlin (1964, p. 231). For the first approximation

$$\sum_{i=1}^{N} c_i^{(1)}\{\mathscr{D}(v_i, v_k) - \lambda_{(1)} H(v_i, v_k)\} = 0. \tag{7.110}$$

For the second eigenvalues we wish to minimize $\mathscr{D}(\Sigma c_i v_i)$ subject to the conditions $H(v) = 1$ and

$$H\Big(v^{(1)}, \sum_{i=1}^{N} c_i v_i\Big) = \sum_{i,j=1}^{N} c_i c_j^{(1)} H(v_j, v_i) = 0. \tag{7.111}$$

Use the method of Lagrange multipliers and minimize (with respect to c_i) the quantity

$$\mathscr{D}(v) - \lambda H(v) - 2\mu H(v, v^{(1)}). \tag{7.112}$$

This gives

$$\sum_{i=1}^{N} \{c_i[\mathscr{D}(v_i, v_k) - \lambda H(v_i, v_k)] - \mu c_i^{(1)} H(v_k, v_i)\} = 0. \tag{7.113}$$

Multiply by $c_k^{(1)}$ and sum over k:

$$\sum_{i=1}^{N} c_i \sum_{k=1}^{N} c_k^{(1)}[\mathscr{D}(v_i, v_k) - \lambda H(v_i, v_k)] = \mu \sum_{i,k=1}^{N} c_i^{(1)} c_k^{(1)} H(v_i, v_k) = \mu. \tag{7.114}$$

Equation (7.107), written for $c_k^{(1)}$ and $\lambda_{(1)}$, can be used to reduce (7.114) to

$$\sum_{i=1}^{N} c_i \sum_{k=1}^{N} c_k^{(1)} H(v_i, v_k)(\lambda_{(1)} - \lambda) = \mu. \tag{7.115}$$

This is zero by (7.111). Consequently (7.113) with $\mu = 0$ is the same as (7.107), and the approximation to the second eigenvalue is therefore the second root of (7.108) for λ. Similar results hold for higher eigenvalues.

The variational method is illustrated by application to problem (3.6).

The Rayleigh quotient (7.100) is

$$\lambda = \int_0^1 (X')^2\,dx \bigg/ \int_0^1 (1 - x^2)X^2\,dx. \tag{7.116}$$

A trial function is taken in the form

$$X = \sum_{i=1}^{N} c_i X_i, \qquad X_i'(1) = X_i(0) = 0, \tag{7.117}$$

and the Rayleigh quotient becomes

$$\lambda = \sum_{i,j=1}^{N} c_i c_j \int_0^1 X_i' X_j'\,dx \bigg/ \sum_{i,j=1}^{N} c_i c_j \int_0^1 (1 - x^2) X_i X_j\,dx. \tag{7.118}$$

The numerator can be integrated by parts to give

$$\int_0^1 X_i' X_j'\,dx = -\int_0^1 X_i X_j''\,dx + [X_i X_j']_0^1 \tag{7.119}$$

and the boundary term vanishes. Equation (7.108) for the approximate solution, obtained by minimizing (7.118) with respect to the c_i, is then

$$-\int_0^1 X_k \left[\sum_{j=1}^{N} c_j X_j'' + \lambda(1 - x^2)\sum_{j=1}^{N} c_j X_j\right] dx = 0. \tag{7.120}$$

The term in the brackets is just the residual, which is made orthogonal to the approximating functions. Thus the variational method is equivalent to the Galerkin method, and this result holds for other problems as well. The Galerkin results in Section 3.1 are thus upper bounds on the exact values, as is shown in Table 3.1. It is in problems such as these that the power of the variational method is evident. The calculations are the same as the Galerkin method, but only the variational principle shows the results to be upper bounds on the solution.

The second approximation gives a good result for the first eigenvalue and a poor result for the second eigenvalue. The reason is clear from the minimax principle. The second eigenfuntion has only one degree of freedom (say c_2) since c_1 is determined by the requirement that the approximate second eigenfunction be orthogonal to the approximate first eigenfunction. Furthermore, the exact second eigenfunction is orthogonal to the exact first eigenfunction, whereas the orthogonality condition (7.111) involves the approximate first eigenfunction. The reduced number of degrees of freedom and the approximate orthogonality condition give rise to the poor approximation of the second eigenvalue. For the third approximation, when these conditions are relaxed, the approximate second eigenvalue is much better, and the third eigenvalue is approximated poorly, for the same reasons.

The variational method can be applied to entry-length and initial-value problems, provided they are linear and separation of variables can be applied to reduce the problem to an eigenvalue problem. Thus the treatment in Section 3.3 applies equally well to variational methods, which are used to solve the eigenvalue problem (3.33). Sparrow and Siegel (1960) have applied the variational method to the entry-length problem with a constant heat flux on the wall in parallel plate and circular pipe geometries. For the circular pipe the eigenvalue problem is a special case of (3.38a) and (3.56a):

$$(1/r)(rX')' + \lambda(1 - r^2)X = 0, \qquad X' = 0 \quad \text{at} \quad r = 0, 1. \tag{7.121}$$

Sparrow and Siegel choose trial functions which are the exact solution when the velocity is uniform:

$$X = \sum_{i=0}^{N} c_i \cos i\pi y. \tag{7.122}$$

These satisfy the natural conditions of the variational principle. The results of the computations are shown in Table 7.2, and can then be substituted

TABLE 7.2

LAMINAR FLOW IN A CIRCULAR TUBE WITH UNIFORM WALL HEAT FLUX[a]

	First approximation ($N=1$)	Second approximation ($N=2$)	Exact
λ_2	28.997	25.6956	25.6796
$R_2(1)$	−0.83911	−0.49262	−0.49252
c_2		0.40680	0.40348

$\lambda_1 = 0, R_1(1) = 1, c_1 = 0$

$R_2^{(1)} = 0.080445 + 0.91956 \cos \pi y$
$R_2^{(2)} = 0.109207 + 0.746309 \cos \pi y + 0.144484 \cos 2\pi y$

[a] Sparrow and Siegel (1960).

into (3.58a) to determine the temperature. The value of $R_2(1)$ is needed to evaluate the wall temperature.

Sparrow and Siegel (1960) also treat the entry-length problem for a square duct and uniform heat flux on the wall. The eigenvalue problem is then

$$\begin{gathered} z_{xx} + z_{yy} + \lambda u(x, y)z = 0, \\ z_x = 0 \quad \text{at } x = \pm 1, \qquad z_y = 0 \quad \text{at } y = \pm 1. \end{gathered} \tag{7.123}$$

The velocity is a function of both x and y, and thus separation of variables cannot be applied. An approximate velocity profile is calculated from the variational principle of Eq. (7.63), Sparrow and Siegel (1959):

$$u = (x^2 - 1)(y^2 - 1)[2.0983 + 0.2918(x^2 + y^2) + 0.8755x^2y^2]. \quad (7.124)$$

Trial functions for temperature are chosen which satisfy the boundary conditions, are symmetric about $x = 0$ and $y = 0$ and are symmetric in x and y:

$$Z_1 = 1, \qquad Z_2 = \cos \pi x \cos \pi y, \qquad Z_3 = \cos \pi x \cos 2\pi y + \cos 2\pi x \cos \pi y. \quad (7.125)$$

The results show that the temperature varies around the periphery, particularly near the corners. Earlier Dennis *et al.* (1959) applied the Galerkin method to the same problem, using the exact infinite series for velocity, trial functions as solutions to $\phi_{xx} + \phi_{yy} + \Lambda\phi = 0$, and taking many terms in the expansion. Due to the equivalence between the Galerkin and Rayleigh–Ritz methods, their calculations were also variational calculations. Schechter (1967, p. 209) treats a similar equation for a rectangular duct under boundary conditions of constant wall temperature.

The fully developed Nusselt number is determined by the first eigenvalue, as discussed in Section 3.3, Eq. (3.50d). To determine it the velocity must first be found by solving (7.62), either exactly or approximately. This velocity is then used in the eigenvalue problem, which is solved for the first eigenvalue. Very good upper bounds can be achieved, as illustrated in Table 3.1. In some cases lower bounds can be derived as well, and these are discussed in Section 7.7.

7.6 Enclosure Theorems

There is another feature of the minimum formulation of eigenvalue problems which is important: it may be possible to predict the effect of parameters on the solution without solving the problem. To do so we use the following Lemma, and follow the treatment of Courant and Hilbert (1953, p. 407).

Let C_u and C_v be two classes of functions and let $C_u \subset C_v$. Then if

$$\lambda_u \equiv \min_u \Phi(u), \qquad \lambda_v \equiv \min_v \Phi(v), \quad (7.126)$$

then

$$\lambda_u \geq \lambda_v. \quad (7.127)$$

The notation $C_u \subset C_v$ means that the class of functions C_u is contained within the class of functions C_v. Since we widen the class of admissible functions when going from C_u to C_v we get a lower value of λ, or at the least the same value. Consider the following applications.

APPLICATION 1. Consider two problems with the same differential equation but different domains V_1 and V_2, with $V_1 \subset V_2$. Define C_i as the class of functions having the appropriate continuity requirements and obeying $v = 0$ on S_i, the boundary of V_i. Then $C_1 \subset C_2$ since a function in C_1 can be extended as zero to the boundary of V_2, and hence is in C_2:

$$\lambda_{(1)} \geq \lambda_{(2)}. \tag{7.128}$$

Thus if the domain is enlarged the eigenvalue is lowered.
This is useful information when treating problems with irregularly shaped domains.

APPLICATION 2. Consider the effect of a boundary condition of the third kind. Take the two problems

$$\lambda = \min_{C_\lambda} \frac{(\nabla u, \nabla u)}{(u, u)} \qquad \mu = \min_{C_\mu} \frac{(\nabla u, \nabla u) + \sigma(u, u)_s}{(u, u)} \tag{7.129}$$

C_λ: all continuous u with piecewise continuous first derivatives and $u = 0$ on S.

C_μ: all continuous u with piecewise continuous first derivatives.

The Euler equations for these two problems are the same [see Eqs. (7.96), (7.97), (7.100), and (7.104)] and the boundary conditions are

$$\begin{aligned} u &= 0 \qquad \text{for } \lambda \quad \text{(essential BC)}, \\ \partial u/\partial n + \sigma u &= 0 \qquad \text{for } \mu \quad \text{(natural BC)}. \end{aligned} \tag{7.130}$$

The two problems correspond to one with boundary conditions of the first or third kind. Because $C_\lambda \subset C_\mu$ it follows from the lemma that

$$\lambda \geq \mu. \tag{7.131}$$

The imposition of a boundary condition of the third kind always lowers the eigenvalue compared to its value for a boundary condition of the first kind. In transient heat transfer problems, the eigenvalues determine the speed of response to a change in the boundary conditions, with a higher eigenvalue indicating a faster speed of response. Thus a system always responds fastest if the temperature remains fixed on the boundary rather than having the heat flux proportional to the temperature there.

APPLICATION 3. It is clear from the definition of $\mathscr{D}$, Eqs. (7.98), that if the function σ is either increased or decreased at every point, then each individual eigenvalue can change only in the same sense. This means that as the heat transfer coefficient is increased the eigenvalues increase, and the time-response becomes faster.

APPLICATION 4. If in (7.96) the function ρ increases (or remains the same) at each point, then the nth eigenvalue decreases (or remains the same). We can use this principle to predict the relationship of the eigenvalues of the following two problems

$$\begin{aligned} X'' + \lambda X &= 0, & X'' + \mu(1 - x^2)X &= 0, \\ X(0) &= 0, & X(0) &= 0, \\ X'(1) &= 0, & X'(1) &= 0, \\ \lambda_k &= k^2\pi^2/4. & & \end{aligned} \tag{7.132}$$

Thus

$$\mu_k \geq \lambda_k. \tag{7.133}$$

The problem with eigenvalues λ corresponds to a diffusion or heat transfer problem with uniform flow rate, whereas the other problem allows a parabolic profile of velocity (see Section 3.4). The principle says the eigenvalue for parabolic flow is above that for uniform flow.

Similar theorems are useful in more complicated situations, such as the uniqueness of steady-state solutions to chemical reaction in catalyst particles (see Section 9.5).

7.7 Least Squares Interpretation of D.H. Weinstein's Method

The variational method developed by Weinstein (1934) can be related to the least squares method. To do so we need to explain the proper way to apply the least squares method to eigenvalue problems.

Consider the least squares method for the problem,

$$y'' + \lambda y = 0, \qquad y(0) = y(1) = 0, \tag{7.134}$$

which has eigenvalues $\lambda_k = k^2\pi^2$. Take $y = c_1 x(1 - x)$ so that the mean-square residual, which is to be minimized, is

$$I = \int_0^1 R^2\, dx = c_1{}^2 \int_0^1 [-2 + \lambda x(1 - x)]^2\, dx = c_1{}^2[4 - \tfrac{2}{3}\lambda + \tfrac{1}{30}\lambda^2]. \tag{7.135}$$

This gives the approximation

$$\lambda = 10 \pm 2\sqrt{5}\, i. \tag{7.136}$$

This is clearly a poor approximation to π^2. A better approach is suggested by Becker (1964). For the first approximation take $\partial I/\partial \lambda = 0$. Then

$$-\tfrac{2}{3} + \tfrac{2}{30}\lambda = 0, \qquad \lambda = 10, \tag{7.137}$$

which is a good approximation. Becker's method depends on the fact that for linear problems one can write

$$I(c_1, c_2, \ldots c_N, \lambda) = c_1{}^2 I(1, \alpha_2, \ldots, \alpha_N, \lambda); \tag{7.138}$$

that is, the mean-square residual is a function of only $\alpha_i = c_i/c_1$, $i = 2, \ldots, N$, so that there is no point in minimizing I with respect to c_1. In the correct application of the least-squares method, then, the weighting functions are $\partial I/\partial \lambda$, $\partial I/\partial \alpha_2, \ldots, \partial I/\partial \alpha_N$.

Now investigate Weinstein's method of obtaining upper and lower bounds for eigenvalue problems. Consider the problem

$$Lu = \lambda u, \tag{7.139}$$

where Lu is a self-adjoint operator, positive bounded below, so that $(u, Lu) = \mathscr{D}(u)$ and $\mathscr{D}(u, v) = \mathscr{D}(v, u)$. The variational principle corresponding to this problem is

$$\text{Minimize} \quad \lambda = (u, Lu) \quad \text{subject to} \quad (u, u) = 1.$$

Define the following integrals:

$$I_1 = (u, Lu), \qquad I_2 = (Lu, Lu), \qquad I_3 = \int (Lu - \Lambda u)^2\, dV, \tag{7.140}$$

and consider $I_3(u)$ under the restriction $(u, u) = 1$. Expand u in terms of the eigenfunctions of the operator (7.139), and assume that the eigenfunctions are orthonormal:

$$u = \sum_{i=1}^{\infty} c_i u_i, \qquad \sum c_i{}^2 = 1,$$

$$Lu - \Lambda u = \sum_{i=1}^{\infty} c_i u_i(\lambda_i - \Lambda), \tag{7.141}$$

$$I_3 = \sum_{i=1}^{\infty} c_i{}^2(\lambda_i - \Lambda)^2,$$

since the eigenfunctions are orthogonal. Call λ_j the eigenvalue closest to Λ. Then

$$\begin{aligned}(\lambda_j - \Lambda)^2 &\le (\lambda_i - \Lambda)^2, \qquad \text{all } i,\\ I_3 &\ge (\lambda_j - \Lambda)^2 \,\Sigma c_i^{\,2} = (\lambda_j - \Lambda)^2.\end{aligned} \tag{7.142}$$

When $\lambda_j - \Lambda \ge 0$,

$$\sqrt{I_3} \ge \lambda_j - \Lambda \ge 0 \qquad \text{or} \qquad \sqrt{I_3} + \Lambda \ge \lambda_j \ge \Lambda, \tag{7.143}$$

whereas when $\lambda_j - \Lambda \le 0$,

$$\sqrt{I_3} \ge -\lambda_j + \Lambda \ge 0 \qquad \text{or} \qquad \Lambda - \sqrt{I_3} \le \lambda_j \le \Lambda. \tag{7.144}$$

Hence

$$\sqrt{I_3} + \Lambda \ge \lambda_j \ge \Lambda - \sqrt{I_3}\,. \tag{7.145}$$

In order to get the best estimate we need to choose Λ to minimize I_3:

$$\begin{aligned}I_3 &= I_2 - 2\Lambda I_1 + \Lambda^2\\ &= (I_2 - I_1^{\,2}) + (I_1 - \Lambda)^2.\end{aligned} \tag{7.146}$$

Clearly this choice is $\Lambda = I_1$. The estimates on λ_j are then

$$I_1 + (I_2 - I_1^{\,2})^{\frac{1}{2}} \ge \lambda_j \ge I_1 - (I_2 - I_1^{\,2})^{\frac{1}{2}}. \tag{7.147}$$

This formula can be used to assess the error of the Rayleigh–Ritz method, or it provides a method to minimize the error in the approximation of the eigenvalue.

This method is just the least squares method, an application of MWR. Consider the functional

$$I(u) = (Lu, Lu) - (u, Lu)^2, \tag{7.148}$$

which is to be minimized subject to $(u, u) = 1$ in the Weinstein method. The mean-square residual

$$J(u) = \int (Lu - \lambda u)^2 \, dV = (Lu, Lu) - 2\lambda(u, Lu) + \lambda^2(u, u) \tag{7.149}$$

is to be minimized in the least squares method:

$$\frac{\partial J}{\partial \lambda} = 0 \quad \text{gives} \quad \lambda = \frac{(u, Lu)}{(u, u)}. \tag{7.150}$$

Thus the functional $J(u)$ takes the value at the minimum

$$J(u) = (Lu, Lu) - \frac{(u, Lu)^2}{(u, u)} = I(u), \tag{7.151}$$

when $(u, u) = 1$. Thus the Weinstein method gives results identical to the least squares method. The error bounds (7.147) can be calculated for any method—variational or MWR—although they are the smallest for the least squares method. There are also ways to improve the bounds (7.147) (Wilson, 1965).

As an example consider

$$y'' + \lambda y = 0, \qquad y(0) = y(1) = 0, \tag{7.152}$$

which has the solution $y_i = \sin i\pi x$, $\lambda_i = i^2\pi^2$. Eigenvalues have been calculated using Eq. (7.147). The trial functions used were

$$y = \sum_{i=1}^{N} c_i x^i(1 - x). \tag{7.153}$$

The eigenfunctions fall into two classes: symmetric about $x = \frac{1}{2}$ and an odd function about $x = \frac{1}{2}$. Both classes are included in (7.153). The results are shown in Table 7.3. The bounds are poor for low-order approximations, but

TABLE 7.3

ERROR BOUNDS FOR EIGENVALUES[a]

Number of terms	Lower bound	Eigenvalue	Upper bound
	$N = 1$, $\lambda = 9.86960440$		
1	5.52	10.00000	14.48
3	9.56	9.86975	10.18
5	9.8621	9.86960443	9.8771
	$N = 2$, $\lambda = 39.47842$		
2	14.50	42.0000	69.50
4	34.81	39.5016	44.19
6	39.12	39.47847	39.84
	$N = 3$, $\lambda = 88.82$		
3	18.80	102.13	185.46
5	67.86	89.17	110.49
	$N = 4$, $\lambda = 157.91$		
4	10.81	200.50	390.18
6	100.53	159.99	219.45

[a] Eigenvalues calculated using Rayleigh–Ritz method. Bounds calculated using (7.147). Value listed under "Eigenvalue" is the Rayleigh–Ritz upper bound.

improve as more terms are added in the trial expansion. Even so the bounds are conservative: for $N = 1$ and 5 terms, the error bounds give a possible error of 0.08% whereas the actual error is 3×10^{-7}%. The Rayleigh–Ritz eigenvalue is actually an upper bound, too. Using the Rayleigh–Ritz result as the upper bound gives an error of 0.04% for $N = 1$ and 5 terms. Another method of finding lower bounds on eigenvalues is suggested in the next section.

7.8 Lower Bounds for Eigenvalues

We consider lower bounds for the eigenvalues of

$$\begin{aligned} Ly + \lambda f(\mathbf{x})y &= 0 \qquad \text{in } V, \\ B_i y &= 0 \qquad \text{on } S, \end{aligned} \tag{7.154}$$

where L is a self-adjoint operator (see Section 9.2) and appropriate homogeneous boundary conditions are satisfied. Weinstein's method does not give lower bounds for this problem unless f is a constant. Although the least squares method gives approximations to the eigenvalue, it does not provide upper and lower bounds. We present here two techniques applicable to (7.154), as well as to more general problems. We limit consideration to (7.154) because of its importance in engineering applications, such as the asymptotic Nusselt or Sherwood number in entry-length problems (Sections 3.3, 3.4, and 7.5). Indeed in these problems often the only information desired is the first eigenvalues to (7.154), which is closely approximated using these methods.

Variational methods for eigenvalue problems, including lower bounds, are treated in the comprehensive treatise by Gould (1966). We present the method due to Weinstein (1963) which constructs comparison equations, which are easier to solve than (7.154), and which are constructed such that the eigenvalues are lower bounds to the eigenvalues of (7.154). The second method we consider relies on the Schwarz quotients and involves an iterative procedure, outlined by Collatz (1948, p. 182) and Crandall (1956, p. 298). Another review, although not applicable directly to (7.154) is by Fox and Rheinboldt (1966). Pnueli (1967) calculates rough lower bounds by finding a function $g(x)$ such that $f(x) \geq g(x)$, and using the enclosure theorems. It is necessary to solve exactly the problem $Ly + g(x)y = 0$. Pnueli (1969) gives a method for obtaining lower bounds to the eigenvalue of Helmholtz equation, $\nabla^4 y + \lambda y = 0$, in arbitrary, plane regions, and McLaurin (1968) gives bounds for the first four eigenvalues corresponding to elliptical clamped plates.

First consider Weinstein's method, following the treatment of Gould (1966, p. 203). Rewrite the problem (7.154) as

$$Ly + \lambda\alpha y - \lambda(\alpha - f)y = 0, \tag{7.155}$$

where $\alpha \geq \max f(\mathbf{x})$ and introduce the intermediate problem,

$$Ly = \mu\left\{\alpha y - (y, y_n)\left(\int \frac{y_n^2\, dV}{\alpha - f}\right)^{-1} y_n\right\} = 0, \tag{7.156}$$

where y_n is any eigenfunction of the problem $Ly + \Lambda y = 0$, normalized so that $(y_n, y_k) = \delta_{kn}$.

We first calculate the Rayleigh quotient for the intermediate problem:

$$R(u) = (-Ly, y)\Big/\left\{\alpha(y, y) - (y, y_n)^2\left(\int \frac{y_n^2\, dV}{\alpha - f}\right)^{-1}\right\}. \tag{7.157}$$

By the Schwarz inequality (p. 227) we have

$$(y, y_n)^2 = \left(\int \frac{(\alpha - f)^{1/2} y y_n\, dV}{(\alpha - f)^{1/2}}\right)^2 \leq \left(\int (\alpha - f) y^2\, dV\right)\left(\int \frac{y_n^2\, dV}{\alpha - f}\right). \tag{7.158}$$

Thus the denominator of (7.157) is greater than or equal to (fy, y) and is less than or equal to $\alpha(y, y)$. This implies

$$\frac{(-Ly, y)}{\alpha(y, y)} \leq R(u) \leq \frac{(-Ly, y)}{(fy, y)}. \tag{7.159}$$

The first intermediate problem is easy to solve. For $k \neq n$ the eigenfunction is y_k and the eigenvalue is $\mu_k = \Lambda_k/\alpha$. For the nth eigenvalue $y = y_n$ is an eigenfunction with

$$\mu D(\alpha) y_n = \Lambda_n y_n, \qquad D(\alpha) = \alpha - \left(\int \frac{y_n^2\, dV}{\alpha - f}\right)^{-1}. \tag{7.160}$$

The eigenvalues then satisfy

$$\frac{\Lambda_1}{\alpha} \leq \frac{\Lambda_2}{\alpha} \leq \cdots \leq \frac{\Lambda_{n-1}}{\alpha} \leq \frac{\Lambda_{n+1}}{\alpha} \leq \cdots \tag{7.161}$$

and the value $\mu_q = \Lambda_n / D(\alpha)$ must be inserted in its proper numerical position in the sequence (7.161). We wish to choose α to obtain the best bound for λ_n. $\Lambda_n / D(\alpha)$ is above Λ_n/α for $\alpha \geq f$ from (7.160). The function $D(\alpha)$ decreases with increasing α. To see this calculate

$$\frac{dD}{d\alpha} = 1 - \left(\int \frac{y_n^2\, dV}{\alpha - f}\right)^{-2}\left(\int \frac{y_n^2\, dV}{(\alpha - f)^2}\right). \tag{7.162}$$

From the Schwarz inequality

$$\left(\int \frac{y_n^2\, dV}{\alpha - f}\right)^2 \le \left(\int \frac{y_n^2\, dV}{(\alpha - f)^2}\right)(y_n, y_n) = \int \frac{y_n^2\, dV}{(\alpha - f)^2} \tag{7.163}$$

so that $dD/d\alpha$ is negative. Thus $\Lambda_n/D(\alpha)$ increases with increasing α. The other eigenvalues in Eq. (7.161) obviously decrease with increasing α. Thus for small α the eigenvalue $\Lambda_n/D(\alpha)$ fits between Λ_n/α and Λ_{n+1}/α. As α is increased we reach a point where $\Lambda_n/D(\alpha) = \Lambda_{n+1}/\alpha$. Any further increase in α gives no improvement, because the lower bound for λ_n is then Λ_{n+1}/α and $\Lambda_n/D(\alpha)$ is a lower bound for λ_{n+1}, and so forth. Thus the best value of α satisfies

$$\Lambda_n/D(\alpha) = \Lambda_{n+1}/\alpha. \tag{7.164}$$

There is one further complication. Consider $\alpha = M$ = positive maximum of f. If

$$\Lambda_{n+1}/M \le \Lambda_n/D(M), \tag{7.165}$$

then the best lower bound for λ_n is Λ_{n+1}/M. The α cannot be taken any lower (the integrals are then undefined) and any larger α gives worse bounds, Λ_{n+1}/α. We thus have the procedure: check (7.165). If not satisfied, choose α to satisfy (7.164), which gives the best lower bound for λ_n.

As an application consider the problem solved in Section 3.4:

$$y'' + \lambda(1 - x^2)y = 0, \qquad y'(1) = 0, \qquad y(0) = 0. \tag{7.166}$$

The base problem and its solution is then

$$\begin{gathered} y'' + \Lambda y = 0, \qquad y'(1) = y(0) = 0, \\ y_k = \sqrt{2}\sin[(2k-1)\pi x/2], \qquad \Lambda_k = (2k-1)^2\pi^2/4. \end{gathered} \tag{7.167}$$

The intermediate problem is (7.156). Calculation of the integrals in (7.165), where $M = 1$, shows that they are not satisfied, so that the best bounds are achieved from (7.164). We thus choose α to satisfy a rearranged form of (7.164).

$$\frac{\Lambda_{n+1}}{\Lambda_{n+1} - \Lambda_n} = 2\alpha \int_0^1 \frac{\sin^2[(2n-1)\pi x/2]}{\alpha - 1 + x^2}\, dx. \tag{7.168}$$

The integrals were evaluated numerically to obtain the bounds shown in Table 7.4. The first eigenvalue is determined within 0.1%, but the lower bounds on the other eigenvalues are not as precise. For heat transfer problems, the first eigenvalue is of primary interest. When higher eigenvalues are important, we need the eigenfunctions and all the quantities given in Section 3.3. Lower bounds on just the eigenvalues are not as useful. The same procedure can be applied to partial differential equations, although the

TABLE 7.4

LOWER BOUNDS FOR EIGENVALUES USING THE WEINSTEIN METHOD

n	Lower bound	Upper bound[a]
1	5.112095	5.121670
2	30.94142	39.660957
3	81.25363	106.25021
4	153.6031	204.85999
5	247.1393	335.48423

[a] Obtained using Rayleigh–Ritz, 10 expansion functions; see Section 3.4.

results would not be expected to be as good as the number of dimensions is increased. This is clearly a very simple method of obtaining a very good estimate of the first eigenvalue. All that is necessary is to solve (7.168) for α.

The second procedure uses Schwarz quotients, as presented by Collatz (1948, p. 182) and Crandall (1956, p. 298). We outline the mechanics without proving the relevant theorems. Define the boundary value problem

$$-Lu_{k+1} = \lambda f(\mathbf{x})u_k, \qquad B_i u_{k+1} = 0, \tag{7.169}$$

which is to be solved iteratively. u_0 is specified, and satisfies the boundary conditions, and Eq. (7.169) is used to calculate u_i, $i \geq 1$. Define the integrals

$$R_k = \frac{-\int u_{k+1} L u_{k+1}\, dV}{\int f u_{k+1}^2\, dV}, \qquad Q_k = \frac{\int f {u_k}^2\, dV}{\int f u_{k+1}^2\, dV}. \tag{7.170}$$

Crandall gives the bounds

$$R_k - \frac{Q_k - {R_k}^2}{1_{p+1} - R_k} \leq \lambda_p \leq R_k + \frac{Q_k - {R_k}^2}{R_k - u_{p-1}}, \tag{7.171}$$

where 1_{p+1} and u_{p-1} are lower and upper bounds on the $p+1$ and $p-1$ eigenvalues, respectively. It is necessary that the operator be self-adjoint and satisfy

$$-\int uLu\, dV \geq 0, \qquad \int fu^2\, dV \geq 0 \tag{7.172}$$

for (7.171) to hold.

To apply this to (7.166) take $u_0 = x - \frac{1}{2}x^2$. A formula for the successive solutions to $-u''_{k+1} = \lambda(1 - x^2)u_k$ is easily generated and the u_k are found conveniently on a computer. Then the integrals (7.170) are evaluated. For the first eigenvalue we need a lower bound for the second eigenvalue. This can be a rough bound, since the numerator of the error term in (7.171) goes to zero as the iteration proceeds. We use the bound in Table 7.4. The lower bounds are listed in Table 7.5 and are very good. After three iterations the

TABLE 7.5

LOWER BOUNDS FOR EIGENVALUES USING SCHWARZ QUOTIENTS

k	Lower bound	Upper bound
1	5.1213174	5.1219856
2	5.1216674	5.1216734
3	5.12166928	5.12166937[a]

[a] The best upper bound by the Rayleigh–Ritz method is 5.121670, for 10 trial functions (see Section 3.4).

lower bound and upper bound differ by only one digit in the eighth significant figure. Note that for $p = 1$, R_k is an upper bound, since it is just the Rayleigh quotient, which is an upper bound for the first eigenvalue. The correction in (7.171) is needed for higher eigenvalues, since then the admissible trial functions must be orthogonal to the other eigenfunctions. This procedure is less suitable for higher eigenvalues because the iterations tend to converge to the first eigenvalue unless the first guess is close to the exact eigenfunction. The technique is suitable for partial differential equations, although the solution of (7.169) may be more difficult.

The two techniques given here give very good lower bounds on the first eigenvalue. The problem (7.154) is important in engineering applications when $f(\mathbf{x})$ represents the velocity distribution. Then for arbitrary geometries we must solve

$$\nabla^2 u = -1 \quad \text{in } A, \qquad u = 0 \quad \text{on } C, \tag{7.173a}$$

$$\nabla^2 y + \lambda u y = 0 \quad \text{in } A, \qquad B_i y = 0 \quad \text{on } C. \tag{7.173b}$$

The exact solution for u may not be known, but bounds can be derived as illustrated in Sections 4.1 and 11.5. Then we can use the monotone property of the eigenvalues (Application 4 in Section 7.6). Use an upper bound for u, which decreases λ, to calculate a lower bound for the exact λ. Use a lower bound for u to calculate the upper bound for λ. The combined results give upper and lower bounds for the eigenvalue of (7.173). Moler and Payne (1968) provide error bounds for (7.173b) when $u = 1$. Their approach, extended to $u(\mathbf{x})$, would be particularly useful for arbitrary geometry.

EXERCISES

7.1. For the functionals listed in Table 7.1, verify the Euler equations, essential and natural boundary conditions.

7.2. Find the Euler–Lagrange equation for $f(x, y, y') = y'^2 + g(x)\ y$ for $y(x)$ subject to $y(x_1) = y_1$, $y(x_2) = y_2$. Show that the Legendre condition is satisfied by any admissible $y(x)$. What are the admissibility conditions? Answer: $g - 2y'' = 0$; $f_{y'y'} = 2 \geq 0$; $y(x)$ continuous, satisfies the boundary conditions, and has continuous second derivatives.

7.3. One of the first problems in the calculus of variations was solved by Euler: find the closed curve of given circumference and greatest area. It can be formulated as the following. Find $y(x) > 0$ which maximizes

$$I(y) = \int_0^1 y\,dx$$

subject to

$$C = \int_0^1 [1 + (y')^2]^{1/2} = \pi/2, \qquad y(0) = y(1) = 0.$$

Find the Euler equations, verify that a semi-circle with center $x = \frac{1}{2}$, $y = 0$ and diameter 1 satisfies the equation, and show the Legendre condition is satisfied by any admissible $y(x)$.

7.4. Consider the variational principle: make $I(y)$ stationary among all $y \in C^1$ and $y(0) = 0$, $y(1) = 0$:

$$I(y) = \int_0^1 (y'^2 - y^2 - 2y)\,dx.$$

(a) Find the Euler–Lagrange equation. Are the boundary conditions essential or natural?

(b) Find the exact solution to the Euler–Lagrange equation in the form

$$y = a \sin x + b \cos x + c.$$

(c) Use the variational method to find an approximate solution of the form

$$y = x(1 - x) \sum_{i=1}^{N} c_i x^{i-1}.$$

Find the first and second approximations and compare to the exact solution: $y(0.25)$, $y(0.5)$, $y(0.75)$, $y'(0)$, $y'(1)$.
Answer: $N = 1: c_1 = 5/9$.

7.5. Consider the problem of heat conduction in a square:

$$T_{xx} + T_{yy} = 0, \qquad 0 < x < 1, \qquad 0 < y < 1,$$
$$T(x, 0) = 1, \qquad T_x(0, y) = T_y(x, 1) = 0,$$
$$T_x + \text{Nu}\, T = 0 \qquad \text{at} \quad x = 1.$$

Derive a variational principle and apply the variational method using the trial function

$$T = 1 + a_1 yx^2(2 - y).$$

Which boundary conditions does this satisfy? Calculate the average heat flux along $x = 1$.

Answer: $a_1 = -15\ \text{Nu}/(22 + 12\ \text{Nu})$; for $\text{Nu} = 1$ average flux is 0.706; exact result is 0.640.

7.6. Use the enclosure theorems to derive upper and lower bounds for the flow rate of a fluid in rectilinear flow through a pipe with the cross section of an equilateral triangle with sides of length a. The flow rate through a square of side-length b or a circle of diameter c are $0.0351b^4$ and $0.0245c^4$, respectively. The flow is governed by (7.173a).

Answer: Using the circle for the lower bound and the square for the upper bound gives $0.00272a^4 < Q < 0.0351a^4$.

7.7. Prove Eq. (7.72).

7.8. Verify Eq. (7.88).

7.9. Are the approximate eigenvalues calculated in Exercises 3.1–3.3 and 3.8–3.10 upper bounds on the exact answer? Why? Would the integral or collocation method have given results known to be upper bounds (i.e., known without knowledge of the exact result)? Formulate the variational principles for the asymptotic Nusselt numbers in Exercises 3.8 and 3.10.

REFERENCES

Atkinson, B., Brocklebank, M. P., Card, C. C. H., and Smith, J. M. (1969). Low Reynolds Number Developing Flows, *AIChE J.* **15**, 548–553.

Becker, M. (1964). "The Principles and Applications of Variational Methods." MIT Press, Cambridge, Massachusetts.

Becker, M. (1969). Nuclear Reactor Systems and Their Associated Variational and Optimization Problems, *in* "Computing Methods in Optimization Problems" (L. A. Zadeh, ed.), pp. 13–24. Academic Press, New York.

Bellman, R. (1962). Quasi-Linearization and Upper and Lower Bounds for Variational Principles, *Quart. Appl. Math.* **19**, 349–350.

Collatz, L. (1948). "Eigenwertprobleme und ihre Numerische Behandlung." Chelsea, New York.

Courant, R. (1943). Variational Methods for the Solution of Problems of Equilibrium and Vibrations, *Bull. Amer. Math. Soc.* **49**, 1–23.

Courant, R., and Hilbert, D. (1953). "Methods of Mathematical Physics," Vol I. Wiley (Interscience), New York.

Crandall, S. H. (1956). "Engineering Analysis." McGraw-Hill, New York.

Delleur, J., and Sooky, A. (1961). Variational Methods in Fluid Dynamics, *Proc. Amer. Soc. Civ. Eng.* **87**, EM6, 57–77.

Denn, M. M. (1969). "Optimization by Variational Methods." McGraw-Hill, New York.

Dennis, S. C. R., Mercer, A. McD., and Poots, G., (1959) Forced Heat Convection in Laminar Flow through Rectangular Ducts, *Quart. Appl. Math.* **17**, 285–297.

Erle, M. A., Dyson, D. C., and Gillette, R. D. (1970). Stability of Interfaces of Revolution with Constant Surface Tension—the Case of the Catenoid, *Chem. Eng. J.* **1**, 97–109.

Erle, M. A., Dyson, D. C., and Morrow, N. R. (1971). Liquid Bridges between Cylinders, in a Torus, and between Spheres, *AIChE J.* **17**, 115–121.

Forray, M. J. (1968). "Variational Calculus in Science and Engineering." McGraw-Hill, New York.

Fox, D. W., Rheinboldt, W. C. (1966). Computational Methods for Determining Lower Bounds for Eigenvalues of Operators in Hilbert Space, *SIAM Rev.* **8**, 427–462.

Galerkin, B. G. (1915). Rods and Plates. Series in Some Problems of Elastic Equilibrium of Rods and Plates, *Vestnik. Inzh. Tech.* (*USSR*) **19**, 897–908. Translation 63–18924 available from Clearinghouse Fed. Sci. Tech. Info., Springfield, Virginia.

Gelfand, I. M., and Fomin, S. V. (1963). "Calculus of Variations." Prentice-Hall, Englewood Cliffs, New Jersey.

Gould, S. H. (1966). "Variational Methods for Eigenvalue Problems." 2nd ed. Univ. of Toronto Press, Toronto, Canada.

Greenspan, D. (1967). On Approximating Extremals of Functionals. II, Theory and Generalizations Related to Boundary-Value Problems for Nonlinear Differential Equations, *Int. J. Eng. Sci.* **5**, 571–588.

Greenspan, D., and Jain, P. (1967). Application of a Method for Approximating Extremals of Functionals to Compressible Subsonic Fluid Flows, *J. Math. Anal. Appl.* **18**, 85–111.

Harrington, R. F. (1968). "Field Computation by Moment Methods." Macmillan, New York.

Herrmann, G. (1963). On Variational Principles in Thermoelasticity and Heat Conduction, *Quart. Appl. Math.* **21**, 151–155.

Hirschfelder, J. O., Bird, R. B., and Curtiss, C. F. (1954). "Molecular Theory of Gases and Liquids." Wiley, New York.

Kaplan, S. (1969). Variational Methods in Nuclear Engineering, *Advan. Nucl. Sci. Tech.* **5**, 185–221.

Kauzmann, W. (1957). "Quantum Chemistry." Academic Press, New York.

Leitmann, G. (ed.) (1962). "Optimization Techniques." Academic Press, New York.

Lewins, J. (1965). "Importance; The Adjoint Function." Pergamon Press, Oxford.

Löwdin, P. O. (ed.) (1966). "Quantum Theory of Atoms, Molecules, and the Solid State." Academic Press, New York.

Lynn, L. L., Parkin, E. S., and Zahradnik, R. L. (1970). Near-Optimal Conrol by Trajectory Approximation, *Ind. Eng. Chem. Fund.* **9**, 58–62.

McLaurin, J. (1968). Bounding Eigenvalues of Clamped Plates, *Z. Angew. Math. Phys.* **19**, 676–681.

Mikhlin, S. G. (1964). "Variational Methods in Mathematical Physics." Pergamon, Oxford.

Moiseiwitsch, B. L. (1966). "Variational Principles." Wiley (Interscience), New York.

Moler, C. B., and Payne, L. E. (1968). Bounds for Eigenvalues and Eigenvectors of Symmetric Operators, *SIAM J. Num. Anal.* **5**, 64–70.

Petrov, I. P. (1968). "Variational Methods in Optimum Control Theory." Academic Press, New York.

Pnueli, D. (1967). A Computational Scheme for the Asymptotic Nusselt Number in Ducts of Arbitrary Cross Section, *Int. J. Heat Mass Transfer* **10**, 1743–1748.

Pnueli, D. (1969). Lower Bound to the *n*th Eigenvalue of the Helmholtz Equation over Two-Dimensional Regions of Arbitrary Shape, *J. Appl. Mech. Trans. ASME, Ser. E* **36**, 630–631.

Rayleigh, Lord (J. W. Strutt) (1871). On the Theory of Resonance, *Phil. Trans. Roy. Soc.* (*London*) **A161**, 77–118. Also (1964). "Scientific Papers," Vol I, §5, 1869–1881, Dover, New York.

Rayleigh, Lord (J. W. Strutt) (1873). Some General Theorems Relating to Vibrations, *Proc. London Math. Soc.* **4**, 357–368, also (1964). "Scientific Papers," Vol I, §21, 1869–1881. Dover, New York.

Rayleigh, Lord (J. W. Strutt) (1896). "The Theory of Sound," 2nd ed. Macmillan, London.

Rayleigh, Lord (1899). On the Calculation of the Frequency of Vibration of a System in Its Greatest Mode, with an Example from Hydrodynamics, *Phil. Mag.* [5] **47**, 566–572.

Rayleigh, Lord (1911). On the Calculation of Chladni's Figures for a Square Plate, *Phil. Mag.* [6] **22**, 225–229.

Ritz, W. (1908). Über eine Neue Methode zur Lösung gewisser Variationsprobleme der Mathematischen Physik, *J. Reine Angew. Math.* **135**, 1–61.

Ritz, W. (1909). Theorie der Transversalschwingungen einer Quadratischen Platte mit Freien Rändern, *Ann. Phys.* **28**, 737–786.

Sani, R. L. (1963). Dual Variational Statements Viewed from Function Space, *AIChE J.* **9**, 277–278.

Schechter, R. S. (1967). "The Variational Method in Engineering." McGraw-Hill, New York.

Slater, J. C. (1960) "Quantum Theory of Atomic Structure." Vol I. McGraw-Hill, New York.

Sparrow, E. M., and Siegel, R. (1959). A Variational Method for Fully Developed Laminar Heat Transfer in Ducts, *J. Heat Transfer, Trans. ASME, Ser. C* **81**, 157–167.

Sparrow, E. M., and Siegel, R. (1960). Applications of Variational Methods to Thermal Entrance Region of Ducts, *Int. J. Heat Mass Transfer* **1**, 161–172.

Stewart, W. E. (1962). Application of Reciprocal Variational Principles to Laminar Flow in Uniform Ducts. *AIChE J.* **8**, 425–428.

Trefftz, E. (1927). Ein Gegenstück zum Ritzschen Verfahren, *Proc. Int. Congr. Appl. Mech.*, 2nd *Zurich* pp. 131–137.

Villadsen, J. V., and Stewart, W. E. (1967). Solution of Boundary-Value Problems by Orthogonal Collocation, *Chem. Eng. Sci.* **22**, 1483–1501.

Weinstein, A. (1963). On the Sturm–Liouville Theory and the Eigenvalues of Intermediate Problems, *Num. Math.* **5**, 238–245.

Weinstein, D. H. (1934). Modified Ritz Method, *Proc. Nat. Acad. Sci.* (*U.S.*) **20**, 529–532.

Weinstock, R. (1952). "Calculus of Variations with Applications to Physics and Engineering." McGraw-Hill, New York.

Wexler, A. (1969). Computation of Electromagnetic Fields, *IEEE Trans. Microwave Theory Tech.* **MT T-17**, 416–439.

Williams, T. I. (ed.) (1969). "A Biographical Dictionary of Scientists." Adams and Charles Black, London.

Wilson, E. B., Jr. (1965). Lower Bounds for Eigenvalues, *J. Chem Phys.* **43**, S172–174.

Yasinsky, J. B. (1966). Extensions and Applications of Synthesis Methods, with Error Bounds, to Problems in Reactor Physics, Heat Conduction and Elasticity, Ph.D. Thesis, Carnegie Inst. Tech, Pittsburgh, Pennsylvania.

Yasinsky, J. B. (1968). Numerical Studies of Combined Space-Time Syntheses, *Nucl. Sci. Eng.* **34**, 158–168.

Yasinsky, J. B., and Kaplan S. (1967). Synthesis of Three-Dimensional Flux Shapes Using Discontinuous Sets of Trial Functions, *Nucl. Sci. Eng.* **28**. 426–437.

Chapter 8

Variational Principles in Fluid Mechanics

The variational principles governing equations in fluid mechanics are summarized below, including perfect fluids, magnetohydrodynamics, non-Newtonian fluids, and flow around suspended drops and particles. In Section 8.6 it is shown that there is no variational principle for the Navier–Stokes equation, and in Section 8.7 a variational principle is presented for the Reynold's number below which any solution of the Navier–Stokes equation is stable. There are many other areas governed by variational principles: the Rayleigh–Taylor instability of two superposed fluids of different densities (Chandrasekhar, 1961, Chapter 10; Selig, 1964). Many problems of hydrodynamic stability are governed by variational principles, specialized for the particular case (see Chandrasekhar, 1961, for example; also see Chapter 9 for the generation of variational principles for new problems). Water waves are governed by variational principles (Luke, 1967; Whitham, 1967; Simmons, 1969). Variational principles have been found for some problems in rarefied gas flow (Cercignani and Pagani, 1966; Su, 1970; Strieder, 1970). Rocket exhaust nozzles are designed for optimum thrust using variational principles (Rao, 1958; Fanselau, 1959; Guderley and Armitage, 1965; Hoffman, 1967; Hoffman and Thompson, 1967; Kraiko, 1964; 1967; Kraiko and Osipov, 1968).

8.1 Basic Equations

The equation underlying fluid mechanics is the Cauchy momentum equation (1.13):

$$\rho \frac{du_i}{dt} \equiv \rho\left(\frac{\partial u_i}{\partial t} + u_j u_{i,j}\right) = \rho \hat{F}_i + T_{ji,j}. \tag{8.1}$$

This vector equation must be augmented by a constitutive relation for the stress tensor, which depends on the material. For one type of non-Newtonian fluid,

$$T_{ji} = -p\,\delta_{ji} + g_{ji}(d), \tag{8.2}$$

where

$$d_{ji} = \tfrac{1}{2}(u_{i,j} + u_{j,i}) \tag{8.3}$$

is the rate of deformation. This simplifies for an incompressible, Newtonian fluid to

$$T_{ji} = -p\,\delta_{ji} + 2\mu\, d_{ji}. \tag{8.4}$$

For a perfect fluid there are no viscous terms,

$$T_{ji} = -p\,\delta_{ji}. \tag{8.5}$$

The equation of continuity is

$$\frac{\partial \rho}{\partial t} + (\rho u_i)_{,i} = 0, \tag{8.6}$$

and when the density is constant (for an incompressible fluid)

$$u_{i,i} = 0. \tag{8.7}$$

When the motion is irrotational,

$$\nabla \times \mathbf{u} = 0, \tag{8.8}$$

then the velocity can be represented in terms of a gradient,

$$\mathbf{u} = \nabla\phi. \tag{8.9}$$

If electromagnetic forces are important the force term in (8.1) becomes (Pai, 1962)

$$\rho\hat{\mathbf{F}} = \rho_e \mathbf{E} + \mu \mathbf{j} \times \mathbf{H}. \tag{8.10}$$

The equations must then be augmented by Maxwell's equations and an equation for the current density,

$$\nabla \times \mathbf{H} = \mathbf{j} + \frac{\partial \mathbf{D}}{\partial t}, \tag{8.11a}$$

$$\nabla \cdot \mathbf{D} = 0, \tag{8.11b}$$

$$\nabla \times \mathbf{E} = -\frac{\partial \mathbf{B}}{\partial t}, \tag{8.11c}$$

$$\nabla \cdot \mathbf{B} = 0, \tag{8.11d}$$

$$\mathbf{j} \equiv \mathbf{i} + \rho_e \mathbf{u}, \tag{8.11e}$$

where ρ_e is the electric charge density, $\mathbf{j}$ is the electric current density, $\mathbf{H}$ is the magnetic field, $\mathbf{B}$ is the magnetic flux density, $\mathbf{D}$ is the electric flux density, and $\mathbf{E}$ is the electric field intensity. Constitutive relations are also needed to relate

$$\mathbf{D} = \mathbf{D}(\mathbf{E}), \qquad \mathbf{B} = \mathbf{B}(\mathbf{H}) \tag{8.12}$$

as well as a relation for the current, $\mathbf{i}$.

The equations are written above in Eulerian form in terms of the velocity $\mathbf{u}(\mathbf{x}, t)$ as a function of position and time. If Lagrangian coordinates are used, the position of a particle is denoted by $\mathbf{x}(\mathbf{a}, t)$, where $\mathbf{a}$ is the position of the particle at $t = 0$. The equation of motion for a perfect fluid is

$$\rho\left(\frac{\partial^2 x_i}{\partial t^2}\right)_{\mathbf{a}} = -\frac{\partial p}{\partial x_i} + \rho \hat{F}_i. \tag{8.13}$$

The conservation of mass is

$$\rho(\mathbf{x}(\mathbf{a}, t))J = \rho_0 \equiv \rho(\mathbf{a}, 0), \qquad J = \frac{\partial(x_1, x_2, x_3)}{\partial(a_1, a_2, a_3)}. \tag{8.14}$$

The equations in this form are particularly useful for deriving variational principles, although both forms are used below.

Some of the variational principles are motivated by Hamilton's principle, which governs the equations of mechanics and dynamics of discrete particles. The variational principle is the following.

Make stationary the variational integral I subject to variations in the generalized coordinates q_r which vanish at t_1 and t_2:

$$I = \int_{t_1}^{t_2} L(q_r, q_r')\, dt, \qquad L = T - V. \tag{8.15}$$

T is the kinetic energy and V is the potential energy, which depends only on the generalized coordinates, q_r, not their derivatives, q_r'. The Euler equation is easily found from Eq. (7.10),

$$\frac{d}{dt}\left(\frac{\partial L}{\partial q_r'}\right) - \frac{\partial L}{\partial q_r} = 0. \tag{8.16}$$

The form of the Lagrangian—the kinetic energy minus the potential energy—is common to many of the variational principles described below. Finally, throughout this chapter the variational shorthand (7.14) is used.

8.2 Variational Principles for Perfect Fluids

Many attempts have been made to obtain the momentum equations from a variational principle patterned after Hamilton's principle, which is so powerful and useful in particle mechanics. These attempts have not all been successful except in the case of perfect fluids. As Truesdell and Toupin (1960, p. 595) point out, "the lines of thought which have led to beautiful variational statements for systems of mass-points have been applied in continuum mechanics also, but only rarely are the results beautiful or useful." We consider first the steady, irrotational flow of an incompressible fluid, then admit compressible fluids, and finally treat unsteady flow of compressible fluids. Some of this treatment is summarized in the elegant treatise by Serrin (1959a).

Consider the steady irrotational flow of an incompressible fluid. Assume the body force is a potential function, $\rho\hat{\mathbf{F}} = -\nabla\Phi$, and rearrange the convective terms in (8.1) to get

$$\begin{aligned} u_j u_{i,j} &= \phi_{,j}\phi_{,ij} = \tfrac{1}{2}(\phi_{,j}\phi_{,j})_{,i}, \\ \nabla(\tfrac{1}{2}\mathbf{u}\cdot\mathbf{u}) &= -\nabla\Phi - \nabla(p/\rho). \end{aligned} \tag{8.17}$$

The last relation gives a form of Bernoulli's theorem,

$$\tfrac{1}{2}\mathbf{u}\cdot\mathbf{u} + \Phi + \frac{p}{\rho} = \text{constant}, \tag{8.18}$$

which determines the pressure once the velocity is found. The velocity equations are then Eqs. (7.7) and (7.9), subject to a specified mass flux on the boundary:

$$\mathbf{u} = \nabla\phi \quad \text{in } V \tag{8.19a}$$

$$\text{or} \quad \nabla\times\mathbf{u} = 0 \quad \text{in } V \tag{8.19b}$$

$$\nabla\cdot\mathbf{u} = 0 \quad \text{in } V \tag{8.19c}$$

$$\rho\mathbf{n}\cdot\mathbf{u} = f \quad \text{on S.} \tag{8.19d}$$

Reciprocal variational principles are found using (8.19a) or (8.19c) as admissibility conditions and the other as the Euler equation and vice versa. They are special cases of Eqs. (7.48) and (7.59).

The first principle is due to Lord Kelvin (Thomson, 1849) and minimizes the kinetic energy, known in his day as *vis-viva*.

Among all motions of an incompressible fluid in V which have a specified mass flux on S, Eq. (8.19d), the irrotational motion has the least kinetic energy, T.

$$T(\mathbf{u}) = \tfrac{1}{2}\rho \int_v \mathbf{u} \cdot \mathbf{u} \, dV. \tag{8.20}$$

The stationary character is proved using a Lagrange multiplier. Consider the variational integral

$$I = \tfrac{1}{2}\rho \int_v \mathbf{u} \cdot \mathbf{u} \, dV + \rho \int \lambda \nabla \cdot \mathbf{u} \, dV. \tag{8.21}$$

Calculation of the first variation, and use of the divergence theorem gives

$$\delta I = \rho \int_v [\delta\mathbf{u} \cdot (\mathbf{u} - \nabla\lambda) + \delta\lambda \, \nabla \cdot \mathbf{u}] \, dV + \int_s \lambda\rho \mathbf{n} \cdot \delta\mathbf{u} \, dS. \tag{8.22}$$

The boundary term vanishes since the variations vanish on S, and the Euler equations are Eqs. (8.19a,c).

The minimum property is proved by considering a motion $\mathbf{u} = \nabla\phi + \mathbf{u}_0$, where ϕ satisfies (8.19) and $\mathbf{u}_0$ satisfies $\nabla \cdot \mathbf{u}_0 = 0$ in V and $\mathbf{n} \cdot \mathbf{u}_0 = 0$ on S. Then

$$T(\nabla\phi + \mathbf{u}) = \tfrac{1}{2}\rho \int_v (\nabla\phi \cdot \nabla\phi + 2\mathbf{u}_0 \cdot \nabla\phi + \mathbf{u}_0 \cdot \mathbf{u}_0) \, dV. \tag{8.23}$$

The divergence theorem is applied to the middle term

$$\int_v \mathbf{u}_0 \cdot \nabla\phi \, dV = \int_s \phi\mathbf{n} \cdot \mathbf{u}_0 \, dV - \int_v \phi\nabla \cdot \mathbf{u}_0 \, dV = 0, \tag{8.24}$$

which is zero by the conditions imposed on $\mathbf{u}_0$. Thus

$$T(\nabla\phi + \mathbf{u}_0) = T(\nabla\phi) + T(\mathbf{u}_0). \tag{8.25}$$

Since T is positive, the kinetic energy of any other permitted motion is greater than that corresponding to ϕ.

This principle is obtained by imbedding the problem in a vector space which is solenoidal but not irrotational. The reciprocal variational principle uses a vector space which is irrotational but not solenoidal [see (7.48)].

Among all irrotational motions in V, the one satisfying Eqs. (8.19c,d) maximizes the functional $J(\phi)$:

$$J(\phi) = -\tfrac{1}{2}\rho \int_v \nabla\phi \cdot \nabla\phi \, dV + \int_s \phi f \, dS. \tag{8.26}$$

The Euler equation is clearly $\nabla^2\phi = 0$ and the natural boundary condition is $\rho\mathbf{n} \cdot \nabla\phi = f$ on S. The maximum nature follows as a special case of Eq.

(7.48). The two variational principles thus give bounds on the kinetic energy:

$$J(\phi) \le \text{kinetic energy} \le T(\mathbf{u}). \tag{8.27}$$

These principles are not widely used, however, since few situations are governed by Eqs. (8.19).† Reciprocal variational principles are also valid for the compressible fluid, which is of more interest.

The first variational principle for the irrotational flow of a compressible fluid is due to Bateman (1929, 1930) [Serrin (1959a) calls it the Bateman–Dirichlet principle]. Hargreaves (1908) first showed that the pressure integral is a potential of the motion, although he did not use this fact to write a variational principle. The formulation presented here is due to Lush and Cherry (1956).

In a compressible fluid the density is no longer constant and an equation of state must be given to relate the pressure and density. The internal energy is a function of entropy and specific volume, but here we consider the case for constant entropy. Then the internal energy and thermodynamic pressure can be written as

$$dU = -p\,dV = \frac{p}{\rho^2}\,d\rho, \qquad p \equiv -\left(\frac{\partial U}{\partial V}\right)_S. \tag{8.28}$$

The U is specific internal energy, V is specific volume, and S is specific entropy, i.e., all quantities are per unit mass. The equation is integrated to give the internal energy function

$$U = \int_{\rho_0}^{\rho} \frac{p\,d\rho}{\rho^2} \tag{8.29}$$

once the $p(\rho)$ relation is specified. The momentum equation takes a particularly simple form—called the Bernoulli theorem. Take $\mathbf{u}\,\cdot$ into Eq. (8.1), assume a conservative force, a perfect fluid, Eq. (8.5), and add to it the energy equation, $\rho\mathbf{u}\cdot\nabla U = -T_{ji}u_{i,j}$:

$$\mathbf{u}\cdot\nabla\left(\tfrac{1}{2}u^2 + U + \Phi + \frac{p}{\rho}\right) = -\frac{p}{\rho^2}\,\mathbf{u}\cdot\nabla\rho - \frac{p}{\rho}\,\nabla\cdot\mathbf{u} = 0 \tag{8.30}$$

by the continuity equation. The term in parentheses is a constant, whose value is taken as zero. If $\Phi = 0$ and we use the thermodynamic relation $p = \rho^2\,dU/d\rho$ this is

$$\tfrac{1}{2}u^2 + \frac{d}{d\rho}(\rho U) = 0. \tag{8.31}$$

† One application is of historical interest, however, since it relates to the historical justification for naming the variational method after Rayleigh. See the footnote on p. 224.

This is the equation of motion for this problem.

Maximize the variational integral

$$J=\int p\,dV+\int \phi f\,dS \tag{8.32}$$

among all continuous functions ϕ satisfying the following equations:

$$p=\rho^2U', \qquad \tfrac{1}{2}\mathbf{u}\cdot\mathbf{u}+(\rho U)'=0, \qquad \mathbf{u}=\nabla\phi. \tag{8.33}$$

The prime refers to differentiation with respect to density, ρ. Here it is more convenient to regard these equations as constraints which the variations must satisfy than to use Lagrange multipliers. The first variation of J is

$$\delta J=\int \delta p\,dV+\int \delta\phi f\,dS. \tag{8.34}$$

The variations in p, ρ, and $\mathbf{u}$ are related by the variations of (8.33).

$$\begin{aligned}\delta p &= 2\rho U'\,\delta\rho+\rho^2U''\,\delta\rho=p'\,\delta\rho\equiv c^2\,\delta\rho,\\ \mathbf{u}\cdot\delta\mathbf{u} &= -2U'\,\delta\rho-\rho U''\,\delta\rho=-\delta p/\rho=-c^2\,\delta\rho/\rho,\end{aligned} \tag{8.35}$$

where $dp/d\rho=c^2$. The first variation is then

$$\delta J=-\int \rho\mathbf{u}\cdot\delta\mathbf{u}\,dV+\int \delta\phi f\,dS. \tag{8.36}$$

Write $\mathbf{u}=\nabla\phi$, $\delta\mathbf{u}=\nabla\delta\phi$, integrate by parts and apply the divergence theorem to obtain

$$\delta J=\int \delta\phi\,\nabla\cdot(\rho\mathbf{u})\,dV+\int \delta\phi(f-\rho\mathbf{n}\cdot\nabla\phi)\,dS, \tag{8.37}$$

which gives the Euler–Lagrange equation and natural boundary condition

$$\nabla\cdot(\rho\mathbf{u})=0 \quad \text{in } V, \qquad f-\rho\mathbf{n}\cdot\nabla\phi=0 \quad \text{on } S. \tag{8.38}$$

To examine the maximum property we calculate the second variation:

$$\delta^2J=-\int \rho\,\delta\mathbf{u}\cdot\delta\mathbf{u}\,dV-\int \delta\rho\mathbf{u}\cdot\delta\mathbf{u}\,dV. \tag{8.39}$$

Combination with (8.35) gives

$$\delta^2J=-\int \frac{\rho}{c^2}\sum_{i,j=1}^{3}(c^2\delta_{ij}-u_iu_j)\,\delta u_i\,\delta u_j\,dV. \tag{8.40}$$

When the flow is everywhere subsonic, that is, when $c^2>\mathbf{u}\cdot\mathbf{u}$, the second variation is negative definite for the extremal and the principle satisfies the necessary condition for a maximum principle.

The reciprocal variational principle is called the Bateman–Kelvin principle by Serrin (1959a). This principle was formulated by Lush and Cherry (1956) in terms of a stream function for two-dimensional flow. Here we present the principle for three-dimensional flow. The proof is given by Serrin (1959a).

The variational principle uses (8.38) as constraints and obtains Eq. (8.33c) as the Euler equation.

Minimize the integral

$$I(\mathbf{u}) = \int_v (p + \rho \mathbf{u} \cdot \mathbf{u})\, dV \tag{8.41}$$

among all velocity fields which satisfy

$$p = \rho^2 U', \qquad \tfrac{1}{2}\mathbf{u} \cdot \mathbf{u} + (\rho U)' = 0, \qquad \nabla \cdot (\rho \mathbf{u}) = 0 \quad \text{in } V,$$
$$\rho \mathbf{n} \cdot \mathbf{u} = f \quad \text{on } S. \tag{8.42}$$

The velocity fields which do so are irrotational Eq. (8.8).

The two variational principles (8.32) and (8.41) are reciprocal to each other and for the extremal the two variational integrals are equal to each other. This can be seen by setting $\mathbf{u} = \nabla \phi$ for one of the $\mathbf{u}$ in (8.41), integrating by parts, applying the divergence theorem and (8.42) to obtain (8.32). The variational principles thus provide upper and lower bounds on the sum of pressure and twice the kinetic energy:

$$J(\phi) \leq \int (p + \rho \mathbf{u} \cdot \mathbf{u})\, dV \leq I(\mathbf{u}). \tag{8.43}$$

Wang (1948, 1961) and Lush and Cherry (1956) have used the principle of Eq. (8.32) to calculate the flow past a circular airfoil and a thin curved surface. In these cases the flow is in an infinite region and the integrals do not exist. It is necessary to reformulate the variational integral as follows (for a two-dimensional region). The proposed extremal flow is taken as $\phi = Vx + \chi$, where Vx is the flow potential at infinity. Lush and Cherry then subtract from Eq. (8.32) an integral whose variation vanishes at infinity:

$$J(\phi) = \iint \left[p - p_\infty + V\rho_\infty \frac{\partial \chi}{\partial x} \right] dx\, dy + \int_{c_0} [\phi f - V\rho_\infty \chi \mathbf{n} \cdot \nabla x]\, ds. \tag{8.44}$$

In the numerical work, Wang (1948, 1961) restrict attention to compressible fluids with the equation of state $p\rho^{-\gamma} = \text{constant}$, with $\gamma/(\gamma - 1)$ an integer. This is done for convenience in calculating the integrals. Lush and Cherry (1956), however, treat more general values of γ using a polynomial approximation to calculate the integrals. This difficulty is avoided in the study by Wang and Rao (1950), who used the linearized equations for flow around a circular cylinder, ellipse, and Kaplan's bump, and by Wang and Brodsky (1949) who used Galerkin's method rather than a variational method. More recently Greenspan and Jain (1967) have applied the variational principle (8.32) and (8.44) using difference expressions for the derivatives appearing in the integrand, and maximizing the integral with respect to the potential at the nodes [see (7.90)].

Shiffman (1952) proves an extremal exists for the infinite region using the modified integral (8.44). Modifications are possible in certain cases: Lin and Rubinov (1948) derive a principle for isoenergetic flows, and this was later modified by Lin (1952) to include the equations written in terms of Crocco's stream function. Giese (1951) and Skobelkin (1957) also present variational principles when the velocity assumes a particular form in terms of stream functions.

Next consider the time-dependent equations, first in the Lagrangian formulation and then in the Eulerian formulation. The results for the Lagrangian form of the equations are due to Herivel (1955) and Eckart (1960). Consider the equation of motion (8.13), the conservation of mass (8.14), and the conservation of the entropy†:

$$(\partial S/\partial t)_{\mathbf{a}} = 0. \tag{8.45}$$

The pressure term in the momentum equation is rewritten as

$$\frac{\partial p}{\partial x_i} = \sum_j \frac{\partial p}{\partial a_j}\frac{\partial a_j}{\partial x_i} = \frac{1}{J}\sum_j J_{ij}\frac{\partial p}{\partial a_j} \tag{8.46}$$

and the forces are assumed to be conservative, $\hat{\mathbf{F}} = -\nabla\Phi$. The momentum equation is then

$$\left(\frac{\partial^2 x_i}{\partial t^2}\right)_{\mathbf{a}} = -\frac{1}{\rho_0}\sum_j J_{ij}\frac{\partial p}{\partial a_j} - \frac{\partial \Phi}{\partial x_i} \tag{8.47}$$

where J_{ij} is the cofactor of x_{ij} in J:

$$J_{ij} = \frac{\partial J}{\partial x_{ij}}, \qquad x_{ij} = \frac{\partial x_i}{\partial a_j}, \qquad \frac{\partial a_j}{\partial x_i} = \frac{J_{ij}}{J}. \tag{8.48}$$

The variational principal uses a Lagrangian which is the kinetic energy minus the potential and internal energy.

Make the variational integral stationary among functions $\mathbf{x}(\mathbf{a}, t)$ which have continuous second derivatives, and $\rho(\mathbf{a}, t)$ and $S(\mathbf{a}, t)$ which have continuous first derivatives, subject to the constraints (8.14) and (8.45).

† S is used to represent both entropy and the boundary of V. The usage is clear from the context.

S and $\mathbf{x}$ take prescribed values in V_a at $t = t_1$ and t_2 and $\mathbf{n}_a \cdot \mathbf{x}$ takes prescribed values on S for all t:

$$I = \int_{t_1}^{t_2} \int_{V_a} [\tfrac{1}{2}\rho\mathbf{u}\cdot\mathbf{u} - \rho(U + \Phi)]J \, d\mathbf{a} \, dt, \tag{8.49}$$

where V_a is a region in $\mathbf{a}$ space (i.e., fixed mass), with

$$\int_{\text{mass}} (\,)\, d\mathbf{x} = \int_{V_a} (\,)J \, d\mathbf{a}. \tag{8.50}$$

The constraints are handled using Lagrange multipliers. The first variation is

$$\delta \int_{t_1}^{t_2} \int_{V_a} \left\{[\tfrac{1}{2}\rho\mathbf{u}\cdot\mathbf{u} - \rho(U + \Phi)]J - \alpha[\rho J - \rho_0] - \beta\left(\frac{\partial S}{\partial t}\right)_{\mathbf{a}}\right\} d\mathbf{a}\, dt = 0. \tag{8.51}$$

The variations in density and entropy are elementary [use the definition of p in Eq. (8.28)]:

$$\begin{aligned} \delta\rho:&\quad \tfrac{1}{2}\mathbf{u}\cdot\mathbf{u} - (U + \Phi) - \frac{p}{\rho} - \alpha = 0, \\ \delta S:&\quad \frac{\partial\beta}{\partial t} = \rho_0\left(\frac{\partial U}{\partial S}\right)_\rho \equiv \rho_0 T. \end{aligned} \tag{8.52}$$

The variation in $\mathbf{x}$ is more complicated since J is an implicit function of $\mathbf{x}$:

$$dJ = \frac{\partial J}{\partial x_{ij}} dx_{ij}. \tag{8.53}$$

The variation is then

$$\delta J = \frac{\partial J}{\partial x_{ij}} \delta x_{ij} = \frac{\partial J}{\partial x_{ij}} \frac{\partial\, \delta x_i}{\partial a_j}. \tag{8.54}$$

In the variational integral this term is integrated by parts and the divergence theorem applied (with respect to $\mathbf{a}$) to give

$$\delta\mathbf{x}:\quad -\rho J\left(\frac{\partial^2 x_i}{\partial t^2}\right)_{\mathbf{a}} - \rho J \frac{\partial\Phi}{\partial x_i} - \frac{\partial}{\partial a_j}\left\{[\tfrac{1}{2}\rho\mathbf{u}\cdot\mathbf{u} - \rho(U + \Phi) - \rho\alpha]\frac{\partial J}{\partial x_{ij}}\right\} = 0. \tag{8.55}$$

Equation (8.52a) can be used to rewrite this as

$$\rho_0\left(\frac{\partial^2 x_i}{\partial t^2}\right)_{\mathbf{a}} + \rho_0 \frac{\partial\Phi}{\partial x_i} = -\frac{\partial}{\partial a_j}(pJ_{ij}) = -J_{ij}\frac{\partial p}{\partial a_j} - p\frac{\partial}{\partial a_j}(J_{ij}). \tag{8.56}$$

The last term vanishes, as may be seen by expanding it, so that we obtain the equation of motion (8.47).

The variational principle for the Eulerian formulation is more difficult and can be approached in two different ways. A Lagrangian may be assumed, as in (8.49), which is varied subject to the constraints: equation of continuity, entropy balance, etc. An alternative is to use a representation for the velocity and choose a Lagrangian which automatically incorporates some of the constraints. We choose the former approach here, while Seliger and Whitham (1968) illustrate the second approach. In the next section on magnetohydrodynamics the representation of magnetic field, electric field, etc. is more important. See Bretherton (1970) for an interpretation of the geometric coordinates, α, β, and γ in (8.60) and the Lin constraint (8.58b).

The variational principle was first presented by Herivel (1955). Herivel's treatment, however, allowed only irrotational flows. A modification, attributed to C. C. Lin, permits more general flows, as shown by Serrin (1959a). Drobot and Rybarski (1959) also use the same modification. A less general principle due to Eckart (1938) applies to irrotational flow when the internal energy depends only on density. Taub (1949) presents a principle (without using Lin's constraint) in which the temperature replaces entropy as a primary variable. Stephens (1967) presents variational principles for the canonical form of the energy-momentum tensor and Penfield (1966) treats the equations in relativistic form.

The equation of motion is (8.1) for a perfect fluid, and we assume the body force is conservative:

$$\frac{du_i}{dt} = \frac{-1}{\rho}\frac{\partial p}{\partial x_i} - \frac{\partial \Phi}{\partial x_i}. \tag{8.57}$$

In addition we have the equation of continuity (8.6), the entropy conservation relation (8.45) and the additional equation, added by Lin,

$$dS/dt = 0, \tag{8.58a}$$

$$d\mathbf{a}/dt = 0. \tag{8.58b}$$

The vector $\mathbf{a}(\mathbf{x}, t)$ establishes the initial position of the particle which is at position $\mathbf{x}$ at time t. The variational principle is as follows.

Make the variational integral stationary among functions $\mathbf{u}$, ρ, S which have continuous first derivatives in space and time, take prescribed values in V for $t = t_1$ and t_2, $\mathbf{n} \cdot \mathbf{u}$, ρ, and S take prescribed values on S for all t, subject to the constraints (8.14) and (8.58):

$$I = \int_{t_1}^{t_2} \int_V [\tfrac{1}{2}\rho \mathbf{u} \cdot \mathbf{u} - \rho(U + \Phi)]\, dV\, dt. \tag{8.59}$$

Introduce Lagrange multipliers and make the integral (8.59) stationary:

$$\delta \int_{t_1}^{t_2} \int_V \left\{ \tfrac{1}{2}\rho \mathbf{u} \cdot \mathbf{u} - \rho(U + \Phi) - \alpha\left[\frac{\partial \rho}{\partial t} + \nabla \cdot (\rho \mathbf{u})\right] - \beta\rho \frac{dS}{dt} - \rho\boldsymbol{\gamma} \cdot \frac{d\mathbf{a}}{dt}\right\} dV\, dt = 0. \tag{8.60}$$

Variations of α, β, and γ give the constraints,

$$\frac{\partial \rho}{\partial t} + \nabla \cdot (\rho \mathbf{u}) = 0, \qquad \frac{dS}{dt} = 0, \qquad \frac{d\mathbf{a}}{dt} = 0. \tag{8.61}$$

Variations of the other quantities leads to

$$\delta \mathbf{u}: \quad \mathbf{u} + \nabla\alpha - \beta\nabla S - \boldsymbol{\gamma} \cdot \nabla \mathbf{a} = 0, \tag{8.62a}$$

$$\delta \rho: \quad \tfrac{1}{2}\mathbf{u} \cdot \mathbf{u} - (U + \Phi) - \frac{p}{\rho} + \frac{d\alpha}{dt} - \beta\frac{dS}{dt} - \boldsymbol{\gamma} \cdot \frac{d\mathbf{a}}{dt} = 0, \tag{8.62b}$$

$$\delta S: \quad -\rho T + \rho \frac{d\beta}{dt} = 0, \tag{8.62c}$$

$$\delta \mathbf{a}: \quad \frac{d\boldsymbol{\gamma}}{dt} = 0, \tag{8.62d}$$

where we remember that

$$\left(\frac{\partial U}{\partial \rho}\right)_S = \frac{p}{\rho^2}, \qquad \left(\frac{\partial U}{\partial S}\right)_\rho = T, \qquad \frac{d}{dt} = \frac{\partial}{\partial t} + \mathbf{u} \cdot \nabla. \tag{8.63}$$

When we ignore the constraint (8.58b), as Herivel did, the velocity must be irrotational in any isentropic motion.

To obtain the equations of motion we write (8.62a) in the form

$$\mathbf{u} = \sum_\alpha \xi_\alpha \nabla \eta_\alpha \tag{8.64}$$

and evaluate $d\mathbf{u}/dt = \partial \mathbf{u}/\partial t + \mathbf{u} \cdot \nabla \mathbf{u}$:

$$\frac{\partial u_i}{\partial t} + u_j u_{i,j} = \sum_\alpha \left\{ \frac{d\xi_\alpha}{dt} \eta_{\alpha,i} + \xi_\alpha \left(\frac{d\eta_\alpha}{dt}\right)_{,i} - \xi_\alpha u_{i,j} \eta_{\alpha,i} \right\}. \tag{8.65}$$

The last term is $u_i u_{i,j} = \frac{1}{2}(u_i u_i)_{,j}$. Substitution for ξ_α and η_α, using (8.61) and (8.62) gives

$$\frac{d\mathbf{u}}{dt} = -\nabla\left(U + \Phi + \frac{p}{\rho}\right) + T\nabla S. \tag{8.66}$$

This becomes the equation of motion (8.57) after using the thermodynamic relation $dU = T\,dS + pd(1/\rho)$. This shows that every extremal satisfies the equations of motion. Serrin (1959) shows that every flow is an extremal as well. As Seliger and Whitham (1968) point out the expanded Lagrangian (8.60) is just the pressure. This can be seen by integrating the α-term by parts and applying (8.62b). The pressure, of course, occurs in the Lagrangian for the steady state (8.32).

We summarize then: A stationary principle (8.59) exists for the most general case of a compressible perfect fluid in unsteady motion. If the motion is steady and irrotational we have reciprocal variational principles, (8.32) and (8.41), for a compressible fluid and (8.20) and (8.26) for incompressible fluids, which give upper and lower bounds to the variational integral.

8.3 Magnetohydrodynamics

We extend the previous section to include electromagnetic phenomena, treating first Maxwell's equations, then magnetohydrodynamics in various simplifications: magnetostatic case, steady state with flow, and unsteady state with flow.

Maxwell's equations for free space are a simplification of (8.11). Write

$$\mathbf{D} = \varepsilon_0\,\mathbf{E}, \tag{8.67a}$$

$$\mathbf{B} = \mu_0\,\mathbf{H}, \tag{8.67b}$$

$$\mathbf{j} = 0, \tag{8.67c}$$

where ε_0 and μ_0 are the dielectric constant and permeability of free space, and $1/\varepsilon_0\,\mu_0 = c^2$, where c is the velocity of light. Then (8.11a–d) can be rearranged to give

$$c^2\,\nabla \times \mathbf{B} = \partial\mathbf{E}/\partial t, \tag{8.68a}$$

$$\nabla \cdot \mathbf{E} = 0, \tag{8.68b}$$

$$\nabla \times \mathbf{E} = -\partial\mathbf{B}/\partial t, \tag{8.68c}$$

$$\nabla \cdot \mathbf{B} = 0. \tag{8.68d}$$

We present a variational principle for these equations following Seliger and Whitham (1968). Penfield and Haus (1966) treat the more general case for time-dependent phenomena with motion and more general relations than (8.67), and Katz (1961) and Seliger and Whitham (1968) treat a collisionless plasma, in which the charged particles do not collide but interact with the electromagnetic fields. See also Merches (1969) who uses different electromagnetic potentials than arise in (8.71a,b). The first variational principle for (8.68) is the following.

Make the variational integral stationary to variations in **E** and **B** which have continuous first time and spatial derivatives, $\delta\mathbf{E} = 0$ in V at $t = t_1$ and t_2, $\mathbf{n} \cdot \delta\mathbf{E} = 0$ and $\mathbf{n} \times \delta\mathbf{B} = 0$ on S for all t, subject to the restrictions (8.68a,b):

$$I(\mathbf{E}, \mathbf{B}) = \int_{t_1}^{t_2} \int_V \frac{\varepsilon_0}{2} (c^2\mathbf{B} \cdot \mathbf{B} - \mathbf{E} \cdot \mathbf{E})\, dV\, dt. \tag{8.69}$$

Use Lagrangian multipliers **A** and ϕ to form the expanded variational integral:

$$\delta \int_{t_1}^{t_2} \int_V \left[\frac{\varepsilon_0}{2} (c^2\mathbf{B} \cdot \mathbf{B} - \mathbf{E} \cdot \mathbf{E}) + \varepsilon_0 \mathbf{A} \cdot \left(\frac{\partial \mathbf{E}}{\partial t} - c^2 \nabla \times \mathbf{B}\right) + \varepsilon_0 \phi \nabla \cdot \mathbf{E}\right] dV\, dt = 0. \tag{8.70}$$

The variations are easily seen to be

$$\delta\mathbf{E}: \quad \mathbf{E} + \partial\mathbf{A}/\partial t + \nabla\phi = 0, \tag{8.71a}$$

$$\delta\mathbf{B}: \quad \mathbf{B} - \nabla \times \mathbf{A} = 0, \tag{8.71b}$$

$$\delta\mathbf{A}: \quad \partial\mathbf{E}/\partial t - c^2 \nabla \times \mathbf{B} = 0, \tag{8.71c}$$

$$\delta\phi: \quad \nabla \cdot \mathbf{E} = 0, \tag{8.71d}$$

where we have used the vector identity

$$\mathbf{A} \cdot \nabla \times \mathbf{B} = -\nabla \cdot (\mathbf{A} \times \mathbf{B}) + \mathbf{B} \cdot \nabla \times \mathbf{A}. \tag{8.72}$$

The Euler equations (8.71a,b) are the usual representation for electric field intensity and magnetic flux density. If this representation is employed from the start we have the variational principle as follows.

Make the variational integral stationary to variations in **A** which have continuous second derivatives, take prescribed values in V at $t = t_1$ and t_2 and on S for all t, and variations in ϕ, which have continuous first time derivatives and second spatial derivatives and vanish on S for all t:

$$I(\mathbf{A}, \phi) = \int_{t_1}^{t_2} \int_V \frac{\varepsilon_0}{2} \left[c^2(\nabla \times \mathbf{A})^2 - \left(\frac{\partial \mathbf{A}}{\partial t} + \nabla\phi\right)^2\right] dV\, dt. \tag{8.73}$$

The Euler equations are

$$\frac{\partial \nabla \cdot \mathbf{A}}{\partial t} + \nabla^2\phi = 0, \qquad \frac{\partial^2 \mathbf{A}}{\partial t^2} + \frac{\partial}{\partial t} \nabla\phi + c^2 \nabla \times (\nabla \times \mathbf{A}) = 0, \tag{8.74}$$

which are the combined form of (8.71).

Next we consider magnetohydrodynamics of plasmas, assuming displacement currents are small [$\mathbf{D} = 0$ in (8.11)], no free charges ($\rho_e = 0$), Eq. (8.67b), and a perfectly conducting fluid. The generalized Ohm's law is

$$\mathbf{i} = \sigma(\mathbf{E} + \mathbf{u} \times \mathbf{B}), \qquad \mathbf{E} + \mathbf{u} \times \mathbf{B} = \frac{\mathbf{i}}{\sigma} = \frac{\mathbf{j}}{\sigma} = \frac{\nabla \times \mathbf{H}}{\sigma}, \tag{8.75}$$

and in a perfectly conducting fluid $\sigma \to \infty$ so that the right-hand side vanishes. Then the equations are

$$\rho \frac{d\mathbf{u}}{dt} = -\nabla p + \mathbf{j} \times \mathbf{B} \tag{8.76a}$$

$$\frac{d\rho}{dt} = -\rho \nabla \cdot \mathbf{u}, \tag{8.76b}$$

$$\frac{dS}{dt} = 0, \tag{8.76c}$$

$$\nabla \times \mathbf{E} + \partial \mathbf{B}/\partial t = 0, \tag{8.76d}$$

$$\nabla \cdot \mathbf{B} = 0, \tag{8.76e}$$

$$\mathbf{E} + \mathbf{u} \times \mathbf{B} = 0, \tag{8.76f}$$

$$\nabla \times \mathbf{B} \equiv \mu_0 \mathbf{j}. \tag{8.76g}$$

For the static case the equations reduce to

$$\nabla p = \mathbf{j} \times \mathbf{B}, \tag{8.77a}$$

$$\nabla \times \mathbf{B} \equiv \mu_0 \mathbf{j}, \tag{8.77b}$$

$$\nabla \cdot \mathbf{B} = 0. \tag{8.77c}$$

We present the treatment due to Grad and Rubin (1958). Other treatments by Kruskal and Kulsrud (1958) and Bernstein *et al.* (1958) were published at the same time. See also Greene *et al.* (1962). Grad and Rubin show that if $\mathbf{B}$ is solenoidal, it can be represented by

$$\mathbf{B} = \nabla \phi \times \nabla \psi \tag{8.78}$$

provided the region of space is simply covered by $\mathbf{B}$ lines which intersect on a transverse surface S and ϕ is constant on the $\mathbf{B}$ lines. Equation (8.77a) also implies that $\mathbf{B} \cdot \nabla p = 0$, so that we can set $\phi = p$. Using this representation we arrive at the variational principle:

Make the variational integral stationary to variations in functions p, ψ which have continuous first derivatives and take prescribed values of the boundary S:

$$I(p, \phi) = \int_V \left[\frac{1}{2\mu_0}(\nabla p \times \nabla\psi)^2 - p\right] dV. \tag{8.79}$$

The variation then gives (using $\mathbf{j} \equiv \nabla \times \mathbf{B}/\mu_0$)

$$\delta I = \int_V [\delta p(\mathbf{j} \cdot \nabla\psi - 1) - \delta\psi\, (\mathbf{j} \cdot \nabla p)]\, dV = 0, \tag{8.80}$$

which gives the Euler equations. These can be combined to give

$$\nabla p = \nabla p \mathbf{j} \cdot \nabla\psi - \nabla\psi \mathbf{j} \cdot \nabla p = \mathbf{j} \times \mathbf{B}. \tag{8.81}$$

For the more general case including flow but still considering steady state the appropriate equations are (8.76) with the convected time derivative, d/dt replaced by $\mathbf{u} \cdot \nabla$. We present a variational principle for these equations following the treatment of Greene and Karlson (1969). See Woltjer (1958) and Frieman and Rotenberg (1960) for other variational principles for the same problem. Equation (8.76d) for steady state and (8.76f) give

$$\mathbf{E} = -\nabla\phi, \tag{8.82a}$$

$$\mathbf{B} \cdot \nabla\phi = 0. \tag{8.82b}$$

This means that the field lines must lie on constant potential surfaces which are also magnetic surfaces and are a set of nested toroids. The centerline of the set is called a magnetic axis. Following Kruskal and Kulsrud (1958), we introduce two new coordinate functions θ and η. The continuous function η increases by unity in one traversal around the (bounded) system along the magnetic axis, and the function θ, which is continuous everywhere except at the magnetic axis, increases by unity in one traversal looping that axis. We then take the coordinates ϕ, θ, and η to be right-handed.

For each magnetic surface define the following integrals, integrated over the interior of the magnetic surface identified by ϕ:

$$\begin{aligned} \psi(\phi) &= \int \mathbf{B} \cdot \nabla\eta\, dV, & \chi(\phi) &= \int \mathbf{B} \cdot \nabla\theta\, dV, \\ V(\phi) &= \int \rho\mathbf{u} \cdot \nabla\eta\, dV, & W(\phi) &= \int \rho\mathbf{u} \cdot \nabla\theta\, dV. \end{aligned} \tag{8.83}$$

The integrals represent the magnetic flux in two directions and the corresponding mass fluxes.

The variational integral is suggested by the form of the Herivel–Lin principle, namely, the integrand should be the kinetic energy minus the sum of internal, potential and magnetic energy. Other equations must be added as constraints. We consider then the variational integral

$$L = \int \left[\frac{1}{2}\rho u^2 - \frac{1}{2\mu_0} B^2 - \rho U\right] dV, \tag{8.84}$$

which is to be made stationary among functions satisfying the following constraints:

$$\begin{gathered} \nabla \cdot \mathbf{B} = 0, \qquad \nabla \cdot (\rho \mathbf{u}) = 0, \\ \phi \text{ has toroidal level surfaces,} \\ \phi = C \text{ at wall,} \qquad \min \phi = 0, \qquad \max \phi = C, \\ \mathbf{u} \times \mathbf{B} = \nabla\phi, \qquad \psi = \psi_0(\phi), \qquad \chi = \chi_0(\phi), \qquad V = V_0(\phi), \\ W = W_0(\phi), \qquad S = S_0(\phi), \end{gathered} \tag{8.85}$$

where the fluxes ψ, χ, V, and W are defined by (8.83) and the subscript zero means they are prescribed functions of the potential. Equation (8.76c) means the entropy is constant along a streamline and (8.76f) and (8.82b) makes the streamlines lie on the magnetic surfaces. Then we can write $S = S_0(\phi)$ which is used in the above constraints.

To verify the variational principle the constraints are absorbed into the notation following Greene and Karlson (1969) and Kruskal and Kulsrud (1958). The most general magnetic field and velocity satisfying conditions (8.85) is

$$\mathbf{B} = \nabla\phi \times \nabla\nu_B, \qquad \rho\mathbf{u} = \nabla\phi \times \nabla\nu_v, \tag{8.86}$$

where $\nu_B = \lambda_B - \eta\chi'(\phi) + \theta\psi'(\phi)$ and $\nu_v = \lambda_v - \eta W'(\phi) + \theta V'(\phi)$. If these equations are substituted into $\mathbf{u} \times \mathbf{B} = \nabla\phi$ we obtain

$$\rho = \nabla\phi \cdot (\nabla\nu_v \times \nabla\nu_B). \tag{8.87}$$

This last equation is included in the Lagrangian by means of a Lagrange multiplier. The variational principle is then to make stationary the Lagrangian

$$\begin{aligned} L^+ = \int \Bigg\{ & \frac{1}{2\rho} (\nabla\phi \times \nabla\nu_v)^2 - \frac{1}{2\mu_0} (\nabla\phi \times \nabla\nu_B)^2 - \rho U(\rho, S) \\ & + G[\rho - \nabla\phi \cdot (\nabla\nu_v \times \nabla\nu_B)]\Bigg\} \, dV, \end{aligned} \tag{8.88}$$

subject to $S = S_0(\phi)$, and $\phi = C$ at the wall, $\min \phi = 0$, $\max \phi = C$. The details of the verification of the variational principle are given by Greene and Karlson (1969).

Wenger (1970) presents a variational principle for magnetohydrodynamic channel flow including viscous effects, and the principle is considerably simpler than the one given here due to the simplifications, which are possible in channel flow.

For the time-dependent problem, the Lagrangian density is similar to that used by Herivel and proper formulations using Lin's constraint have been given by Lundgren (1963), and Penfield and Haus (1966), Cotsaftis (1962), and Newcomb (1962). Lundgren gives a variational principle for the equations in Lagrangian form when a perfectly conducting, inviscid, compressible fluid is bounded partly by a stationary, rigid, perfect conductor and partly by a vacuum. When he treats the Eulerian form of the equations he considers a simpler boundary condition, that of a fixed boundary on which all variations vanish. The variational principle is the following.

Make I stationary to variations in ρ, $\mathbf{u}$, U, $\mathbf{B}$ subject to the constraints (8.76b,c) and

$$\frac{d\mathbf{a}}{dt} = 0, \qquad \nabla \cdot \mathbf{B} = 0,$$

$$\frac{\partial \mathbf{B}}{\partial t} - \nabla \times (\mathbf{u} \times \mathbf{B}) = 0, \tag{8.89}$$

$$I \equiv \int_{t_1}^{t_2} \int [\tfrac{1}{2}\rho u^2 - \rho U - B^2/2\mu_0]\, dV\, dt.$$

If Lagrange multipliers are introduced [adding $-\mathbf{h} \cdot (\partial \mathbf{B}/\partial t - \nabla \times (\mathbf{u} \times \mathbf{B})) - B^2/2\mu_0$ to (8.60)] the variations are given by

$$\begin{aligned}
\delta\rho&: \quad \tfrac{1}{2}u^2 - (U + p/\rho) + d\alpha/dt - \beta\, dS/dt - \gamma \cdot d\mathbf{a}/dt = 0, \\
\delta S&: \quad d\beta/dt = T, \\
\delta \mathbf{a}&: \quad d\gamma/dt = 0, \\
\delta \mathbf{u}&: \quad \mathbf{u} = -\nabla\alpha + \beta\, \nabla S + \gamma \cdot \nabla \mathbf{a} - (\mathbf{B}/\rho) \times (\nabla \times \mathbf{h}), \\
\delta \mathbf{B}&: \quad \partial \mathbf{h}/\partial t = \mathbf{u} \times (\nabla \times \mathbf{h}) + \mathbf{B}/\mu_0 .
\end{aligned} \tag{8.90}$$

These equations can be rearranged to give the correct momentum balance, (8.76a); see the article by Lundgren for details.

8.4 Non-Newtonian Fluids

The equations for a perfect fluid have no viscous terms but include inertial terms. Next we include the viscous terms, omitting the inertial terms. Consider non-Newtonian fluid in steady flow†:

$$T_{ji,j} - p_{,i} + \rho F_i = 0, \qquad u_{i,i} = 0. \tag{8.91}$$

† Note that now, in contrast to (8.1), the symbol T_{ji} does not include the pressure term $-p\, \delta_{ji}$.

The constitutive relation for the stress tensor is taken as

$$T_{ji} = \partial\Gamma(d_{pq}, x_r)/\partial d_{ij}, \tag{8.92}$$

where Γ is a given, symmetric function of the rate of deformation tensor d_{pq} and x_r, and the stress tensor is taken as symmetric throughout, even when (8.91) is an Euler equation. The most general relation between the stress and rate of deformation is given by the Reiner–Rivlin constitutive relation

$$T_{ji} = G_1 d_{ij} + G_2 d_{ik} d_{kj}, \tag{8.93}$$

where G_1 and G_2 are arbitrary functions of the invariants

$$\text{II} \equiv d_{ij}d_{ij}, \qquad \text{III} \equiv d_{ij}d_{jk}d_{ki}. \tag{8.94}$$

The boundary conditions are: the velocity is specified on S_u and the surface traction is specified on S_t, where $S = S_u \cup S_t$:

$$u_i = f_i \quad \text{on } S_u, \qquad T_{ji}n_j - pn_i = g_i \quad \text{on } S_t. \tag{8.95}$$

The variational principle is presented in a form similar to that used by Johnson (1960, 1961; see also Hill, 1956). Consider the functional

$$J_0 = \int \{[\tfrac{1}{2}(u_{i,j} + u_{j,i}) - d_{ij}]T_{ji} - \rho F_i u_i + \Gamma - pu_{i,i}\}\, dV$$
$$- \int_{S_u} (T_{ji}n_j - pn_i)(u_i - f_i)\, dS - \int_{S_t} g_i u_i\, dS. \tag{8.96}$$

The variation of J_0 with respect to all variables, u_i, d_{ij}, T_{ij}, p, gives as Euler equations the terms in brackets, which are just (8.91)–(8.92), along with natural boundary conditions (8.95):

$$\delta J_0 = \int \{[\tfrac{1}{2}(u_{i,j} + u_{j,i}) - d_{ij}]\, \delta T_{ji} - [T_{ji,j} + \rho F_i - p_{,i}]\, \delta u_i$$
$$+ (-T_{ji} + \partial\Gamma/\partial d_{ij})\, \delta d_{ij} - u_{i,i}\, \delta p\}\, dV$$
$$- \int_{S_u} (\delta T_{ji}\, n_j - \delta p\, n_i)(u_i - f_i)\, dS + \int_{S_t} \delta u_i [T_{ji}n_j - pn_i - g_i]\, dS. \tag{8.97}$$

An alternate variational principle can be obtained by inverting (8.92):

$$d_{ij} = \partial\hat{\Gamma}(T_{pq}, x_r)/\partial T_{ji}. \tag{8.98}$$

Then the Legendre transformation exists

$$\Gamma(d_{pq}, x_r) - d_{ij}T_{ji} = -\hat{\Gamma}(T_{pq}, x_r) \tag{8.99}$$

and the variational integral takes the form

$$I_0 = \int \{\tfrac{1}{2}(u_{i,j} + u_{j,i})T_{ji} - \rho F_i u_i - \hat{\Gamma} - p u_{i,i}\}\, dV$$

$$- \int_{S_u} (T_{ji} n_j - p n_i)(u_i - f_i)\, dS - \int_{S_t} g_i u_i\, dS. \tag{8.100}$$

The variation of I gives the Euler equations and natural boundary conditions in brackets:

$$\delta I_0 = \int \{[\tfrac{1}{2}(u_{i,j} + u_{j,i}) - \partial\hat{\Gamma}/\partial T_{ji}]\, \delta T_{ji} - [T_{ji,j} + \rho F_i - p_{,i}]\, \delta u_i - u_{i,i}\, \delta p\}$$

$$\times\, dV - \int_{S_u} (\delta T_{ji}\, n_j - \delta p n_i)(u_i - f_i)\, dS + \int_{S_t} \delta u_i [T_{ji} n_j - p n_i - g_i]\, dS. \tag{8.101}$$

A variational principle in terms of the velocity is obtained by imposing certain equations as admissible conditions.

Make $J(u)$ stationary among all vector functions $\mathbf{u} \in C^2$ subject to the restrictions (8.91b), (8.92), (8.93), and (8.95a):

$$J(\mathbf{u}) = \int \{\Gamma - \rho F_i u_i\}\, dV - \int_{S_t} g_i u_i\, dS. \tag{8.102}$$

The first variation gives the Euler equation, (8.91a) and the natural boundary conditions (8.95b). The second variation is

$$\delta^2 J = \frac{\varepsilon^2}{2} \int \frac{d^2\Gamma}{d\varepsilon^2}\bigg|_{\varepsilon=0} dV = \frac{1}{2} \int \frac{\partial^2 \Gamma(\bar{d}_{pq}, x_r)}{\partial d_{ij}\, \partial d_{kl}}\, \delta d_{ij}\, \delta d_{kl}\, dV. \tag{8.103}$$

When the quadratic form in the integral is positive at the extremum $\bar{\mathbf{u}}$, the principle exhibits a local minimum. When the quadratic form is positive for all arguments then the functional $J(\mathbf{u})$ attains an absolute minimum for the extremal.

A reciprocal variational principle can be obtained from the alternate functional I_0.

Make $I(\mathbf{T})$ stationary to tensor functions $\mathbf{T} \in C^1$ subject to the restrictions (8.91a) and (8.95b):

$$I(\mathbf{T}) = -\int \hat{\Gamma}\, dV + \int_{S_u} (T_{ji} n_j - p n_i) f_i\, dS. \tag{8.104}$$

The first variation is found using a Lagrange multiplier:

$$I = -\int \hat{\Gamma}\, dV + \int \gamma_i [T_{ji,j} - p_{,i} + \rho F_i]\, dV + \int_{S_u} (T_{ij} n_j - p n_i) f_i\, dS. \tag{8.105}$$

The first variation then gives Euler equations (8.91a) and

$$-\tfrac{1}{2}(\gamma_{i,j}+\gamma_{j,i})=\frac{\partial\hat{\Gamma}}{\partial T_{ji}},\qquad \gamma_i+f_i=0\quad\text{on } S_v. \tag{8.106}$$

The second variation gives, for the extremal,

$$\delta^2 I=-\frac{1}{2}\int\frac{\partial^2\hat{\Gamma}}{\partial T_{ji}\,\partial T_{kl}}\,\delta T_{ji}\,\delta T_{kl}\,dV. \tag{8.107}$$

When the term in the integral is positive then the functional achieves a local maximum. When the quadratic form is positive for all T_{pq} then the principle is an absolute maximum principle. Under conditions that J and I give minimum and maximum principles, respectively, upper and lower bounds are obtained,

$$I(\mathbf{T})\le I(\bar{\mathbf{T}})=J(\bar{\mathbf{u}})\le J(\mathbf{u}). \tag{8.108}$$

The equality holds for the extremals, because then $I_0=J_0$ [Eqs. (8.96), (8.99), and (8.100)].

Consider next several specific constitutive relations, Eq. (8.92). For the Newtonian fluid

$$T_{ij}=2\mu d_{ij}. \tag{8.109}$$

The functions Γ and $\hat{\Gamma}$ and their derivatives are given by

$$\Gamma=\mu d_{ij}d_{ij},\qquad \hat{\Gamma}=\frac{1}{4\mu}T_{ij}T_{ij},$$
$$\frac{\partial^2\Gamma}{\partial d_{ij}\,\partial d_{kl}}=2\mu\,\delta_{ik}\,\delta_{jl},\qquad \frac{\partial^2\hat{\Gamma}}{\partial T_{ij}\,\partial T_{kl}}=\frac{1}{2\mu}\delta_{ik}\,\delta_{jl}. \tag{8.110}$$

Thus the principles given above represent maximum and minimum principles for a Newtonian fluid. When the forces are derivable from a potential, $F_i=\Phi_{,i}$ and $S_t=0$, the functional J can be given a special interpretation:

$$J(\mathbf{u})=\int\mu\,d_{ij}\,d_{ij}\,dV-\int_S\rho u_i n_i\Phi\,dS. \tag{8.111}$$

The boundary term is a constant since $u_i=f_i$ on S_u, so that it can be dropped from the definition of J without altering the Euler equations or second variation. Then the functional $J(\mathbf{u})$ is one-half of the viscous dissipation. The variational principle says that the velocity minimizes the viscous dissipation among all possible velocity functions satisfying the boundary conditions. Unfortunately this physical principle is not valid in more general situations. It was first stated as a minimum principle by Helmholtz in 1882 (see Lamb, 1945, p. 617) and proved to be an absolute minimum by Korteweg

(1883). The reciprocal principle for Newtonian fluids was presented by Hill and Power (1956) and Lippmann (1963).

For the generalized Newtonian fluid the constitutive relation is of the form

$$T_{ij} = 2\eta(\mathrm{II})\, d_{ij}, \tag{8.112}$$

where the viscosity is a function of the second invariant. The function Γ is

$$\Gamma = \int^{\mathrm{II}} \eta(\mathrm{II}')\, d\mathrm{II}' \tag{8.113}$$

and

$$\mathrm{II}_t \equiv T_{ij}T_{ij} = 4\eta^2(\mathrm{II})\mathrm{II}, \tag{8.114}$$

which is inverted to give $\mathrm{II} = g(\mathrm{II}_t)$. Then $d_{ij} = T_{ij}/2\eta(g)$. The function $\hat{\Gamma}$ is then

$$\hat{\Gamma} = \int^{\mathrm{II}_t} \frac{d\mathrm{II}'}{4\eta[g(\mathrm{II}')]}. \tag{8.115}$$

The constants of integration are chosen so that (8.99) is satisfied. To examine the possibility of a maximum or minimum principle the second derivatives are

$$\begin{gathered}\frac{\partial^2 \Gamma}{\partial d_{ij}\, \partial d_{kl}} = 2\eta\, \delta_{ik}\, \delta_{jl} + 4\frac{d\eta}{d\mathrm{II}}\, d_{ij}\, d_{kl}, \\ \delta^2 J = \int \{\eta(\mathrm{II})\, \delta d_{ij}\, \delta d_{ij} + 2\frac{d\eta}{d\mathrm{II}}\, (\bar{d}_{ij}\, \delta d_{ij})^2\}\, dV.\end{gathered} \tag{8.116}$$

Clearly if $\eta > 0$, $d\eta/d\mathrm{II} > 0$, which is the case for a dilatant fluid, the variational principle for velocity is a minimum principle. For a pseudoplastic fluid ($\eta > 0$, $d\eta/d\mathrm{II} < 0$) the velocity principle is a minimum principle if

$$\eta + 2(d\eta/d\mathrm{II})\mathrm{II} > 0 \tag{8.117}$$

holds throughout the region of integration. This follows by means of the inequality†

$$-\int \left(-\frac{d\eta}{d\mathrm{II}}\right)(\bar{d}_{ij}\, \delta d_{ij})^2\, dV \geq -\int \left(-\frac{d\eta}{d\mathrm{II}}\right) \bar{d}_{kl}\bar{d}_{kl}\, \delta d_{ij}\, \delta d_{ij}\, dV. \tag{8.118}$$

The velocity variational principle for a generalized fluid was first given by Pawlowski (1954) and was stated as a minimum principle, although the minimum nature was not proved. Later Bird (1960) independently presented

† See Exercise 8.2.

an identical principle. The variational integral is no longer the viscous dissipation for the generalized Newtonian fluid.

For the power-law fluid the viscosity function is

$$\eta = m\mathrm{II}^{(n-1)/2}, \tag{8.119}$$

where n and m are given constants. Then the stress functions are

$$\Gamma = \frac{2m}{n+1}\,\mathrm{II}^{(n+1)/2}, \qquad \hat{\Gamma} = \frac{n}{n+1}\left(\frac{1}{2m}\right)^{1/n}\mathrm{II}_t^{(n+1)/2n}, \tag{8.120}$$

$$T_{ij} = 2m(\mathrm{II})^{(n\pm 1)/2}\, d_{ij}.$$

Values of $n > 1$ correspond to dilatant behavior and the velocity principle is a minimum principle and the functional is $2/(n+1)$ times the viscous dissipation. Values of n below unity correspond to pseudoplastic flow and the condition (8.117) becomes simply $n > 0$. The first minimum principle specifically for a power-law fluid is due to Tomita (1959).

For the more general case of a Reiner–Rivlin fluid, the stress function Γ depends upon III as well as II. Then (8.92) becomes

$$T_{ij} = 2\frac{\partial\Gamma}{\partial\mathrm{II}}\, d_{ij} + 3\frac{\partial\Gamma}{\partial\mathrm{III}}\, d_{ik}\, d_{kj}. \tag{8.121}$$

Equation (8.93), compared to Eq. (8.121) requires that

$$\frac{\partial\Gamma}{\partial\mathrm{II}} = \tfrac{1}{2}G_1(\mathrm{II}, \mathrm{III}), \qquad \frac{\partial\Gamma}{\partial\mathrm{III}} = \tfrac{1}{3}G_2(\mathrm{II}, \mathrm{III}). \tag{8.122}$$

The total differential for $\Gamma(\mathrm{II}, \mathrm{III})$ is

$$d\Gamma = \frac{\partial\Gamma}{\partial\mathrm{II}}\, d\mathrm{II} + \frac{\partial\Gamma}{d\mathrm{III}}\, d\mathrm{III}. \tag{8.123}$$

If Γ is a continuous function of its arguments then

$$\frac{\partial^2\Gamma}{\partial\mathrm{III}\,\partial\mathrm{II}} = \frac{\partial^2\Gamma}{\partial\mathrm{II}\,\partial\mathrm{III}} \quad \text{or} \quad \frac{1}{2}\frac{\partial G_1}{\partial\mathrm{III}} = \frac{1}{3}\frac{\partial G_2}{\partial\mathrm{II}}. \tag{8.124}$$

Then the function Γ is given by the line integral

$$\Gamma = \int^{\mathrm{II},\mathrm{III}} \left(\tfrac{1}{2}G_1\, d\mathrm{II} + \tfrac{1}{3}G_2\, d\mathrm{III}\right). \tag{8.125}$$

It is clear that in general the minimum or maximum conditions are not satisfied for the Reiner–Rivlin fluid.

The type of non-Newtonian fluids considered here are not the most general type, since the stress can depend on the deformation as well as its rate (e.g., visco-elastic materials) or on the entire previous history of the fluid (see Fredrickson, 1964, for example). Variational principles have not been developed for all these cases, however (see Hill, 1956). Prager (1954) presents both maximum and minimum principles for a Bingham plastic, which has the constitutive relation

$$2\mu\, d_{ij} = \begin{cases} 0 & \text{if} \quad J < k, \\ \left(1 - \dfrac{k}{J}\right) T_{ij} & \text{if} \quad J \geq k, \end{cases} \tag{8.126}$$

where $J = (\frac{1}{2}T_{ij}T_{ij})^{1/2}$ and k is a yield stress which must be exceeded before the rate of deformation differs from zero. One fluid which is modeled well by the Bingham plastic model is drilling mud, which is a water-based suspension used to remove chips of rock from the bottom of an oil well.

Two problems are frequently of interest: flow past a sphere or other body (see Section 8.5) and rectilinear flow in ducts. For rectilinear flow of a Newtonian fluid, the reciprocal variational principles (7.63) and (7.67) give upper and lower bounds on the flow rate. A similar theorem applies to some non-Newtonian rectilinear flows. Consider the momentum equation in the absence of body forces:

$$0 = -p_{,i} + T_{ji,j}. \tag{8.127}$$

Take the scalar product with $\mathbf{u}$, integrate by parts and apply the divergence theorem to obtain

$$0 = -\int p u_i n_i \, dS + \int [p u_{i,i} - u_{i,j} T_{ji}] \, dV + \int u_i T_{ji} n_j \, dS. \tag{8.128}$$

For an incompressible fluid $u_{i,i} = 0$ and since T_{ij} is symmetric $u_{i,j}T_{ij} = d_{ij}T_{ij}$. On the sides of the duct $\mathbf{u} = 0$ so that the surface integrals vanish there. On the ends the last surface integral involves T_{zz} which vanishes, at least for the Newtonian and power law fluids, when $d_{zz} = 0$. On the ends $\mathbf{n} = \pm\mathbf{e}_z$, the pressure is constant, and

$$\int_{\text{ends}} p u_i n_i \, dS = -p_0 \int_{S_0} u_z \, dS + p_L \int_{S_L} u_z \, dS \equiv (p_L - p_0) Q, \tag{8.129}$$

where S_0 and S_L are the surfaces at $z = 0$ and $z = L$, respectively. Thus

$$\int d_{ij} T_{ij} \, dV = (p_0 - p_L) \int u_z \, dS = Q \, \Delta p. \tag{8.130}$$

For Newtonian fluids

$$d_{ij}\, T_{ij} = 2\mu\, d_{ij}\, d_{ij} = 2\Gamma \tag{8.131}$$

and the variational integral (8.102) is proportional to flow rate Q. Thus the two variational principles (8.102) and (8.104) provide upper and lower bounds on the flow rate, as developed in Section 7.3. The same result applies to the non-Newtonian fluid provided $d_{ij} T_{ij}$ is proportional to Γ. This is true for a power-law fluid, but not for the generalized Newtonian fluid:

$$\begin{aligned} d_{ij} T_{ij} &= 2m(\mathrm{II})^{(n+1)/2} = (n+1)\Gamma, \qquad \text{power law}, \\ d_{ij} T_{ij} &= 2\mathrm{II}\eta(\mathrm{II}) \neq 2 \int^{\mathrm{II}} \eta(\mathrm{II}')\, d\mathrm{II}', \qquad \text{generalized Newtonian}. \end{aligned} \tag{8.132}$$

These results can be used to provide *a priori* inequalities on the flow rate. Denote by U and U' the class of vector functions which vanish on S and S', respectively. Suppose that S is contained within S'. Then $U \subset U'$, that is any function in U (vanishes on S) can also be extended to be zero on the boundary S', and hence is in U'. Thus as the cross section of the duct increases, the admissible class of functions increases, and this decreases the minimum $J(\mathbf{u})$. Thus we have the following theorem.

For rectilinear flow in ducts of a Newtonian or power-law fluid, the product of pressure drop per unit length times mass flow rate is decreased when the cross section of the duct is enlarged.

Calculations for Newtonian fluids are reported in Section 7.3, and Schechter (1961) treats a power-law fluid in rectilinear flow between infinite flat plates (to check with the known exact solution) and flow through a rectangular duct. He uses the trial function

$$u = \sum_{i,j} c_{ij} \sin \frac{(2i-1)\pi x}{2a} \sin \frac{(2j-1)\pi x}{2b}, \tag{8.133}$$

which is the form of the solution for a Newtonian fluid. In this way Schechter obtains upper bounds on the flow rate, and the product $f\,\mathrm{Re}$. For small values of the power-law model parameter n low-order results can possibly be improved using a different trial function (see Section 4.3). For the Newtonian fluid between flat plates the solution is $u = 1 - (x/a)^2$ and for flow in a rectangular duct good results were achieved (Sections 7.3, 4.1) using the trial function $u = u_{\max}[1 - (x/a)^2][1 - (y/b)^2]$. For the power-law model, the solution to flow between flat plates is $u = (1 - x^{(n+1)/n})$. Thus a suggested profile for the power-law fluid is

$$u = u_{\max}[1 - (x/a)^{(n+1)/n}][1 - (y/b)^{(n+1)/n}]. \tag{8.134}$$

Calculations have not been made with this profile. Mitsuishi and Aoyagi (1969) present calculations for both power law and Sutterby fluids flowing through noncircular ducts, in particular, rectangular and isosceles triangular ducts. Thompson *et al.* (1969) use the variational principle (8.102) in a finite element calculation for a non-Newtonian fluid squeezed between two plates.

8.5 Slow Flow past Drops and Particles

A useful application of the principles in Section 8.4 is to the slow flow past drops or particles. The inertial terms are assumed zero or negligible, which is valid for small Reynold's number, $\mathrm{Re} = dU/\nu$. We consider the flow of a fluid containing solid particles, liquid drops or gaseous bubbles. Keller *et al.* (1967) presented the first variational principle which determined the equations of motion of the particles as well as the equations for the basic flow. Skalak (1970) extended the results to include deformable drops, so that the shape can change during the motion. These authors used Newtonian fluids in the main stream and in the drops. We generalize their results to include non-Newtonian fluids in the main stream and drop, using the constitutive relations developed in Section 8.4.

Consider first a finite domain, or an infinite domain in which all integrals exist. Take N fluid drops and M solid particles, of arbitrary shape, and the fluid drops can be deformed. The configuration of the particles and drops is specified. The main fluid is in volume V, enclosed by the surface $S_u \cup S_t$, and the particles and drops are in nonintersecting volumes V_k, which are bounded by the surface S_k, with normal pointing outward from the drop $\mathbf{n}^{(k)}$. The subscript k takes values from 1 to N, referring to drops, and $N + 1$ to $N + M$, referring to solid particles. A superscript k denotes a function defined in V_k or on S_k. All fluids are incompressible and non-Newtonian, with a constitutive relation given by (8.92). The ΔT_{ji} denotes a jump condition at a boundary and is the value outside minus the value inside the boundary. The equations are collected into two groups in anticipation that some will be Euler equations and the rest admissibility conditions for a variational principle, with the reverse interpretation applying to the reciprocal principle. The equations are

$$u_{i,i} = 0, \qquad u_{i,i}^{(k)} = 0, \tag{8.135a}$$

$$u_i = f_i \quad \text{on } S_u, \tag{8.135b}$$

$$u_i = U_i^{(k)} + \varepsilon_{imj}\,\omega_m^{(k)} r_j^{(k)} \quad \text{on } S_k, \qquad k = N+1, \ldots, N+M, \tag{8.135c}$$

$$u_i = u_i^{(k)} \quad \text{on } S_k, \qquad k = 1, \ldots, N, \tag{8.135d}$$

$$n_i^{(k)} u_i = n_i^{(k)} u_i^{(k)} = n_i^{(k)}[U_i^{(k)} + \varepsilon_{imj}\,\omega_m^{(k)} r_j^{(k)}] \quad \text{on } S_k, \qquad k = 1, \ldots, N \tag{8.135e}$$

The terms $\mathbf{U}^{(k)}$ is a rigid body motion of a reference point in V_k and $\boldsymbol{\omega}^{(k)}$ is an angular velocity about an axis through this point. The surface tension is $\sigma^{(k)}$ and R_1 and R_2 are the two principal radii of curvature (positive into drop). The remaining equations are

$$T_{ji,j} - p_{,i} + \rho F_i = 0 \quad \text{in } V, \tag{8.136a}$$

$$\left.\begin{aligned} &T^{(k)}_{ji,j} - p^{(k)}_{,i} + \rho^{(k)}F_i = 0 \quad \text{in } V_k, && (8.136\text{b})\\ &(T_{ji} - p\,\delta_{ji})n_j = g_i \quad \text{on } S_t, && (8.136\text{c})\\ &\Delta(T_{ji} - p\,\delta_{ji})n_j\,n_i = \sigma^{(k)}\left(\frac{1}{R_1} + \frac{1}{R_2}\right), && (8.136\text{d})\\ &\Delta(T_{ji} - p\,\delta_{ji})n_j - \Delta(T_{pq} - p\,\delta_{pq})n_p\,n_q\,n_i = 0, && (8.136\text{e}) \end{aligned}\right\} \quad k = 1, \ldots, N$$

$$F^{(k)}_i + \int_{S_k} n^{(k)}_j(T_{ji} - p\,\delta_{ji})\,dS = 0, \qquad k = N+1, \ldots, N+M, \quad (8.136\text{f})$$

$$N^{(k)}_i + \int_{S_k} \varepsilon_{iqj}\,r^{(k)}_q n^{(k)}_m(T_{jm} - p\,\delta_{jm})\,dS = 0. \quad (8.136\text{g})$$

$F^{(k)}_i$ and $N^{(k)}_i$ are the external force and torque on particle k. The following two variational principles are reciprocal to each other.

Minimize the functional among functions $\mathbf{u}$, $\mathbf{u}^{(k)}$, $\mathbf{U}^{(k)}$, $\boldsymbol{\omega}^{(k)}$ which satisfy Eqs. (8.135) and $\mathbf{u}$ and $\mathbf{u}^{(k)}$ are continuous and piecewise continuously differentiable:

$$\begin{aligned} J(\mathbf{u}, \mathbf{u}^{(k)}, \mathbf{U}^{(k)}, \boldsymbol{\omega}^{(k)}) = &\int \{\Gamma - \rho F_i u_i\}\,dV - \int_{S_t} g_i u_i\,dS \\ &+ \sum_{k=1}^{N} \left\{\int_{v_k} [\Gamma^{(k)} - \rho^{(k)} F_i u^{(k)}_i]\,dV \right.\\ &\left. + \sigma^{(k)} \int_{S_k} u_i n^{(k)}_i \left(\frac{1}{R_1} + \frac{1}{R_2}\right) dS\right\} \\ &- \sum_{k=N+1}^{N+M} \{F^{(k)}_i U^{(k)}_i + N^{(k)}_i \omega^{(k)}_i\}. \end{aligned} \quad (8.137)$$

The details of the proof are sketched as follows. Introduce the incompressibility constraint using Lagrange multipliers $-p$ in V and $-p^{(k)}$ in V_k. The first variation is

$$\begin{aligned} \delta J = &\int_{v} \delta u_i[-T_{ji,j} - \rho F_i + p_{,i}]\,dV + \int_{S_t} \delta u_i[T_{ji}n_j - pn_i - g_i]\,dS \\ &+ \sum_{k=1}^{N} \int_{v_k} \delta u^{(k)}_i[-T^{(k)}_{ji,j} - \rho^{(k)}F_i + p^{(k)}_{,i}]\,dV \\ &+ \sum_{k=1}^{N+M} \int_{S_k} \delta u_i[T_{ji} - p\,\delta_{ji}](-n^{(k)}_j)\,dS \\ &+ \sum_{k=1}^{N} \int_{S_k} \delta u^{(k)}_i[T^{(k)}_{ji} - p^{(k)}\,\delta_{ji}]n^{(k)}_j\,dS - \sum_{k=N+1}^{N+M} \{F^{(k)}_i\,\delta U^{(k)}_i + N^{(k)}_i\,\delta\omega^{(k)}_i\} \\ &+ \sum_{k=1}^{N} \sigma^{(k)} \int_{S_k} \delta u_i\,n^{(k)}_i \left(\frac{1}{R_1} + \frac{1}{R_2}\right) dS. \end{aligned} \quad (8.138)$$

The Euler equations in V and V_k are easily identified. Remember that the outward pointing normal to V on S_k is the inward pointing normal to V_k, $\mathbf{n} = -\mathbf{n}^{(k)}$. The variations on S_k are not independent, however. Consider one drop surface. By Eq. (8.135d) $\delta\mathbf{u} = \delta\mathbf{u}^{(k)}$ on S_k. The surface terms then become (for one drop)

$$\int_{S_k} \delta u_i[-\Delta(T_{ji} - p\,\delta_{ji})]n_j^{(k)}\,dS + \int_{S_k} \delta u_i\,\sigma^{(k)}\left(\frac{1}{R_1} + \frac{1}{R_2}\right)n_i^{(k)}\,dS. \quad (8.139)$$

The velocity variation and the vector $\Delta T_{ji}n^{(k)}$ can be split into their tangential and normal components. For example,

$$\delta u_i = (\delta u_i - n_l^{(k)}\,\delta u_l\,n_i^{(k)}) + n_l^{(k)}\,\delta u_l\,n_i^{(k)}$$

and the surface terms in Eq. (8.139) become

$$\int_{S_k} \delta u_i^{\text{II}}\{-\Delta(T_{ji} - p\,\delta_{ji})n_i^{(k)} + \Delta(T_{pq} - p\,\delta_{pq})n_p\,n_q\,n_i^{(k)}\}\,dS$$

$$+ \int_{S_k} \delta u_l\,n_l^{(k)}\left[-\Delta(T_{ji} - p\,\delta_{ji})n_j^{(k)}n_i^{(k)} + \sigma^{(k)}\left(\frac{1}{R_1} + \frac{1}{R_2}\right)\right]dS. \quad (8.140)$$

Since the variations in tangential and normal components are independent we obtain Eqs. (8.136d,e) as natural boundary conditions. For the solid particles we substitute for δu_i from the variation of (8.135c) and obtain (8.136f,g) directly, since $\delta\mathbf{U}^{(k)}$ and $\delta\boldsymbol{\omega}^{(k)}$ are independent variations. The minimum nature follows, of course, only for non-Newtonian fluids for which (8.103) is positive. Keller *et al.* (1967) demonstrate for the Newtonian fluid that the principle is a minimum principle. The reciprocal principle is the following.

Maximize the functional among tensor functions $\mathbf{T}$ and $\mathbf{T}^{(k)}$ which satisfy (8.136) and are symmetric, piecewise continuous and piecewise continuously differentiable in V and $V^{(k)}$ respectively, and on surfaces of discontinuity of $\mathbf{T}$ (other than surfaces of drops) the traction $\mathbf{n}' \cdot \mathbf{T}$ is continuous, where $\mathbf{n}'$ is the normal to the surface of discontinuity:

$$I(\mathbf{T}, \mathbf{T}^{(k)}) = -\int_v \hat{\Gamma}\,dV + \int_{S_u} (T_{ji}n_j - pn_i)f_i\,dS - \sum_{k=1}^{N} \int \hat{\Gamma}^{(k)}\,dV. \quad (8.141)$$

The principle is verified using the Lagrange multipliers $-\mathbf{u}$, $-\mathbf{u}^{(k)}$, $-\mathbf{U}^{(k)}$, $-\boldsymbol{\omega}^{(k)}$, and $-n_i^{(k)}(U_i^{(k)} + \varepsilon_{imj}\omega_m^{(k)}r_j^{(k)})$ for (8.136a,b,d,f,g), respectively.

For an infinite domain Skalak (1970) gives additional conditions needed at infinity. The variations must satisfy

$$\delta u = O(r^{-1}), \qquad \delta T_{ji} - p\,\delta_{ji} = O(r^{-2}) \qquad \text{as } r \to \infty, \qquad (8.142)$$

while the limit

$$p^* = \lim_{r \to \infty} (p + \Phi) \tag{8.143}$$

must exist, the forces are derivable from a potential, and the flow is zero at infinity. Skalak (1970) also extends his results to include spatially periodic flows. Both Skalak and Keller *et al.* prove uniqueness theorems, which follow immediately from the minimum principles.

The solution $\mathbf{u}$ of (8.135) and (8.136) is unique within a rigid body motion and the stress is unique within a uniform pressure.

Consider the minimum principle for a Newtonian fluid and assume there are two solutions. Then they satisfy

$$J(\mathbf{u}_1, \mathbf{u}_1^{(k)}, \mathbf{U}_1^{(k)}, \boldsymbol{\omega}_1^{(k)}) \le J(\mathbf{u}_2, \mathbf{u}_2^{(k)}, \mathbf{U}_2^{(k)}, \boldsymbol{\omega}_2^{(k)}) \tag{8.144}$$

as well as the reverse inequality. Thus the equality must hold. Integration by parts and application of the differential equations and boundary conditions shows that the equality holds only for rigid body motion. The same results apply to non-Newtonian fluids when the term Γ can be integrated by parts, as is the case for the power law fluid.

Keller *et al.* (1967) use these variational principles to obtain bounds for the viscosity of a suspension. They construct admissible functions as follows: around each sphere construct a sphere of radius b_k, where $2b_k$ is the distance to the nearest particle. Assume each particle is within one such sphere, and they do not overlap. In the region outside the spheres the trial functions are taken as either the velocity for a uniform shear flow or the stress for the same flow. Within the sphere of radius b_k a flow problem is solved exactly satisfying a boundary condition which makes the velocity or stress match continuously with the flow exterior to the sphere. Then the particles are assumed to be on a simple cubic lattice (this assumption weakens the bounds) and the viscosity of the suspension is calculated to be

$$\frac{\mu_s}{\mu} \le 1 - \frac{\pi\lambda^3[5(1-\eta)\lambda^7 + (5\eta+2)]}{3[4(1-\eta)\lambda^{10} + 5(5\eta-2)\lambda^7 - 42\eta\lambda^5 + 5(5\eta+2)\lambda^3 - 4(1+\eta)]},$$

$$\frac{\mu_s}{\mu} \ge 1 + \frac{\pi\lambda^3[-80(1-\eta)\lambda^7 + 19(2+5\eta)]}{6[-48(1-\eta)\lambda^{10} - 40(2-5\eta)\lambda^7 - 336\eta\lambda^5 + 45(2+5\eta)\lambda^3 + 38(1+\eta)]}, \tag{8.145}$$

where $\eta = \mu_1/\mu$, $\lambda = a/b$, μ_1 = viscosity of fluid in a spherical drop of radius a, in a sphere of radius b. The volume concentration is $c = \pi\lambda^3/6$ and (8.145) provides good bounds for $c \gtrsim 0.1$.

Variational principles of the type (8.137) and (8.141) have been applied mainly to non-Newtonian flow past a single spherical particle. For power law fluids, the functional is proportional to the drag. Thus upper and lower bounds are achieved for a parameter of interest, and the bounds give a measure of the accuracy. Consider the calculations by Wasserman and Slattery (1964) for the flow of a power-law fluid past a sphere (see also Slattery, 1962). The velocity is represented by a stream function, as in the case of a Newtonian fluid. The continuity equation is then satisfied identically:

$$u_r = -\frac{1}{r^2 \sin\theta}\frac{\partial\psi}{\partial\theta}, \qquad u_\theta = \frac{1}{r\sin\theta}\frac{\partial\psi}{\partial r}, \qquad u_\phi = 0. \tag{8.146}$$

The trial function is a generalization of the exact solution for a Newtonian fluid, and satisfies the boundary conditions:

$$\begin{aligned} \psi &= -\tfrac{1}{2}Ur^2\sin^2\theta\,[1-(R/r)^a]^2, \\ u_r &= u_0 = 0 \quad \text{at} \quad r = R, \\ u_z &= U \qquad\quad \text{at} \quad r\to\infty. \end{aligned} \tag{8.147}$$

The parameter a is determined by minimizing the functional (8.137). Both the evaluation of J and the minimization are done numerically.

In this problem $S = S_v$ and $F_i = 0$. Then the functional is related to the drag coefficient:

$$\begin{aligned} c_D\,\mathrm{Re} &= 24Q, \\ Q &= J(u)/\alpha, \qquad \alpha = 3\pi m U^{n+1} D^{2-n}/(n+1), \\ \mathrm{Re} &= \rho U^{2-n} D^n/m, \qquad c_D = 8F/(\rho U^2 \pi D^2), \end{aligned} \tag{8.148}$$

where Q depends on n, D is the diameter of the sphere, F is the force on the sphere, and c_D is the drag coefficient.

To determine a lower bound it is necessary to find stress functions satisfying the equations of motion, Eq. (8.136a). For flow past a sphere the equations reduce to

$$\begin{aligned} \frac{\partial(p+\phi\rho)}{\partial r} &= \frac{1}{r^2}\frac{\partial}{\partial r}(r^2 T_{rr}) + \frac{1}{r\sin\theta}\frac{\partial}{\partial\theta}(T_{r\theta}\sin\theta) - \frac{T_{\phi\phi}+T_{\theta\theta}}{r}, \\ \frac{\partial(p+\phi\rho)}{\partial\theta} &= \frac{1}{r}\frac{\partial}{\partial r}(r^2 T_{r\theta}) + \frac{1}{\sin\theta}\frac{\partial}{\partial\theta}(T_{\theta\theta}\sin\theta) + T_{r\theta} - T_{\phi\phi}\cot\theta. \end{aligned} \tag{8.149}$$

These can be combined by taking $\partial/\partial\theta$ of the first, $\partial/\partial r$ of the second and subtracting. The stress distributions are taken in the following form, arrived at through some trial and error and knowledge of their form for the Newtonian fluid:

$$
\begin{aligned}
T_{r\theta} &= -Ax^B \sin\theta\, m(2V/D)^n, x \equiv R/r, \\
T_{rr} &= -(Cx^D + C'x^B)\cos\theta\, m(2V/D)^n, \\
T_{\theta\theta} &= -(Fx^D + F'x^B)\cos\theta\, m(2V/D)^n, \\
T_{\phi\phi} &= -(Ex^D + E'x^B)\cos\theta\, m(2V/D)^n.
\end{aligned} \tag{8.150}
$$

Equation (8.149) then gives an expression involving powers of x, the radial coordinate, R/r. The coefficients of different powers of x are set to zero in order that the equation be satisfied identically, giving the following relations:

$$
\begin{gathered}
E = F, \qquad E' = F', \qquad D = 2, \\
AB^2 - B(3A + C' - F') + 2(A + C' - F') = 0.
\end{gathered} \tag{8.151}
$$

Wasserman and Slattery arbitrarily assume that

$$
B = 4, \qquad C' = -C, \qquad F' = -F, \tag{8.152}
$$

which requires that $C' = 3A + F'$. The functional $I(T)$ becomes a function of A and F'. It is minimized analytically to obtain $A = -F = (2/K)^n$, where K is an integral evaluated numerically for each parameter n. The value of $I(T)$ gives the lower bound for the constant in (8.148). The results are shown in Table 8.1

TABLE 8.1

UPPER AND LOWER BOUNDS ON Q[a]

n	Q_{upper}	Q_{lower}	n	Q_{upper}	Q_{lower}
1.0	1.005	1.006	0.7	1.366	1.173
0.9	1.140	1.141	0.5	1.499	0.916
0.8	1.260	1.199	0.3	1.617	0.527

[a] Wasserman and Slattery (1964). Used by permission of the copyright owner, the American Institute of Chemical Engineers.

as values of the coefficient Q. For $n > 0.8$ the upper and lower bound are within 5% of each other. The values for $n \approx 1$ are in error because the lower bound is above the upper bound, which is impossible and probably is due to the numerical integration. For the lower bound and $n = 1$ an analytic result is possible which gives a value of 1.0 instead of 1.006 found numerically. These results can be compared to experimental measurements on power-law fluids which should fall between the bounds. Unfortunately, this is not the case, and Wasserman and Slattery conclude that the power-law model, using constants obtained from viscometric measurements, is not adequate to describe these fluids. This is an illustration of the power of the error bounds provided by the variational principle. Since the error bounds are rigorous, the lack of agreement with experiment is due to the model, not to the approximate solution of the equation. If there were no error bounds, the analyst would have difficulty deciding if the discrepancy was due to the approximate solution or the model (constitutive relation).

Tomita (1959) has solved the same problem, using a less general trial function: Eqs. (8.147) with $a = n$. His results were later corrected (Slattery and Bird, 1961; Wallick *et al.*, 1962) and the upper bound is considerable above that listed in Table 8.1. This illustrates the general guideline that the trial function should have some free parameters in order to achieve good results. Other non-Newtonian fluids have been treated as well. Ziegenhagen *et al.* (1961) treat the generalized Newtonian fluid with the constitutive relation

$$\eta = \eta_0 - \eta_1 \mathrm{II}. \tag{8.153}$$

They provide bounds for the constant $Q = F_{\mathrm{drag}}/(6\pi\eta_0 UR)$. For a trial function appropriate to a Newtonian fluid they obtain

$$Q = 1 - 0.414H, \qquad H = (\eta_1/\eta_0)(U/R)^2, \tag{8.154}$$

and with one free parameter the result is

$$Q = 1 - 0.376\, H - 1.508\, H^2. \tag{8.155}$$

Ziegenhagen (1965) treat a Powell–Eyring fluid with a generalized viscosity

$$\eta = \eta_N + \frac{x\beta}{\alpha} \frac{\sinh^{-1}[\beta(\frac{1}{2}\mathrm{II})^{1/2}]}{\beta|(\frac{1}{2}\mathrm{II})^{1/2}|} \tag{8.156}$$

and Slattery (1961) studies the flow of a non-Newtonian fluid whose viscosity changes abruptly at a certain shear rate. Ehrlich and Slattery (1968) study power-law fluids as lubricants and obtain upper and lower bounds for the torque exerted by a journal bearing. Hopke and Slattery (1970) compute the drag on a sphere for slow flow of an Ellis-model fluid. The bounds are not as good as for a power-law fluid, since the variational integral gives bounds on the drag, but is not proportional to it. Even when the exact solution is known, so that the exact value of the variational integral is known, the drag is not determined exactly (in terms of the variational integral). It could be determined exactly, of course, by a direct calculation of the viscous dissipation or drag. Flumerfelt and Slattery (1965) use a more complicated variational principle (similar to an adjoint variational principle) for flow of a Newtonian fluid past a sphere.

Nakano and Tien (1968) report a calculation for the creeping flow of a power-law fluid past a Newtonian fluid sphere. Inside the sphere they apply a method of moments [see Eqs. (4.76) and (4.77)], while outside they apply a variational principle. The integral used by Nakano and Tien is the first term of Eq. (8.137). Thus they obtain the correct Euler equation, but the boundary term is

$$\int_{S_k} \delta u_i (T_{ji} n_j - p n_i)\, dS = 0 \tag{8.157}$$

The boundary conditions on velocity are

$$\mathbf{e}_r \cdot \mathbf{u} = 0, \qquad \mathbf{e}_r \cdot \mathbf{u}^{(k)} = 0, \qquad \mathbf{e}_\theta \cdot \mathbf{u} = \mathbf{e}_\theta \cdot \mathbf{u}^{(k)} \quad \text{on } S_k. \tag{8.158}$$

The radial components clearly satisfy $\delta u_r = 0$ but the azimuthal components do not. The natural boundary condition is then

$$\mathbf{e}_\theta \cdot [\mathbf{T} \cdot \mathbf{e}_r - p\mathbf{e}_r] = 0 \qquad \text{or} \qquad T_{r\theta} = 0. \tag{8.159}$$

This is not the correct condition. In fact, the correct condition is a jump condition (8.136e), so that their results are in error.

A priori bounds can be obtained for the drag on a particle just as was done for the pressure drop through ducts (Section 8.4). For the power-law fluid the variational integral is proportional to the drag (8.148) when the forces are conservative and the solid particle is stationary. Employing the same reasoning used in Section 8.4 we denote by U and U' the class of vector functions which vanish on S and S' respectively. If S is contained within S' then $U' \subset U$. Thus as the size of the particle is increased the class of admissible functions decreases, so that the minimum is increased. Thus we have the following theorem.

For creeping flow of a power-law (or Newtonian) fluid past a solid particle, the drag is increased when the particle size is increased.

This theorem can be used to find bounds on the drag for irregularly-shaped particles, provided we know the solution for a particle which either contains or is contained by the irregularly-shaped particle. Spheres are useful, but other shapes have known solutions (disks, ellipses, etc; see Happel and Brenner, 1965). In addition, the effect of a wall is immediately evident. Since the velocity is constrained to be zero at a wall, the class of admissible functions is decreased, and the minimum increases. Thus the drag on the sphere increases as well. Hill and Power (1956) obtain bounds in this fashion for a torus.

8.6 Variational Principles for the Navier–Stokes Equations

In Section 8.2 we obtained variational principles for the equation of motion for perfect fluids, including the inertial terms but in the absence of viscous terms. In Section 8.4 we obtained variational principles for the equations of motion including viscous terms but not inertial terms. We next consider *both* viscous and inertial terms, but find that there is no variational principle which gives the equation of motion as an Euler equation.

We assume the simplest type of viscous terms, those appropriate to a Newtonian fluid, take an incompressible fluid, constant physical properties and a conservative body force field. The equation of motion is then

$$\mathbf{u} \cdot \nabla \mathbf{u} = -(1/\rho)\nabla p - \nabla\Phi + \nu \nabla^2 \mathbf{u}. \tag{8.160}$$

Using the vector identity

$$\mathbf{u} \cdot \nabla \mathbf{u} = \nabla(\tfrac{1}{2}\mathbf{u} \cdot \mathbf{u}) - \mathbf{u} \times (\nabla \times \mathbf{u}). \tag{8.161}$$

Equation (8.160) is rewritten in the form

$$\nabla(\tfrac{1}{2}\mathbf{u} \cdot \mathbf{u}) - \mathbf{u} \times (\nabla \times \mathbf{u}) = -(1/\rho)\nabla p - \nabla\Phi + \nu \nabla^2 \mathbf{u}. \tag{8.162}$$

The Helmholtz–Korteweg principle, Eqs. (8.102) and (8.111), gives a variational principle for Eq. (8.162) when the inertial terms are absent. Brill (1895) was the first to recognize that the Helmholtz–Korteweg principle could be generalized to include the term $\nabla(\tfrac{1}{2}\mathbf{u} \cdot \mathbf{u})$.

Minimize the functional among vector functions $\mathbf{u} \in C^2$ which satisfy $u = f$ on S and $\nabla \cdot \mathbf{u} = 0$ in V:

$$J(\mathbf{u}) = \int \nu \, d_{ij} d_{ij} \, dV. \tag{8.163}$$

The Euler equation is derived using a Lagrange multiplier, adding $\int \lambda \nabla \cdot \mathbf{u} \, dV$ to the variational integral. The Euler equations are derived in the usual fashion, and interpretation of the Lagrange multiplier $\lambda = -u^2 - 2(p/\rho) - 2\Phi$ gives Eq. (8.162) without the $\mathbf{u} \times (\nabla \times \mathbf{u})$ terms. A similar generalization obviously applies to the principles in Section 8.4 for non-Newtonian fluids. Lord Rayleigh (1913) also noted that the viscous dissipation (8.163) was a minimum whenever the velocity satisfied

$$\nabla^2 \mathbf{u} = \nabla H, \qquad \nabla^2 H = 0. \tag{8.164}$$

This is of course the same as (8.162) without the $\mathbf{u} \times (\nabla \times \mathbf{u})$ term.

It was Millikan (1929), however, who gave a definitive treatment of the question of the existence of a variational principle for (8.162). He concluded that a principle existed only if $\mathbf{u} \times (\nabla \times \mathbf{u}) = 0$. Because of the widespread interest in deriving the equations of motion from a variational principle, we outline his arguments here.

Millikan asked the question: is there a Lagrangian L such that the variational integral $J = \int L \, dV$ gives as Euler equations the Navier–Stokes equations for the steady motion of an incompressible fluid? The answer he obtained was no. Millikan proved the following theorem.

If L is restricted to be a non-singular function of the velocity and its first-order space derivatives only, then it is impossible to find a Lagrangian L which will give (8.162) as the Euler–Lagrange equations.

His argument goes as follows. He first tries a Lagrangian of the form

$$J = (\tfrac{1}{2}\mu\, d_{ij}\, d_{ij} + \tfrac{1}{2}\rho\mathbf{u} \cdot \nabla u^2 - \lambda\nabla \cdot \mathbf{u})\, dV, \tag{8.165}$$

where λ is a Lagrange multiplier. The corresponding Euler equations are

$$\tfrac{1}{2}\rho\, \nabla u^2 + \nabla\lambda = \mu\, \nabla^2\mathbf{u}. \tag{8.166}$$

This means that a variational principle exists whenever $\mathbf{u} \times (\nabla \times \mathbf{u}) = 0$.

In the next step of the proof he examines the term $\mathbf{e}_x \cdot [\mathbf{u} \cdot \nabla\mathbf{u}] = uu_x + vu_y + wu_z$ and asks what L gives vu_y in the Euler equation. The general expression for the Euler equation is

$$\frac{\partial}{\partial x}\left(\frac{\partial L}{\partial u_x}\right) + \frac{\partial}{\partial y}\left(\frac{\partial L}{\partial u_y}\right) + \frac{\partial}{\partial z}\left(\frac{\partial L}{\partial u_z}\right) - \frac{\partial L}{\partial u} = 0. \tag{8.167}$$

Terms of the form $\partial f/\partial x$ or $\partial g/\partial z$ cannot give rise to vu_y, however, so that the important terms in the Euler equation reduce to

$$\frac{\partial}{\partial y}\left(\frac{\partial L}{\partial u_y}\right) - \frac{\partial L}{\partial u} = 0. \tag{8.168}$$

Millikan then tries a power series for L,

$$L = u^\alpha v^\beta w^\gamma u_x^{\alpha_1} u_y^{\alpha_2} \cdots w_z^{\gamma_3} \tag{8.169}$$

and through a detailed examination of the resulting Euler equation concludes that such a Lagrangian will not give rise to a term vu_y alone. In cases where the term vu_y appeared in conjunction with other terms, the other terms could not be cancelled as required to obtain the correct Euler equation. In addition a sum of power series will not give rise to vu_y alone. He then assumes that the Lagrangian has no singularities in the volume V, so that any power series

$$S = \sum_{-\infty}^{\infty} c_i z_i \tag{8.170}$$

converges uniformly to L inside V, where z^i represents one term of the form (8.169). Consequently, any possible L can be represented as a power series in the form (8.170), which cannot give rise to the term vu_y in the Euler equation. Thus there exists no Lagrangian (without singularities) which gives the inertial terms $\mathbf{u} \cdot \nabla\mathbf{u}$ in the Euler equation. Gerber (1949) reached the same conclusion for the equation in Lagrangian form.

Millikan's analysis is tedious to apply and we would like a simpler way to decide whether there is a variational principle for (8.162). Fréchet differentials can be used to do this, and they are introduced in Chapter 9. We use them here, with the understanding that the reader unfamiliar with them will read Sections 9.1–9.3 first. Consider the four-vector $\mathbf{w} = (\mathbf{u}, p)$ and the four equations

$$f^\alpha = -w_{4,\alpha} + \mu w_{\alpha,\beta\beta}, \qquad f^4 = w_{\beta,\beta} = 0. \tag{8.171}$$

Greek indices run from one to three and Latin indices go from one to four; repeated indices are summed over their range. If a Fréchet differential is to exist for the set (8.171) it is necessary that Eqs. (C) of Table 9.1 be satisfied. The derivatives are

$$\frac{\partial f^\alpha}{\partial w_{\beta,\gamma\delta}} = \mu\delta_{\alpha\beta}\delta_{\gamma\delta}; \qquad \text{zero otherwise,}$$

$$\frac{\partial f^4}{\partial w_{\alpha,\beta}} = \delta_{\alpha\beta}; \qquad \frac{\partial f^\alpha}{\partial w_{4,\beta}} = -\delta_{\alpha\beta}; \qquad \text{zero otherwise,} \tag{8.172}$$

$$\frac{\partial f^l}{\partial w_s} = 0.$$

Equation (9.C1) is easily satisfied. Equation (9.C2) is

$$\frac{\partial f^4}{\partial w_{\alpha,\beta}} = \delta_{\alpha\beta} = -\frac{\partial f^\alpha}{\partial w_{4,\beta}} = -(-\delta_{\alpha\beta}) \tag{8.173}$$

and is zero for other indices; it is thus satisfied. Equation (9.C3) is satisfied since all terms are zero. The functional is determined from (9.19) to be $\frac{1}{2}w_i f^i$, which gives

$$J = -\tfrac{1}{2}\int \mu u_{\alpha,\beta}u_{\alpha,\beta}\,dV + \int p u_{\gamma,\gamma}\,dV. \tag{8.174}$$

This is the variational integral (8.163) with the Lagrange multiplier introduced. To derive a variational principle for (8.166) we use the same equations but with $w_4 = p + \frac{1}{2}u^2$.

Consider the full equations (8.162):

$$f^\alpha = -w_{4,\alpha} + \mu w_{\alpha,\beta\beta} + A\rho\varepsilon_{\alpha\beta\gamma}\varepsilon_{\gamma\delta\varepsilon}w_\beta w_{\varepsilon,\delta}, \qquad f^4 = w_{\beta,\beta} = 0. \tag{8.175}$$

We take $w_4 = p + \frac{1}{2}u^2$ and A is a constant, introduced for convenience, which takes the values 0 or 1. The various derivatives are

$$\frac{\partial f^\alpha}{\partial w_{\beta,\gamma\delta}} = \mu\delta_{\alpha\beta}\delta_{\gamma\delta}; \qquad \text{zero otherwise}$$

$$\frac{\partial f^\alpha}{\partial w_{\beta,\gamma}} = + A\rho\varepsilon_{\alpha\delta\varepsilon}\varepsilon_{\varepsilon\gamma\beta}w_\delta$$

$$\frac{\partial f^\alpha}{\partial w_{4,\gamma}} = -\delta_{\alpha\gamma}; \qquad \frac{\partial f^4}{\partial w_{\beta,\gamma}} = \delta_{\beta\gamma}; \qquad \frac{\partial f^4}{\partial w_{4,\gamma}} = 0 \tag{8.176}$$

$$\frac{\partial f^\alpha}{\partial w_\beta} = A\rho\varepsilon_{\alpha\beta\gamma}\varepsilon_{\gamma\delta\varepsilon}w_{\varepsilon,\delta}; \qquad \text{zero otherwise.}$$

Equation (9.C2) gives for l, $s = 1, 2$, or 3

$$A\rho(\varepsilon_{\alpha\delta\varepsilon}\varepsilon_{\varepsilon\gamma\beta} + \varepsilon_{\beta\delta\varepsilon}\varepsilon_{\varepsilon\gamma\alpha})w_\delta = 0. \tag{8.177}$$

Taking $\alpha = 1$ and $\beta = 2$ gives

$$A\rho(\varepsilon_{1\delta 3}\varepsilon_{3\gamma 2}w_\delta + \varepsilon_{2\delta 3}\varepsilon_{3\gamma 1}w_\delta) = 0 \tag{8.178}$$

or $Aw_1 = Aw_2 = 0$. $\beta = 3$ gives $Aw_3 = 0$. Thus we obtain

$$A\mathbf{u} = 0 \tag{8.179}$$

as the equation which must be satisfied if the Fréchet differential is symmetric and a variational principle exists. This gives only a trivial result. We conclude then that a variational principle exists for the steady-state Navier–Stokes equations (8.162) only if $\mathbf{u} \times (\nabla \times \mathbf{u}) = 0$ or $\mathbf{u} \cdot \nabla\mathbf{u} = 0$.

The problem (8.162) can be expanded to include "adjoint" equations and a variational principle exists for this expanded system.

Make $I(\mathbf{w}, \mathbf{w}^*)$ stationary among vector functions $\mathbf{w}, \mathbf{w}^* \in C^2$, $\mathbf{w} = (\mathbf{u}, p)$, $w_\alpha = g_\alpha$, $w_\alpha^* = g_\alpha^*$ on S;

$$I(\mathbf{w}, \mathbf{w}^*) = \int \mathbf{w}^* \cdot \mathbf{f}\, dV$$

$$= \int_v [w_\alpha^*(-\rho w_\beta w_{\alpha,\beta} - w_{4,\alpha} + \mu w_{\alpha,\gamma\gamma}) + p^* w_{\alpha,\alpha}]\, dV. \tag{8.180}$$

The variational integral is deduced from (9.75). A principle applicable to more general boundary conditions is given by Finlayson (1972). The first variation gives

$$\delta I = \int_v \delta w_\alpha^*(-\rho w_\beta w_{\alpha,\beta} - w_{4,\alpha} + \mu w_{\alpha,\gamma\gamma})\, dV + \int_v \delta p^* w_{\alpha,\alpha}\, dV$$

$$+ \int_v \delta w_\alpha(-\rho w_\beta^* w_{\beta,\alpha} + \rho w_\beta w_{\alpha,\beta}^* + \rho w_\alpha^* w_{\beta,\beta} - p_{,\alpha}^* + \mu w_{\alpha,\gamma\gamma}^*)\, dV$$

$$+ \int_v \delta w_4 w_{\alpha,\alpha}^*\, dV. \tag{8.181}$$

The "adjoint" equations are thus

$$\begin{aligned} \rho(w_\beta w_\alpha{}^*)_{,\beta} - \rho w_\beta{}^* w_{\beta,\alpha} - w^*_{4,\alpha} + \mu w^*_{\alpha,\gamma\gamma} &= 0, \\ w^*_{\alpha,\alpha} &= 0. \end{aligned} \tag{8.182}$$

We have found a variational principle for the system of equations (8.7), (8.160), and (8.182), and it is similar to that suggested by Dryden *et al.* (1956, p. 156). The variational integral has no apparent physical interpretation, and the variational method is MWR. When the same trial functions are used for **w** and **w***, the variational method is a Galerkin method. Since these computational techniques are well-developed, there appears no important use for the variational principle (8.180).

We conclude by summarizing the equations for which variational principles have been found. For perfect fluids there are stationary principles for unsteady motion, (8.49). The steady flow of perfect fluids is governed by minimum and maximum principles (8.32) and (8.41) for compressible fluids and (8.20) and (8.26) for incompressible fluids. For viscous fluids when $\mathbf{u} \times (\nabla \times \mathbf{u}) = 0$ or $\mathbf{u} \cdot \nabla\mathbf{u} = 0$ stationary principles exist for certain types of non-Newtonian fluids (8.102) and (8.104). These reduce to minimum and maximum principles for Newtonian, power-law and certain types of generalized Newtonian fluids. A variational principle does not exist for (8.162) although it does for (8.162) and its adjoint. The lesson to be learned is: in many (but not all) cases stationary principles have been found, and under additional restrictions maximum and minimum principles hold. Under even more restrictive conditions we can give the variational principles a physical interpretation, such as minimizing the kinetic energy or viscous dissipation. The final conclusion is, however, that variational principles do not exist for all the equations of interest in fluid mechanics. We return to this point in Chapter 10, after looking at variational principles in heat and mass transfer in the next chapter.

8.7 Energy Methods for Stability of Fluid Motion

Although the full Navier–Stokes equations cannot be derived from a variational principle, there are variational principles associated with certain stability theorems applicable to the equations. Serrin (1959b) has developed an energy method for deducing the stability, with respect to *arbitrary* disturbances, of solutions to the Navier–Stokes equations, and the best possible bounds are deduced from a variational principle. We present Serrin's treatment.

Consider two solutions, $\mathbf{v}$ and $\mathbf{v}'$, to the Navier–Stokes equations, (1.18):

$$\rho\left(\frac{\partial \mathbf{v}}{\partial t} + \mathbf{v} \cdot \nabla \mathbf{v}\right) = \rho \hat{\mathbf{F}} - \nabla p + \mu \nabla^2 \mathbf{v}. \tag{8.183}$$

The perturbation motion, $\mathbf{u} \equiv \mathbf{v}' - \mathbf{v}$, then satisfies

$$\frac{\partial \mathbf{u}}{\partial t} + \mathbf{u} \cdot \nabla \mathbf{v} + \mathbf{v}' \cdot \nabla \mathbf{u} = \nabla\left(\frac{p - p'}{\rho}\right) + \nu \nabla^2 \mathbf{u} \tag{8.184}$$

Take the scalar product of (8.184) with $\mathbf{u}$, integrate over a bounded region V, apply $\nabla \cdot \mathbf{v} = \nabla \cdot \mathbf{v}' = 0$ and $\mathbf{u} = \mathbf{v} = \mathbf{v}' = 0$ on S to obtain

$$\frac{d}{dt} \int \tfrac{1}{2} u^2 \, dV = -\int (\mathbf{u} \cdot \mathbf{D} \cdot \mathbf{u} + \nu \nabla \mathbf{u} : (\nabla \mathbf{u})^T) \, dV, \tag{8.185}$$

where $D_{ij} = \frac{1}{2}(v_{i,j} + v_{j,i})$ is the deformation matrix of the basic motion. Clearly viscous terms tend to damp the motion, whereas, for high enough rates of shear in the basic motion ($\mathbf{D}$) the disturbance can grow. Equation (8.135) can be rewritten using the incompressibility condition,

$$u_i \tfrac{1}{2}(v_{i,k} + v_{k,i}) u_k = (v_i u_i u_k)_{,k} - u_k u_{i,k} v_k, \tag{8.186}$$

and the divergence theorem:

$$\frac{d}{dt} \int \tfrac{1}{2} u^2 \, dV = \int [\mathbf{u} \cdot (\nabla \mathbf{u}) \cdot \mathbf{v} - \nu \nabla \mathbf{u} : (\nabla \mathbf{u})^T] \, dV. \tag{8.187}$$

The stability theorem depends on the following inequality, proved by Serrin (1959b):

$$\frac{\alpha}{d^2} \int u^2 \, dV \leq \int \nabla \mathbf{u} : (\nabla \mathbf{u})^T \, dV \tag{8.188}$$

for any $\mathbf{u}$ satisfying $\mathbf{u} = 0$ on S and $\nabla \cdot \mathbf{u} = 0$ in V. The symbol d stands for the edge of a cube which encloses V. Serrin proves (8.188) for $\alpha = (3 + \sqrt{13})\pi^2/2$. In addition we need

$$\mathbf{u} \cdot (\nabla \mathbf{u}) \cdot \mathbf{v} \leq \tfrac{1}{2}\left(\nu \nabla \mathbf{u} : (\nabla \mathbf{u})^T + \frac{u^2 v^2}{\nu}\right). \tag{8.189}$$

This follows from

$$(\mathbf{A} - \mathbf{u}\mathbf{v}) : (\mathbf{A} - \mathbf{u}\mathbf{v}) = \mathbf{A} : \mathbf{A} - 2\mathbf{u} \cdot \mathbf{A} \cdot \mathbf{v} + u^2 v^2 \geq 0 \tag{8.190}$$

with $\mathbf{A} = \nu \nabla \mathbf{u}$. Combining the results gives

$$\frac{d}{dt} \int \tfrac{1}{2} u^2 \, dV \leq (1/\nu)(V_m{}^2 - \alpha \nu^2 d^{-2}) \int \tfrac{1}{2} u^2 \, dV, \tag{8.191}$$

where V_m is the maximum velocity of the base flow. Clearly the flow is stable if

$$V_m d/\nu \leq \sqrt{\alpha} < 8.98, \tag{8.192}$$

where the last value is an improved estimate by Payne and Weinberger (1963), for a sphere of diameter d. For different geometries better values of α are determined by Velte (1962), who obtains upper and lower bounds:

$$\begin{array}{ll}
\text{Straight channel, width } d & 3.74\pi^2 \leq \alpha \leq 3.78\pi^2. \\
\text{Straight pipe, diameter } d & 4.70\pi^2 \leq \alpha \leq 4.78\pi^2. \\
\multicolumn{2}{l}{\text{Plane flow in bounded region, maximum diameter } d} \\
\text{(i) Three-dimensional disturbances} & 4.70\pi^2 \leq \alpha \leq 4.78\pi^2 \\
\text{(ii) Two-dimensional disturbances} & 5.302\pi^2 \leq \alpha \leq 5.312\pi^2 \\
\text{Cube with width } d & 6\pi^2 \leq \alpha \leq 6.33\pi^2.
\end{array} \tag{8.193}$$

The goal of the variational principle (8.198) is to improve these results.

Write (8.185) and (8.188) in dimensionless notation, using the standards V_m for velocity and d^2/ν for time. We desire the maximum coefficient α such that

$$\alpha \int u^2 \, dV \leq \int \nabla \mathbf{u} : (\nabla \mathbf{u})^T \, dV, \tag{8.194}$$

which is Eq. (8.188) in dimensionless notation, for any $\mathbf{u}$ satisfying

$$\nabla \cdot \mathbf{u} = 0 \quad \text{in} \quad V, \tag{8.195a}$$

$$\mathbf{u} = 0 \quad \text{on} \quad S. \tag{8.195b}$$

Similarly for Eq. (8.185) we want the largest $\widetilde{\text{Re}}$ such that

$$\frac{d}{dt} \int \tfrac{1}{2} u^2 \, dV = -\widetilde{\text{Re}} \int [\mathbf{u} \cdot \mathbf{D} \cdot \mathbf{u} + \widetilde{\text{Re}}^{-1} \nabla \mathbf{u} : (\nabla \mathbf{u})^T] \, dV \leq 0 \tag{8.196}$$

subject to the same restrictions (8.185). Consider the following variational principle.

Maximize $I(u)$ among all vector fields $\mathbf{u}$ which have continuous second derivatives and satisfy Eqs. (8.195) and (8.197):

$$\int \nabla \mathbf{u} : (\nabla \mathbf{u})^T \, dV = 1, \tag{8.197}$$

$$I(u) \equiv -\int \mathbf{u} \cdot \mathbf{D} \cdot \mathbf{u} \, dV. \tag{8.198}$$

The solution gives Eq. (8.196) and for $\mathbf{D} = -\mathbf{I}$ we get Eq. (8.194). The Euler equations are derived using a Lagrange multiplier:

$$\delta \int [\mathbf{u} \cdot \mathbf{D} \cdot \mathbf{u} + \mathrm{Re}^{*-1} \nabla \mathbf{u} : (\nabla \mathbf{u})^T - 2\lambda \nabla \cdot \mathbf{u}] \, dV = 0. \tag{8.199}$$

Variations with respect to $\mathbf{u}$, and p give Eq. (8.195a), and

$$2 \int \delta \mathbf{u} \cdot [\mathbf{D} \cdot \mathbf{u} + \nabla \lambda - \mathrm{Re}^{*-1} \nabla^2 \mathbf{u}] \, dV = 0. \tag{8.200}$$

When $\mathbf{D} = -\mathbf{I}$ the Euler equation is

$$\mathbf{u} - \nabla \lambda + \mathrm{Re}^{*-1} \nabla^2 \mathbf{u} = 0. \tag{8.201}$$

A function satisfying the Euler equation (8.201) gives $I(u^*) = 1/\mathrm{Re}^*$, as may be seen by substituting for $\mathbf{D} \cdot \mathbf{u}$ in Eq. (8.198) and integrating by parts, applying the conditions (8.195) and (8.197).

The stability theorem which uses these results is due to Serrin and Joseph (see Joseph, 1966). Equation (8.196) can be rewritten using $K = \int \frac{1}{2} u^2 \, dV$, Eq. (8.194), and the maximum nature of the eigenvalue $1/\mathrm{Re}^*$:

$$\begin{aligned} \frac{dK}{dt} &= -\int \nabla \mathbf{u} : (\nabla \mathbf{u})^T \, dV \left\{ 1 - \mathrm{Re} \frac{-\int \mathbf{u} \cdot \mathbf{D} \cdot \mathbf{u} \, dV}{\int \nabla \mathbf{u} : (\nabla \mathbf{u})^T \, dV} \right\} \\ &\leq -\alpha K \left\{ 1 - \frac{\mathrm{Re}}{\mathrm{Re}^*} \right\}. \end{aligned} \tag{8.202}$$

THEOREM. Let Eq. (8.191) and $\mathrm{Re} < \mathrm{Re}^*$ hold. Then

$$K(t) \leq K(0) \exp\{-(1 - \mathrm{Re}/\mathrm{Re}^*)\alpha t\} \tag{8.203}$$

and $K(t) \to 0$ and the flow is asymptotically stable in the mean.

Thus the flow is stable for Re less than the Re* determined from the variational principle (8.198).

Serrin (1959b) applies these results to Couette flow between two rotating concentric cylinders. He solves the Euler equation (8.200) exactly, for the small gap approximation, and obtains

$$\frac{|\Omega_2 - \Omega_1|}{\nu} R_2 R_1 \leq 41.2 \frac{R_2 + R_1}{R_2 - R_1}, \tag{8.204}$$

where Ω and R denote the angular velocity and radius, with the subscript one referring to the inner cylinder and two to the outer cylinder, $\Omega_1 > 0$. For $R_1 = 3.55$ cm and $R_2 = 4.03$ cm this gives

$$\frac{|\Omega_2 - \Omega_1|}{\nu} \leq 45.5, \tag{8.205}$$

whereas experiments show instability if the number on the right-hand side is greater than 50. Serrin proves these results assuming the normal modes are complete, but this restriction has been removed (see Theorem 11.29 and the references listed there).

Bounds on the transport of momentum in turbulent shear flow are derived by Busse (1969, 1970) using variational methods. Averaged equations of motion are the Euler equations of variational principles applied to plane Couette flow, channel flow and pipe flow. The technique is very similar to that used by Howard (see Section 9.6) to derive bounds on the heat transport by turbulent convection.

EXERCISES

8.1. Use the definition of the variation, Eq. (7.3), to prove that the operations differentiation and variation commute, that is, $\delta(dy/dx) = d(\delta y)/dx$.

8.2. Verify the inequality (8.118). First the tensor sums can be transformed to vector sums, $\sum_{i,j=1}^{3} \bar{d}_{ij}\,\delta d_{ij} = \sum_{i=1}^{9} a_i b_i$, $a_1 = \bar{d}_{11}$, $a_2 = \bar{d}_{12}$, etc. Then consider the quantity $\sum_{i=1}^{9} (a_i + kb_i)^2 \geq 0$ for all real k (see p. 227).

8.3. Verify the variational principle (8.141).

8.4. Show that Eq. (8.145) yields the Einstein approximation for the apparent viscosity as $\lambda^3 \to 0$.

Answer: $\mu_s = \mu[1 + c(2 + 5\eta)/2(1 + \eta)]$; as $\eta \to \infty$, $\mu_s = \mu(1 + 2.5c)$.

REFERENCES

Bateman, H. (1929). Notes on a Differential Equation Which Occurs in the Two-Dimensional Motion of a Compressible Fluid and the Associated Variational Problems, *Proc. Roy. Soc. (London)* **A125**, 598–618.

Bateman, H. (1930). Irrotational Motion of a Compressible Inviscid Fluid, *Proc. Nat. Acad. Sci. (U.S.)* **16**, 816–825.

Bernstein, I. B., Frieman, E. A., Kruskal. M. D., and Kulsrud, R. M. (1958). An Energy Principle for Hydromagnetic Stability Problems, *Proc. Roy. Soc. (London)* **A244**, 17–40.

Bird, R. B. (1960) New Variational Principle for Incompressible Non-Newtonian Flow, *Phys. Fluids.* **3**, 539–541.

Bretherton, F. P. (1970). A Note on Hamilton's Principle for Perfect Fluids, *J. Fluid Mech.* **44**, 19–31.

Brill, J. (1895). Note on the Steady Motion of a Viscous Incompressible Fluid, *Proc. Camb. Phil. Soc.* **8**, 313–322.

Busse, F. H. (1969). Bounds on the Transport of Mass and Momentum by Turbulent Flow between Parallel Plates, *Z. Angew. Math. Phys.* **20**, 1–14.

Busse, F. H. (1970). Bounds for Turbulent Shear Flow, *J. Fluid Mech.* **41**, 219–240.

Cercignani, C., and Pagani, C. D. (1966). Variational Approach to Boundary-Value Problems in Kinetic Theory, *Phys. Fluids* **9**, 1167–1173.

Chandrasekhar, S. (1961). "Hydrodynamic and Hydromagnetic Stability." Oxford Univ. Press (Clarendon), London and New York.

Cotsaftis, M. (1962). Formulation Lagrangienne des Équations de la Magnéto-Hydrodynamique Appliquée à l'Étude de la Stabilitié, *Nucl. Fusion Suppl.* Pt. 2, 447–450.

Drobot, S., and Rybarski, A. (1959). A Variational Principle of Hydromechanics, *Arch. Rat. Mech. Anal.* **2**, 393–410.

Dryden, H. L., Murnaghan, F. P., and Bateman, H. (1956). "Hydrodynamics." Dover, New York.

Eckart, C. (1938). The Electrodynamics of Material Media, *Phys. Rev.* **54**, 920–923.

Eckart, C. (1960). Variation Principles of Hydrodynamics, *Phys. Fluids* **3**, 421–427.

Ehrlich, R. and Slattery, J. C. (1968). Evaluation of Power-Model Lubricants in an Infinite Journal Bearing, *Ind. Eng. Chem. Fund.* **7**, 239–246.

Fanselau, R. W. (1959). Comments on "Exhaust Nozzle Contour for Maximum Thrust," *ARS J.* **29**, 456–457.

Finlayson, B. A. (1972). On the Existence of Variational Principles for the Navier–Stokes Equation. *Phys. Fluids* **15**, to be published.

Flumerfelt, R. W., and Slattery, J. C. (1965). A Widely Applicable Type of Variational Integral. II, *Chem. Eng. Sci.* **20**, 157–163.

Fredrickson, A. G. (1964). "Principles and Applications of Rheology." Prentice-Hall, Englewood Cliffs, New Jersey.

Frieman, E. A., and Rotenberg, M. (1960). On Hydromagnetic Stability of Stationary Equilibria, *Rev. Mod. Phys.* **32**, 898–902.

Gerber, R. (1949). Sur la Reduction a un Principe Variationnel des Equations du Mouvement d'un Fluide Visqueux Incompressible, *Ann. Inst. Fourier (Grenoble)* **1**, 157–166.

Giese, J. (1951). Stream Functions for Three-Dimensional Flows, *J. Math. Phys.* **30**, 31–35.

Grad, H., and Rubin, H. (1958). Hydromagnetic Equilibria and Force-Free Fields, *Proc. Int. Conf. Peaceful Uses At. Energy, 2nd* **31**, 190.

Greene, J. M., and Karlson, E. T. (1969). Variational Principle for Stationary Magnetohydrodynamic Equilibria, *Phys. Fluids* **12**, 561–567.

Greene, J. M., Johnson, J. L., Kruskal, M. D., and Wilets, L. (1962). Equilibrium and Stability of Helical Hydromagnetic Systems, *Phys. Fluids* **5**, 1063–1069.

Greenspan, D., and Jain, P. (1967). Application of a Method for Approximating Extremals of Functionals to Compressible Subsonic Fluid Flow, *J. Math. Anal. Appl.* **18**, 85–111.

Guderley, K. C., and Armitage, J. V. (1965). General Approach to Optimum Rocket Nozzles, *in* "Theory of Optimum Aerodynamic Shapes," (A. Miele, ed.), Chapter 11. Academic Press, New York.

Happel, J., and Brenner, H. (1965). "Low-Reynolds Number Hydrodynamics with Special Applications to Particulate Media." Prentice-Hall, Englewood Cliffs, New Jersey.

Hargreaves, P. (1908). A Pressure-Integral as Kinetic Potential, *Phil Mag.* [6] **16**, 436–444.

Herivel, J. W. (1955). The Derivation of the Equations of Motion of an Ideal Fluid by Hamilton's Principle, *Proc. Cambridge Phil. Soc.* **51**, 344–349.

Hill, R. (1956). New Horizons in the Mechanics of Solids, *J. Mech. Phys. Solids* **5**, 66–74.

Hill, R., and Power, G. (1956). Extremum Principles for Slow Flow and the Approximate Calculation of Drag, *Quart. J. Mech. Appl. Math.* **9**, 313–319.

Hoffman, J. D. (1967). A General Method of Determining Optimum Thrust Nozzle Contour for Chemically Reacting Gas Flows, *AIAA J.* **5**, 670–676.

Hoffman, J. D., and Thompson, H. D. (1967). Optimum Thrust Nozzle Contours for Gas Particle Flows, *AIAA J.* **5**, 1886–1887.

Hopke, S. W., and Slattery, J. C. (1970). Upper and Lower Bounds on the Drag Coefficient of a Sphere in an Ellis Model Fluid, *AIChE J.* **16**, 224–229.

Joseph, D. D. (1966). Nonlinear Stability of the Boussinesq Equations, *Arch. Rat. Mech. Anal.* **22**, 163–184.

Johnson, M. W., Jr. (1960). Some Variational Theorems for Non-Newtonian Flow, *Phys. Fluids* **3**, 871–878.

Johnson, M. W., Jr. (1961). On Variational Principles for Non-Newtonian Fluids, *Trans. Soc. Rheol.* **5**, 9–21.

Katz, S. (1961). Lagrangian Density for an Inviscid, Perfect, Compressible Plasma, *Phys. Fluid.* **4**, 345–348.

Keller, J. B., Rubenfeld, L. A., and Molyneux, J. E. (1967). Extremum Principles for Slow Viscous Flows with Applications to Suspensions, *J. Fluid Mech.* **30**, 97–125.

Korteweg, D. G. (1883). On a General Theorem of the Stability of the Motion of a Viscous Fluid, *Phil. Mag.* [5] **16**, 112–120.

Kraiko, A. N., and Osipov, A. A. (1968). On the Solution of Variational Problems of Supersonic Flows of Gas with Foreign Particles, *J. Appl. Math. Mech.* **32**, 617–627.

Kraiko, A. N. (1964). Variational Problems of Gas Dynamics of Nonequilibrium and Equilibrium Flows, *J. Appl. Math. Mech.* **28**, 348–360.

Kraiko, A. N. (1967). On the Solution of Variational Problems of Supersonic Gas Dynamics, *J. Appl. Math. Mech.* **30**, 381–391.

Kruskal, M. D., and Kulsrud, R. M. (1958). Equilibrium of a Magnetically Confined Plasma in a Toroid, *Phys. Fluids.* **1**, 265–274.

Lamb, H. (1945). "Hydrodynamics," 6th ed. Dover, New York.

Lin, C. C. (1952). A New Variational Principle for Isenergetic Flows, *Quart. Appl. Math.* **9**, 421–423.

Lin, C. C., and Rubinov, S. I. (1948). On the Flow of Curved Shocks, *J. Math. Phys.* **27**, 105–129.

Lippmann, H. (1963). Inclusion and Variation Laws for Quasi-Static Incompressible Viscous Flow (in Ger.), *Ing. Arch.* **32**, 347–359.

Luke, J. C. (1967). A Variational Principle for a Fluid with a Free Surface, *J. Fluid Mech.* **27**, 395–397.

Lundgren, T. S. (1963). Hamilton's Variational Principle for a Perfectly Conducting Plasma Continuum, *Phys. Fluids* **6**, 898–904.

Lush, P. E., and Cherry, T. M. (1956). The Variational Method in Hydrodynamics, *Quart. J. Mech. Appl. Math.* **9**, 6–21.

Merches, I. (1969). Variational Principle in Magnetohydrodynamics, *Phys. Fluids* **12**, 2225–2227.

Millikan, C. B. (1929). On the Steady Motion of Viscous Incompressible Fluids; with Particular Reference to a Variation Principle, *Phil. Mag.* [7] **7**, 641–662.

Mitsuishi, N., and Aoyagi, Y. (1969). Non-Newtonian Flow in Noncircular Ducts, *Mem. Fac. Eng., Kyushu Univ.* **28**, 223–241.

Nakano, Y., and Tien, C. (1968). Creeping Flow of Power-Law Fluid over Newtonian Fluid Sphere, *AIChE J.* **14**, 145–151.

Newcomb, W. A. (1962). Lagrangian and Hamiltonian Methods in Magnetohydrodynamics, *Nucl. Fusion Suppl.* Pt. 2, 451–463.

Pai, S. I. (1962). "Magnetogasdynamics and Plasma Dynamics." Springer-Verlag, Berlin,

Pawlowski, J. (1954). Über eine Erweiterung des Helmholtzschen Prinzips. *Kolloid Z.* **138**, 6–11.

Payne, L., and Weinberger, H. (1963). An Exact Stability Bound for Navier–Stokes Flow in a Sphere, *in* "Nonlinear Problems" (R. Langer, ed.), Univ. of Wisconsin Press, Madison, Wisconsin.

Penfield, P., Jr. (1966). Hamilton's Principle for Fluids, *Phys. Fluids* **9**, 1184–1194.

Penfield, P., Jr., and Haus, H. A. (1966). Hamilton's Principle for Electromagnetic Fluids, *Phys. Fluids* **9**, 1195–1204.

Prager, W. (1954). On Slow Viscoplastic Flow, *in* "Studies in Mathematics and Mechanics Presented to R. von Mises." Academic Press, New York.

Rao, G. V. R. (1958). Exhaust Nozzle Contour for Optimum Thrust, *Jet Propulsion* **28**, 377–382.

Rayleigh, Lord (1913). On the Motion of a Viscous Fluid, *Phil. Mag.* [6] **26**, 776–786.

Schechter, R. S. (1961). On the Steady Flow of a Non-Newtonian Fluid in Cylinder Ducts, *AIChE J.* **7**, 445–447.

Selig, F. (1964). Variational Principle for Rayleigh–Taylor Instability, *Phys. Fluids* **7**, 1114–1116.

Seliger, R. L., and Whitham, G. B. (1968). Variational Principles in Continuum Mechanics, *Proc. Roy. Soc. (London)* **A305**, 1–25.

Serrin, J. (1959a). On the Stability of Viscous Fluid Motion, *Arch. Rath. Mech. Anal.* **3**, 1–13.

Serrin, J. (1959b). "Handbuch der Physik," Vol. 8/1. Springer-Verlag, Berlin.

Shiffman, M. (1952). On the Existence of Subsonic Flows of a Compressible Fluid, *J. Rat. Mech. Anal.* **1**, 605–652.

Simmons, W. F. (1969). A Variational Method for Weak Resonant Wave Interactions, *Proc. Roy. Soc. (London)* **A309**, 551–575.

Skalak, R. (1970). Extensions of Extremum Principles for Slow Viscous Flows, *J. Fluid Mech.* **42**, 527–548.

Skobelkin, V. I. (1957). Variational Principles in Hydrodynamics, *Sov. Phys. JETP* **4**, 68–73.

Slattery, J. (1961). Flow of a Simple Non-Newtonian Fluid past a Sphere, *Appl. Sci. Res.* **A10**, 286–296.

Slattery, J. C. (1962). Approximations to the Drag Force on a Sphere Moving Slowly through Either an Ostwald–DeWaele or a Sisko Fluid, *AIChE J.* **8**, 663–667.

Slattery, J. C., and Bird, R. B. (1961). Non-Newtonian Flow past a Sphere, *Chem. Eng. Sci.* **16**, 231–241.

Stephens, J. J. (1967). Alternate Forms of the Herivel–Lin Variational Principle, *Phys. Fluids* **10**, 76–77.

Strieder, W. (1970). Upper and Lower Bounds on the Shear Stress of a Rarefied Gas, *Phys. Fluids* **13**, 206–208.

Su, C. L. (1970). Variational Principles for a Rarefied Gas Flow past an Arbitrary Three-Dimensional Body, *Phys. Fluids* **13**, 71–78.

Taub, A. H. (1949). On Hamilton's Principle for Perfect Compressible Fluids, *in* "Non-linear Problems in Mechanics of Continua," *Proc. Symp. Appl. Math.* Vol. I, pp. 143–157. McGraw-Hill, New York.

Thompson, E. G., Mack, L. R., and Lin, F. S. (1969). Finite-Element Method for Incompressible Slow Viscous Flow with a Free Surface, *in* "Development in Mechanics" (H. J. Weiss, ed.). Vol. 5, Iowa State Univ. Press, Ames, Iowa.

Thomson, W. (1849). On the *Vis-Viva* of a Liquid in Motion, *Cambridge Dub. Math. J.* **4**, 90–94.

Tomita, Y. (1959). On the Fundamental Formula of Non-Newtonian Flow, *Bull. Jap. Soc. Mech. Eng.* **2**, 469–474.

Truesdell, C., and Toupin, R. A. (1960). The Classical Field Theories, *in* "Handbuch der Physik" (S. Flügge, ed.), Vol. III/1. Springer-Verlag, Berlin.

Velte, W. (1962). Über ein Stabilitätskriterium der Hydrodynamik, *Arch. Rat. Mech. Anal.* **9**, 9–20.

Wang, C. T. (1948). Variational Method in the Theory of Compressible Fluid, *J. Aero. Sci.* **15**, 675–685.

Wang, C. (1961). Two Dimensional Subsonic Compressible Flow past Arbitrary Bodies by the Variational Method, NACA TN 2326.

Wang, C. T., and Brodsky, R. F. (1949). Application of Galerkin's Method to Compressible Fluid Flow Problems, *J. Appl. Phys.* **20**, 1255–1256.

Wang, C. T., and Rao, G. V. R. (1950). A Study of the Nonlinear Characteristics of Compressible Flow Equations by Means of Variational Methods, *J. Aeronaut. Sci.* **17**, 343–348.

Wallick, G. C., Savins, J. G., and Arterburn, D. R. (1962). Tomita Solution for the Motion of a Sphere in a Power-Law Fluid, *Phys. Fluids.* **5**, 367–368.

Wasserman, M. L., and Slattery, J. C. (1964). Upper and Lower Bounds on the Drag Coefficient of a Sphere in a Power-Model Fluid, *AIChE J.* **10**, 383–388.

Wenger, N. C. (1970). A Variational Principle for Magnetohydrodynamic Channel Flow, *J. Fluid Mech.* **43**, 211–224.

Whitham, G. B. (1967). Variational Methods and Applications to Water Waves, *Proc. Roy. Soc.* (*London*) **A299**, 6–25.

Woltjer, L. (1958). A Theorem on Force-Free Magnetic Fluids on Hydromagnetic Equilibrium, *Proc. Nat. Acad. Sci.* (*U.S.*) **44**, 489–491.

Ziegenhagen, A. J., Bird, R. B., and Johnson, M. W., Jr. (1961). Non-Newtonian Flow Around a Sphere, *Trans. Soc. Rheol.* **5**, 47–49.

Ziegenhagen, A. (1965). The Very Slow Flow of a Powell–Eyring Fluid around a Sphere, *Appl. Sci. Res.* **A14**, 43–56.

Chapter

9

Variational Principles for Heat and Mass Transfer Problems

Variational principles do not exist for many heat and mass transfer problems of interest. The variety of problems, and hence the number of different differential equations involved, makes it necessary to examine the question of the existence of a variational principle in a more systematic fashion than was done in Chapter 8. We introduce Fréchet differentials and Fréchet derivatives in order to give a general treatment of variational principles for nonlinear differential equations, followed by a discussion of applications of variational principles to heat and mass transfer problems.

9.1 Fréchet Derivatives

As an analogy, consider the vector field, $\mathbf{v}$. Can $\mathbf{v}$ be derived from a potential? If the curl of the vector is zero,

$$\nabla \times \mathbf{v} = 0, \tag{9.1}$$

then the vector can be represented as a potential

$$\mathbf{v} = \nabla \phi. \tag{9.2}$$

When Eq. (9.2) holds, the integral of $\mathbf{v}$ along a path in position space $\mathbf{x}$ is independent of the actual path taken and depends only on the end points $\mathbf{x}_1$ and $\mathbf{x}_2$,

$$\int_1^2 \mathbf{v} \cdot d\mathbf{x} = \phi_2 - \phi_1. \tag{9.3}$$

This provides an equivalent method of determining whether a vector field can be derived from a potential. Calculate the integral along two paths starting from $\mathbf{x}$ and each ending at $\mathbf{x}' = \mathbf{x} + \varepsilon\boldsymbol{\phi} + \nu\boldsymbol{\psi}$. If the integral is the same for every $\mathbf{x}$ and $\mathbf{x}'$ then the vector can be derived from a potential.

Consider the two paths

$$\begin{aligned} \text{I:}\quad & \mathbf{x} \to \mathbf{x} + \varepsilon\boldsymbol{\phi} \to \mathbf{x} + \varepsilon\boldsymbol{\phi} + \nu\boldsymbol{\psi}, \\ \text{II:}\quad & \mathbf{x} \to \mathbf{x} + \nu\boldsymbol{\psi} \to \mathbf{x} + \varepsilon\boldsymbol{\phi} + \nu\boldsymbol{\psi}, \end{aligned} \tag{9.4}$$

where $\varepsilon\boldsymbol{\phi}$ and $\nu\boldsymbol{\psi}$ are infinitesimal vectors. The integrand is

$$\mathbf{v}(\mathbf{x}) \cdot \varepsilon\boldsymbol{\phi} + \mathbf{v}(\mathbf{x} + \varepsilon\boldsymbol{\phi}) \cdot \nu\boldsymbol{\psi} = \mathbf{v}(\mathbf{x}) \cdot \nu\boldsymbol{\psi} + \mathbf{v}(\mathbf{x} + \nu\boldsymbol{\psi}) \cdot \varepsilon\boldsymbol{\phi} \tag{9.5}$$

and this can be rearranged to give

$$\frac{\mathbf{v}(\mathbf{x} + \varepsilon\boldsymbol{\phi}) - \mathbf{v}(\mathbf{x})}{\varepsilon} \cdot \boldsymbol{\psi} = \frac{\mathbf{v}(\mathbf{x} + \nu\boldsymbol{\psi}) - \mathbf{v}(\mathbf{x})}{\nu} \cdot \boldsymbol{\phi}. \tag{9.6}$$

In terms of a Cartesian coordinate system this can be written as

$$\sum_{k,l} \psi_k \frac{\partial v_k}{\partial x_l} \phi_l = \sum_{k,l} \phi_k \frac{\partial v_k}{\partial x_l} \psi_l. \tag{9.7}$$

Equation (9.7) represents a condition equivalent to Eq. (9.1) for the existence of a potential for the vector field $\mathbf{v}$.

It is clear from these concepts in vector calculus that if we regard the Euler equation in a variational principle as the gradient of a functional, analogous to a potential, then we should not expect every differential equation to be derivable from a potential, because every vector field is not derivable from a potential. In order to make the concepts concrete we must define the gradient of a functional and the derivative of a differential operator. The discussion follows closely that of Tonti (1969), although the basic theorem was proved by Vainberg and is available in his book (1964).

Consider the differential equation

$$N(u) = 0, \tag{9.8}$$

which can be nonlinear. We consider here the differential equations and ignore complications introduced by boundary conditions, which are best treated in the context of specific applications. Systems of equations are treated below, but here consider a single equation. The derivative of the operator Eq. (9.8) in the direction ϕ is defined

$$N_u'\phi \equiv \lim_{\varepsilon\to 0} \frac{N(u+\varepsilon\phi)-N(u)}{\varepsilon} = \left[\frac{\partial}{\partial\varepsilon} N(u+\varepsilon\phi)\right]_{\varepsilon=0}. \tag{9.9}$$

The result, $N_u'\phi$ is called the Fréchet differential of the operator in the direction ϕ while N_u' is called the Fréchet derivative of the operator N. The subscript u in N_u' means that the differentiation of the operator is done with respect to the argument u.

The gradient of a functional is defined similarly. Given a functional

$$F(u) = \int L(u)\, dV, \tag{9.10}$$

the Fréchet differential in the direction ϕ is given by

$$\lim_{\varepsilon\to 0} \frac{F(u+\varepsilon\phi)-F(u)}{\varepsilon} = \int \lim_{\varepsilon\to 0} \frac{L(u+\varepsilon\phi)-L(u)}{\varepsilon}\, dV = \int L_u'\phi\, dV. \tag{9.11}$$

The Fréchet differential $L_u'\phi$ depends on both u and ϕ. Integrate by parts to remove the derivative operating on ϕ to obtain

$$\lim_{\varepsilon\to 0} \frac{\Delta F}{\varepsilon} = \int L_u'\phi\, dV = \int \phi N(u)\, dV + \text{boundary terms}, \tag{9.12}$$

and the operator $N(u)$ is the gradient of the functional $F(u)$.

To test if an operator $N(u)$ is the gradient of a functional we must see if the path integral, Eq. (9.12), depends on the path of integration. Consider two paths

$$\begin{aligned} &\text{I:} \quad u \to u+\varepsilon\phi \to u+\varepsilon\phi+\nu\psi,\\ &\text{II:} \quad u \to u+\nu\psi \to u+\nu\psi+\varepsilon\phi. \end{aligned} \tag{9.13}$$

If the path integral is independent of the path taken then the following equation must hold:

$$\int N(u)\varepsilon\phi\, dV + \int N(u+\varepsilon\phi)\nu\psi\, dV = \int N(u)\nu\psi\, dV + \int N(u+\nu\psi)\varepsilon\phi\, dV. \tag{9.14}$$

This can be rearranged to give

$$\int \frac{N(u+\varepsilon\phi)-N(u)}{\varepsilon}\psi\, dV = \int \frac{N(u+\nu\psi)-N(u)}{\nu}\phi\, dV. \tag{9.15}$$

In the limit as $\varepsilon \to 0$ and $\nu \to 0$,

$$\int \psi N_u'\phi\, dV = \int \phi N_u'\psi\, dV. \tag{9.16}$$

This is the extension of Eq. (9.7) to function spaces and expresses the fact that the operator N_u' is symmetric. The fact that Eq. (9.16) is the condition for existence of a functional giving the operator $N(u)$ as its gradient follows from the following theorem given by Vainberg (1964).

THEOREM 9.1. Suppose that the following conditions are fulfilled:

(1) N is an operator from E into the conjugate space E^*.
(2) N has a linear Gateaux differential $D(x, h)$ at every point of the ball D: $\|x - x_0\| < r$.
(3) The functional $(DN(x, h_1), h_2)$ is continuous in x at every point of D.

Then, in order that the operator N be potential in the ball D, it is necessary and sufficient that the bilinear functional $(DN(x, h_1), h_2)$ be symmetric for every x in D, i.e. that

$$(DN(x, h_1), h_2) = (DN(x, h_2), h_1) \tag{9.17}$$

for every h_1, h_2 in E and every x in D.

If a Fréchet differential exists then so does a Gateaux differential. Equation (9.17) is just the symmetry condition (9.16). Under the conditions that an operator is the gradient of a functional, i.e.,

$$N(u) = \text{grad } F(u), \tag{9.18}$$

the functional itself can be written as

$$F(u) = \int u \int_0^1 N(\lambda u)\, d\lambda\, dV. \tag{9.19}$$

The variation of the functional $F(u)$ due to a variation in u is shown below to be

$$\delta F = \int N(u)\, \delta u\, dV. \tag{9.20}$$

Clearly if F is the functional in a variational principle, then $N(u)$ is the Euler equation. Consequently, the question of whether a variational principle exists for a given operator depends on whether the operator has a symmetric Fréchet differential, Eq. (9.16).

We turn next to concrete examples. Suppose the differential operator is of the form

$$f(u; u_{,j}; u_{,jk}) = 0, \tag{9.21}$$

then from Eq. (9.9) we have

$$N_u'\phi = \left[\frac{\partial}{\partial\varepsilon} f(u + \varepsilon\phi;\, u_{,j} + \varepsilon\phi_{,j}\,;\, u_{,jk} + \varepsilon\phi_{,jk})\right]_{\varepsilon=0}$$

$$= \frac{\partial f}{\partial u}\phi + \frac{\partial f}{\partial u_{,j}}\phi_{,j} + \frac{\partial f}{\partial u_{,jk}}\phi_{,jk}. \tag{9.22}$$

Test the symmetry requirement of Eq. (9.16) by integrating by parts

$$\begin{aligned}
\int \psi N_u'\phi\, dV &= \int \psi\left[\frac{\partial f}{\partial u} + \frac{\partial f}{\partial u_{,j}}\nabla_j + \frac{\partial f}{\partial u_{,jk}}\nabla_j\nabla_k\right]\phi\, dV \\
&= \int \phi\left\{\left[\frac{\partial f}{\partial u} - \nabla_j\left(\frac{\partial f}{\partial u_{,j}}\right) + \nabla_k\nabla_j\left(\frac{\partial f}{\partial u_{,jk}}\right)\right]\psi\right. \\
&\quad + \left[-\frac{\partial f}{\partial u_{,j}} + 2\,\nabla_k\left(\frac{\partial f}{\partial u_{,jk}}\right)\right]\nabla_j\psi \\
&\quad \left. + \frac{\partial f}{\partial u_{,jk}}\nabla_k\nabla_j\,\psi\right\} dV + \text{boundary terms} \\
&\equiv \int \phi\tilde{N}_u'\psi\, dV.
\end{aligned} \tag{9.23}$$

The last equation is the definition of the Fréchet derivative $\tilde{N}_u'$ which is "adjoint"† to N_u'. The symmetry requirement is then

$$\int \phi\tilde{N}_u'\psi\, dV = \int \phi N_u'\psi\, dV,$$

$$\int \phi\tilde{N}_u'\psi\, dV = \int \phi\left[\frac{\partial f}{\partial u} + \frac{\partial f}{\partial u_{,j}}\nabla_j + \frac{\partial f}{\partial u_{,jk}}\nabla_j\nabla_k\right]\psi\, dV. \tag{9.24}$$

If this is to hold for arbitrary ϕ nd ψ, then

$$\begin{aligned}
\frac{\partial f}{\partial u} - \nabla_j\left(\frac{\partial f}{\partial u_{,j}}\right) + \nabla_k\,\nabla_j\left(\frac{\partial f}{\partial u_{,jk}}\right) &= \frac{\partial f}{\partial u}, \\
-\frac{\partial f}{\partial u_{,j}} + 2\,\nabla_k\left(\frac{\partial f}{\partial u_{,jk}}\right) &= \frac{\partial f}{\partial u_{,j}}.
\end{aligned} \tag{9.25}$$

These are equivalent to

$$\frac{\partial f}{\partial u_{,j}} - \nabla_k\left(\frac{\partial f}{\partial u_{,jk}}\right) = 0. \tag{9.26}$$

† The adjoint is usually defined for linear operators (see Section 9.2) but the concept is useful for nonlinear operators as well.

Equation (9.26) is then the condition for Eq. (9.21) to be derivable from a potential, and hence have a variational principle. Table 9.1 presents similar conditions for other forms of the operators. Applications of this table for transport problems are discussed in other sections of the chapter.

TABLE 9.1

SUMMARY OF CONDITIONS FOR AN OPERATOR TO BE DERIVABLE FROM A POTENTIAL

Function	Operator	Conditions	
$u(x)$	$f(u,u',u'',u''',u^{\text{iv}})$	$\dfrac{\partial f}{\partial u'} - \dfrac{d}{dx}\dfrac{\partial f}{\partial u''} + \dfrac{d^3}{dx^3}\dfrac{\partial f}{\partial u^{\text{iv}}} = 0;\quad 2\dfrac{d}{dx}\dfrac{\partial f}{\partial u^{\text{iv}}} = \dfrac{\partial f}{\partial u'''}$	(9.A)
$u(x_1, \ldots, x_n)$	$f(u;u_{,j};u_{,jk})$	$\dfrac{\partial f}{\partial u_{,j}} - \nabla_k \dfrac{\partial f}{\partial u_{,jk}} = 0$	(9.B)
$u_s(x_1, \ldots, x_n)$	$f^l(u_s;u_{s,j};u_{s,jk})$	$\dfrac{\partial f^l}{\partial u_{s,jk}} = \dfrac{\partial f^s}{\partial u_{l,jk}}$	(9.C1)
		$\dfrac{\partial f^l}{\partial u_{s,j}} = -\dfrac{\partial f^s}{\partial u_{l,j}} + 2\nabla_k \dfrac{\partial f^s}{\partial u_{l,jk}}$	(9.C2)
		$\dfrac{\partial f^l}{\partial u_s} = \dfrac{\partial f^s}{\partial u_l} - \nabla_j \dfrac{\partial f^s}{\partial u_{l,j}} + \nabla_j \nabla_k \dfrac{\partial f^s}{\partial u_{l,jk}}$	(9.C3)
$u(x_1, \ldots, x_n)$	$f(u;v;u_{,k};v_{,k};u_{,kj})$	$\dfrac{\partial g}{\partial v_{,j}} - \nabla_k \dfrac{\partial g}{\partial v_{,jk}} = 0;\quad \dfrac{\partial f}{\partial u_{,j}} - \nabla_k \dfrac{\partial f}{\partial u_{,jk}} = 0.$	(9.D)
$v(x_1, \ldots, x_n)$	$g(u;v;u_{,k};v_{,k};v_{,kj})$	$\dfrac{\partial g}{\partial u_{,k}} = -\dfrac{\partial f}{\partial v_{,k}};\quad \dfrac{\partial g}{\partial u} = \dfrac{\partial f}{\partial v} - \nabla_k \dfrac{\partial f}{\partial v_{,k}}$	

The variation of the functional is evaluated as follows:

$$F(u + \varepsilon\phi) = \int (u + \varepsilon\phi) \int_0^1 N(\lambda u + \lambda\varepsilon\phi)\, d\lambda\, dV,$$

$$\left.\frac{dF}{d\varepsilon}\right|_{\varepsilon=0} = \int \left[\phi \int_0^1 N(\lambda u)\, d\lambda + u \int_0^1 N'_{\lambda u} \lambda\phi\, d\lambda\right] dV$$

$$= \int \phi \left[\int_0^1 N(\lambda u)\, d\lambda + \int_0^1 \tilde{N}'_{\lambda u} u\lambda\, d\lambda\right] dV$$

(equation continues)

$$= \int \phi \left\{ \int_0^1 N(\lambda u)\, d\lambda - \int_0^1 N(\lambda u)\, d\lambda + \Big[N(\lambda u)\lambda \Big]_0^1 \right\} dV$$

$$= \int \phi N(u)\, dV, \tag{9.27}$$

where we have used a symmetric Fréchet differential and

$$\tilde{N}'_{\lambda u} u\lambda = N'_{\lambda u} u\lambda = \lambda \frac{d}{d\lambda} N(\lambda u) = \frac{d}{d\lambda}[\lambda N(\lambda u)] - N(\lambda u). \tag{9.28}$$

It is clear that Eq. (9.20) provides the Euler equation.

We next consider systems of equations but limit consideration to two equations for illustrative purposes:

$$M(u, v) = 0, \qquad N(u, v) = 0. \tag{9.29}$$

Represent this system as a vector system operating on a vector function

$$R(z) = 0, \tag{9.30}$$

then the condition for the existence of a potential is Eq. (9.16),

$$\int \psi R_z' \phi \, dV = \int \phi R_z' \psi \, dV. \tag{9.31}$$

Written in full this is

$$\begin{aligned} &\int [\psi_1(M_u'\phi_1 + M_v'\phi_2) + \psi_2(N_u'\phi_1 + N_v'\phi_2)]\, dV \\ &\qquad = [\phi_1(M_u'\psi_1 + M_v'\psi_2) + \phi_2(N_u'\psi_1 + N_v'\psi_2)]\, dV. \end{aligned} \tag{9.32}$$

The symmetry condition is then that N is symmetric in v and M is symmetric in u,

$$N_v' = \tilde{N}_v', \qquad M_u' = \tilde{M}_u', \tag{9.33}$$

and that

$$N_u' = \tilde{M}_v'. \tag{9.34}$$

Evaluate these conditions for the set of equations

$$f(u; v; u_{,k}; v_{,k}; u_{,jk}) = 0, \qquad g(u; v; u_{,k}; v_{,k}; v_{,jk}) = 0. \tag{9.35}$$

Equation (9.33) can be tested for the individual operators using Table 9.1. To test Eq. (9.34) we evaluate the Fréchet differentials in the direction ϕ_1 and ψ_2.

$$\begin{aligned} M_v' \psi_2 &= \frac{\partial f}{\partial v} \psi_2 + \frac{\partial f}{\partial v_{,k}} \psi_{2,k}, \\ N_u' \phi_1 &= \frac{\partial g}{\partial u} \phi_1 + \frac{\partial g}{\partial u_{,k}} \phi_{1,k}. \end{aligned} \tag{9.36}$$

The "adjoint" operator is

$$\int \phi_1 M_v{}'\psi_2 \, dV = \int \phi_1 \left[\frac{\partial f}{\partial v} \psi_2 + \frac{\partial f}{\partial v_{,k}} \psi_{2,k} \right] dV$$

$$= \int \psi_2 \left[\frac{\partial f}{\partial v} \phi_1 - \nabla_k \left(\frac{\partial f}{\partial v_{,k}} \right) \phi_1 - \frac{\partial f}{\partial v_{,k}} \phi_{1,k} \right] dV$$

$$+ \text{ boundary terms}$$

$$= \int \psi_2 \tilde{M}_v{}' \phi_1 \, dV. \tag{9.37}$$

Equation (9.34) then requires

$$\frac{\partial g}{\partial u_{,k}} = - \frac{\partial f}{\partial v_{,k}}, \tag{9.38a}$$

$$\frac{\partial g}{\partial u} = \frac{\partial f}{\partial v} - \nabla_k \frac{\partial f}{\partial v_{,k}} \tag{9.38b}$$

in addition to the restrictions imposed by Eq. (9.33) as evaluated in Table 9.1. The functional is

$$F(u, v) = \int \left[u \int_0^1 M(\lambda u, \lambda v) \, d\lambda + v \int_0^1 N(\lambda u, \lambda v) \, d\lambda \right] dV. \tag{9.39}$$

If N is a linear operator, that is, $N = L$,

$$L(au + bv) = aLu + bLv, \tag{9.40}$$

then the Fréchet derivative in the direction ϕ is given by

$$L_u{}'\phi = \lim_{\varepsilon \to 0} \frac{L(u + \varepsilon\phi) - L(u)}{\varepsilon} = L\phi. \tag{9.41}$$

Consequently, $L_u{}' = L$. Condition (9.16) then reduces to

$$\int \psi L\phi \, dV = \int \phi L \psi \, dV \tag{9.42}$$

and the functional (9.19) is

$$F(u) = \tfrac{1}{2} \int uLu \, dV. \tag{9.43}$$

If (9.42) holds the operator is self-adjoint, and this concept is discussed in a more direct fashion below.

Note that we have an easy method of determining if a variational principle exists for a given operator. When the Fréchet derivative is symmetric, (9.16), a variational principle exists and the functional is given by (9.19). When the Fréchet derivative is not symmetric, no variational principal exists, at least for the equation in the form assumed. We see below that sometimes the equation can be transformed into a form which has a symmetric Fréchet differential, or a variational principle applies to the equation and its "adjoint."

9.2 Variational Principles for Non-Self-Adjoint Equations

Given a linear differential operator, we define the adjoint operator (Courant and Hilbert, 1953). When the adjoint operator is the same as the original operator the operator is self-adjoint, and (9.42) holds. A variational principle exists, using (9.43), provided the boundary conditions are appropriate. When the operator is non-self-adjoint and (9.42) does not hold, we formulate a variational principle for the original operator and its adjoint, as has been done by Roussopolos (1953) and Morse and Feshbach (1953). Finally, we extend the concepts to nonlinear equations to illustrate how to derive a variational principle for *any* nonlinear equation and its adjoint.

Consider the linear boundary-value problem,

$$Lu = f \qquad \text{in } V, \tag{9.44a}$$

$$B_i u = 0 \qquad \text{on } S, \tag{9.44b}$$

where $L(u)$ is a linear differential operator in the domain V with a piecewise smooth boundary S. Note that the boundary conditions are homogeneous. If the problem is originally stated in the form

$$Lw = h \quad \text{in } V, \qquad B_i w = g_i \quad \text{on } S, \tag{9.45}$$

we assume that it is possible to find a function v which satisfies the non-homogeneous boundary conditions $B_i v = g_i$ and which can be extended into the region V. Then the new function $u = w - v$ satisfies equations of the form of Eq. (9.44). Such a transformation implies certain conditions of smoothness on the boundary conditions which we assume are satisfied.

For the problem (9.44) define the adjoint operator in the following way. The inner product is

$$(u, v) \equiv \int uv \, dV \tag{9.46}$$

for any piecewise continuous functions u and v. To obtain the adjoint operator take the inner product of v and Lu; successive integration by parts yields

$$(v, Lu) = (u, L^* v) + B(u, v), \tag{9.47}$$

where L^* is called the adjoint operator to L. The term $B(u, v)$ is a boundary term, defined by the integration by parts used to obtain (9.47), and has the following property. $B(u, v) = 0$ for all u satisfying $B_i u = 0$ on S implies

$B_j{}^*v = 0$ on S. Comparison of this equation to (9.16) shows a variational principle exists if $L = L^*$ and $B_i = B_i{}^*$. For example, if $Lu = du/dx$, then

$$\int_a^b vu'\,dx = \int_a^b (uv)'\,dx - \int_a^b uv'\,dx$$

$$= -\int_a^b uv'\,dx + \Big[uv\Big]_a^b. \tag{9.48}$$

Consequently, $L^* = -d/dx$. The following pairs of operators are adjoints.

$$L(u) = \sum_{i=0}^{n} a_i(x)\frac{d^i u}{dx^i},$$

$$L^*(v) = \sum_{i=0}^{n} (-1)^i \frac{d^i(a_i v)}{dx^i}, \tag{9.49}$$

$$L(u) = \frac{\partial u}{\partial t} + \mathbf{u}\cdot\nabla u - \nabla^2 u - Ku,$$

$$L^*(v) = -\frac{\partial v}{\partial t} - \nabla\cdot(\mathbf{u}v) - \nabla^2 v - Kv.$$

To determine the last equation we must integrate over V and t.

The adjoint boundary conditions are determined from Eq. (9.47). For example, in (9.48) when the boundary condition is $u(a) = 0$, the adjoint boundary condition is $v(b) = 0$. A general formula for the boundary terms for partial differential equations is given by Morse and Feshbach (1953, p. 874). The inner product can be defined in a more general way than (9.46) and this proves useful for boundary conditions of the third kind for diffusion problems, as is illustrated in Section 7.2.

When the adjoint operator and boundary conditions are the same as the original operator and boundary conditions, the system is said to be self-adjoint. For a self-adjoint system (9.47) yields

$$(v, Lu) = (u, Lv) \tag{9.50}$$

for all functions u and v satisfying the boundary conditions $B_i u = 0$. A physical interpretation of the self-adjoint property is given by Courant and Hilbert (1953, p. 354). If the force A applied at the point x produces a result B at the point y, then in a self-adjoint system the force A applied at y produces the result B at the point x. The physical situation has a certain symmetry which is retained in the mathematical representation, exhibited in (9.50) as well as the Green's function, which is symmetric. In matrix theory, the analog of a self-adjoint differential operator is a symmetric matrix, and this analogy is exploited particularly well by Lanczos (1961).

Consider the possibility of transforming an apparently non-self-adjoint equation to a self-adjoint form. There are three ways to transform a linear second-order ordinary differential equation of the form

$$Lu = a_2 u'' + a_1 u' + a_0 u, \tag{9.51}$$

where $a_i = a_i(x)$. Table 9.1 says the operator is symmetric and has a variational principle if

$$a_1 - \frac{da_2}{dx} = 0. \tag{9.52}$$

If this condition does not hold, we try to find an integrating factor and write

$$f = g(x)Lu. \tag{9.53}$$

Then Table 9.1 gives the condition

$$ga_1 - \frac{d}{dx}(ga_2) = 0 \tag{9.54}$$

for a symmetric, or self-adjoint, form. Solving for g we obtain

$$g(x) = \exp\left\{\int \frac{a_1 - a_2'}{a_2}\, dx\right\}. \tag{9.55}$$

The operator (9.53) is then self-adjoint. This transformation can be affected in several ways (Courant and Hilbert, 1953, p. 279): (1) multiply Lu by $g(x)$, as in Eq. (9.53), (2) introduce the new independent variable $x' = \int g(x)\, dx$, or (3) introduce a new dependent variable $v = ug(x)$. The resulting self-adjoint form is

$$Lw = (p(x)w')' + q(x)w. \tag{9.56}$$

For equations with constant coefficients, even derivatives are self-adjoint whereas odd derivatives are non-self-adjoint.

Next we derive a variational principle for (9.44) and its adjoint:

$$L^*v = g \qquad \text{in } V, \tag{9.57a}$$

$$B_j^*v = 0 \qquad \text{on } S, \tag{9.57b}$$

where g is, for the present, an arbitrary function. By the definition of the adjoint, (9.47),

$$(v, Lu) = (u, L^*v) \tag{9.58}$$

for functions u and v satisfying $B_i u = 0$, $B_j^*v = 0$. The variational principle is then the following.

Make the functional $I(u, v)$ stationary among functions u and v satisfying $B_i u = 0$ and $B_j^* v = 0$:

$$I(u, v) = (u, L^*v - g) - (v, f). \tag{9.59}$$

To determine the Euler equations we consider $I(u + \varepsilon\eta, v + \kappa\xi)$, where η and ξ satisfy the boundary conditions (9.44b) and (9.57b), respectively. Then the Euler equations are determined from

$$\begin{aligned} \left.\frac{\partial I}{\partial \varepsilon}\right|_{\kappa=\varepsilon=0} &= 0 \rightarrow (\delta u, L^*v - g) = 0, \\ \left.\frac{\partial I}{\partial \kappa}\right|_{\kappa=\varepsilon=0} &= 0 \rightarrow (u, L^*\delta v) - (\delta v, f) = (\delta v, Lu - f) = 0. \end{aligned} \tag{9.60}$$

The last relation holds because of Eq. (9.58). Thus when the functional is stationary to variations in u and v the Euler equations are (9.44) and (9.57). The second variation is

$$\delta^2 I = \frac{\varepsilon^2}{2}\left.\frac{\partial^2 I}{\partial \varepsilon^2}\right|_0 + \varepsilon\kappa\left.\frac{\partial^2 I}{\partial \varepsilon\,\partial \kappa}\right|_0 + \frac{\kappa^2}{2}\left.\frac{\partial^2 I}{\partial \kappa^2}\right|_0 = (\delta u, L^*\,\delta v), \tag{9.61}$$

which can be either positive or negative; consequently the variational principle is not a minimum principle.

Suppose the system is self-adjoint. Then the above relations collapse into the following form, taking $g = f$,

$$\begin{aligned} I(u, v) &\rightarrow I(u) = (u, Lu - 2f), \\ \left.\frac{dI(u + \varepsilon\eta)}{d\varepsilon}\right|_{\varepsilon=0} &= 0 \rightarrow 2(\delta u, Lu - f) = 0. \end{aligned} \tag{9.62}$$

A variational principle thus exists for a linear boundary-value problem if the system is self-adjoint. For non-self-adjoint systems a variational principle exists for the system and its adjoint. The necessary condition that (9.62) give a minimum principle is

$$(\delta u, L\,\delta u) \geq 0 \tag{9.63}$$

and apparently a sufficient condition is

$$(\delta u, L\,\delta u) \geq \gamma^2(\delta u, \delta u), \qquad \gamma^2 > 0. \tag{9.64}$$

This is the requirement at least for second-order differential operators.

We have formulated a variational principle for a general linear boundary-value problem. One of the advantages of variational principles is that the functional is approximated more closely than is the solution. In the case of adjoint variational principles the functional can sometimes be related to a quantity of interest physically, as suggested by Lewins (1965). Suppose we wish to calculate an average of the solution

$$J = (g, u), \tag{9.65}$$

where g is known, The problem is Eq. (9.44) and the adjoint problem is Eq. (9.57). Then the variational integral

$$\begin{aligned} I(u, v) &= (u, L^*v - g) - (v, f) \\ &= (v, Lu - f) - (u, g). \end{aligned} \tag{9.66}$$

Thus when u satisfies Eq. (9.44), $I = -J$. Furthermore first-order errors in u contribute second-order errors in I.

Another approach to finding a variational principle for Eq. (9.44) is closely related to the least-squares method. The least-squares method minimizes the mean-square residual

$$J(u) = \int_v (Lu - f)^2 \, dV \tag{9.67}$$

among functions satisfying $B_i u = 0$. The variation is [see Eq. (9.47)]

$$\delta J = 2 \int_v (Lu - f) L \, \delta u \, dV = 2 \int_v \delta u \, (L^*Lu - L^*f) \, dV + B(\delta u, Lu - f) \tag{9.68}$$

Thus the Euler equation and natural boundary conditions are

$$L^*Lu = L^*f \quad \text{in } V, \qquad B_j^*Lu = B_j^*f \quad \text{on } S. \tag{9.69}$$

Given the problem (9.44), we can form the self-adjoint, higher-order system

$$\begin{aligned} L^*Lu &= L^*f \quad \text{in V}, \\ \left.\begin{aligned} B_i u &= 0 \qquad \text{(essential BC)} \\ B_j^*Lu &= B_j^*f \quad \text{(natural BC)} \end{aligned}\right\} &\text{ on } S. \end{aligned} \tag{9.70}$$

Mikhlin (1964, p. 495) discusses the conditions under which the solution of this problem reduces to the solution to (9.44). If (9.44) is soluble and the inequality

$$(u, u) \le K(Lu, Lu) \tag{9.71}$$

holds, then the least-squares method, using (9.67), is equivalent to the variational method applied to (9.69). If the adjoint problem is unique, that is

$$L^*v = 0, \qquad B_j^*v = 0 \tag{9.72}$$

implies $v = 0$, then solving (9.70) is equivalent to solving (9.44). Using the adjoint operator to change the problem to a form easier to solve has been used by Southwell (1956) in relaxation methods. To change $u_t = u_{xx}$ to a jury problem, he operates on it with the adjoint $u_t + u_{xx}$ to obtain $w_{tt} = w_{xxxx}$, a jury problem.

In Section 9.1 we determined that if a linear operator is self-adjoint, Eq. (9.42), then a variational principle exists. In this section we considered linear, non-self-adjoint operators and constructed variational principles for the system of equations: the original and adjoint equations. For nonlinear operators, we determined that a variational principle exists if the Fréchet differential is symmetric, Eq. (9.16). The corresponding generalization—a variational principle for a nonlinear operator and its "adjoint"—has apparently not been developed. We do that here in a formal fashion.

For the problem

$$N(u) = f, \qquad B_i(u) = 0, \tag{9.73}$$

we define the "adjoint" by integration by parts, as in (9.23),

$$N^*(u, v) = \tilde{N}_u{}'v. \tag{9.74}$$

Assume that the boundary terms in (9.23) vanish whenever $B_i u = B_i \phi = 0$ and $B_j{}^* v = B_j{}^* \psi = 0$. Then the variational principle is the following.

Make stationary the functional $I(u, v)$ among all functions u, v satisfying the appropriate continuity requirements and $B_i u = 0$, $B_j{}^* v = 0$:

$$I(u, v) = \int [vN(u) - ug - vf]\, dV. \tag{9.75}$$

The Euler equations are

$$\begin{aligned} \delta v&: N(u) - f = 0, \\ \delta u&: \tilde{N}_u{}'v - g \equiv N^*(u, v) - g = 0. \end{aligned} \tag{9.76}$$

Consequently we can derive a variational principle for any operator, linear or nonlinear, provided we are willing to treat a system of equations and allow operators of the form (9.74). These are purely mathematical constructs and may have no physical meaning. Example applications are in Sections 8.6 and 9.5. The variational principle (9.75) is very similar to that presented by Slattery (1964) for the equations of motion and energy, although (9.73)–(9.76) give a systematic way to derive the variational principle. In this book whenever we say a variational principle does not exist, we mean that one does not exist for the problem as stated, usually in the form of conservation equations of mass, momentum, and energy. If the problem is expanded to include its "adjoint" a variational principle exists.

9.3 Variational Principles for the Transport Equation

The formalism of Fréchet differentials and non-self-adjoint operators can be applied to the transport equations (1.5) and (1.6), to see when variational principles exist. Consider in turn the various complications: unsteady state, convection, nonlinear, diffusivity and chemical reaction.

Take first the unsteady state equation with no convection and constant diffusivity. Let $x_4 = t$:

$$f(c_{,4}; c_{,jj}) = c_{,4} - \mathscr{D}c_{,jj}. \tag{9.77}$$

Table 9.1, Eq. (9.B), gives $\partial f/\partial c_{,4} = 1 \neq 0$, so that a variational principle does not exist. Consider the possibility of an integrating factor

$$f = g(c_{,4}; c_{,jj})(c_{,4} - \mathscr{D}c_{,jj}). \tag{9.78}$$

Table 9.1 gives the restrictions for a variational principle:

$$\frac{\partial g}{\partial c_{,4}}(c_{,4} - \mathscr{D}c_{,jj}) + g = 0, \qquad \frac{\partial g}{\partial c_{,jk}}(c_{,4} - \mathscr{D}c_{,ll}) - \mathscr{D}g\,\delta_{jk} = q(t) \tag{9.79}$$

with q an arbitrary function of time. The solution is

$$g = (c_{,4} - \mathscr{D}c_{,jj})^{-1}, \tag{9.80}$$

which makes $f = 1$ and gives no useful result. Thus no integrating factor exists.

Consider next the convective term,

$$f(c_{,j}; c_{,ll}) = u_j c_{,j} - \mathscr{D}c_{,ll}. \tag{9.81}$$

Table 9.1 gives $\partial f/\partial c_{,j} = u_j \neq 0$ so a variational principle does not exist. Assume an integrating factor of the form $g(c; c_{,k})$:

$$f = g(c; c_{,k})(u_j c_{,j} - \mathscr{D}c_{,ll}). \tag{9.82}$$

Table 9.1 gives the following condition:

$$\frac{\partial g}{\partial c_{,k}}(u_j c_{,j} - \mathscr{D}c_{,ll}) + g u_k + \mathscr{D}g_{,k} = 0. \tag{9.83}$$

Since g does not depend on $c_{,jj}$ this reduces to

$$\frac{\partial g}{\partial c_{,k}} = 0, \tag{9.84a}$$

$$g u_k + \mathscr{D}g_{,k} = 0. \tag{9.84b}$$

Equation (9.84b) can be solved to give $u_k = -(\mathscr{D}\ln g)_{,k}$ which means the velocity is a potential function. A variational principle exists only when the velocity is

$$\mathbf{u} = -\nabla\Omega. \tag{9.85}$$

The integrating factor is

$$g = \exp(\Omega/\mathscr{D}). \tag{9.86}$$

The fact that the problem can be cast into a self-adjoint form when the velocity is of the form (9.85) was pointed out by Ames and de la Cuesta (1963). The variational integral is given by (9.19):

$$F(c) = \int c\left\{\int_0^1 \exp(\Omega/\mathscr{D})[\Omega_{,l}\lambda c_{,l} + \mathscr{D}\lambda c_{,ll}]\, d\lambda\right\} dV$$

$$= \tfrac{1}{2}\int c \exp(\Omega/\mathscr{D})[\Omega_{,l} c_{,l} + \mathscr{D} c_{,ll}]\, dV. \tag{9.87}$$

This is integrated by parts and a boundary term is added to give

$$F(c) = -\tfrac{1}{2}\int \mathscr{D} \exp(\Omega/\mathscr{D})\, \nabla c \cdot \nabla c\, dV - \tfrac{1}{2}\int \exp(\Omega/\mathscr{D}) k(c - c_s)^2\, dS. \tag{9.88}$$

The Euler equation and natural boundary conditions are

$$\nabla \cdot (\exp(\Omega/\mathscr{D})\, \nabla c) = 0 \quad \text{in } V,$$

$$e^{\Omega/D}[\mathscr{D}\mathbf{n} \cdot \nabla c + k(c - c_s)] = 0 \quad \text{on } S. \tag{9.89}$$

A variational principle similar to this has been presented by Krajewski (1964).

Consider next the nonlinear steady-state heat conduction equation:

$$f(T, T_{,ll} T_{,kk}) = (k(T)T_{,l})_{,l} = \frac{dk}{dT} T_{,l} T_{,l} + kT_{,kk}. \tag{9.90}$$

Table 9.1 gives

$$2T_{,l}\frac{dk}{dT} - \frac{dk}{dT} T_{,l} = 0, \tag{9.91}$$

which is satisfied only if $dk/dT = 0$. Next try an integrating factor

$$f = g(T)\left[\frac{dk}{dT} T_{,l} T_{,l} + kT_{,kk}\right]. \tag{9.92}$$

The condition for a variational principle is then

$$2T_{,l}\, g \frac{dk}{dT} - \nabla_l(kg) = 0, \tag{9.93}$$

which is satisfied if $g = k$. The variational principle for $f = k\ (kT_{,k})_{,k}$ is the same as presented in Section 7.2. The use of an integrating factor to find variational principles for several $k(\mathbf{x}, T)$ relationships was first illustrated by Li (1964), and the treatment above, Eqs. (9.92) and (9.93), was given by Tonti (1969). Wang (1970) derives conditions similar to these for the existence of a "thermokinetic potential" and hence a variational principle.

For the combined problem take

$$f = g(T, T_{,k})(\rho C_p u_j T_{,j} - \frac{dk}{dT} T_{,l} T_{,l} - kT_{,kk}) \tag{9.94}$$

with ρC_p constant. Table 9.1 gives

$$\frac{\partial g}{\partial T_{,k}}\left(\rho C_p u_j T_{,j} - \frac{dk}{dT} T_{,l} T_{,l} - k T_{,kk}\right) + g\left(\rho C_p u_k - 2\frac{dk}{dT} T_{,k}\right) + (kg)_{,k} = 0. \tag{9.95}$$

This requires

$$\frac{\partial g}{\partial T_{,k}} \rho C_p u_j T_{,j} + \frac{\partial g}{\partial T} k T_{,k} + g \rho C_p u_k - \frac{dk}{dT}\left(g T_{,k} + \frac{\partial g}{\partial T_{,k}} T_{,l} T_{,l}\right) = 0. \tag{9.96}$$

Clearly Eq. (9.96) is seldom satisfied so that no variational principle exists for the combined problem (9.94).

Consider the steady-state diffusion equation with a chemical reaction term:

$$\nabla^2 c + R(c) = 0. \tag{9.97}$$

Thus with $f = c_{,jj} + R(c)$, conditions (9.B) in Table 9.1 are satisfied. The variational integral is

$$F(c) = \int \left[-\tfrac{1}{2}\nabla c \cdot \nabla c + c \int_0^1 R(\lambda c)\, d\lambda \right] dV. \tag{9.98}$$

The last term can be rearranged to give

$$c \int_0^1 R(\lambda c)\, d\lambda = \int_0^c R(c')\, dc' \tag{9.99}$$

in which case

$$F(c) = \int \left[-\tfrac{1}{2}\nabla c \cdot \nabla c + \int_0^c R(c')\, dc' \right] dV. \tag{9.100}$$

When the reaction rate is a polynomial expression, $R = c^n$, this becomes

$$F(c) = \int \left[-\tfrac{1}{2}\nabla c \cdot \nabla c + \frac{c^{n+1}}{n+1} \right] dV. \tag{9.101}$$

In summary, for the transport equation, a variational principle exists for steady state when k = constant and $\mathbf{u} = -\nabla\Omega$ or for steady state with no motion and $k = k(T)$.

For more general situations we define variational principles using the adjoint operators, restricting attention here to the linear transport equation:

$$\frac{\partial c}{\partial t} + \mathbf{u} \cdot \nabla c - \nabla \cdot \mathcal{D}\, \nabla c = 0. \tag{9.102}$$

The adjoint variational principles presented here were first derived by Morse and Feshbach (1953, p. 313), Washizu (1955) and Schmit (1956) for the unsteady-state heat conduction equation, and Nichols and Bankoff (1965) for the same equation including convection terms. The variational principle is the following.

Make stationary $I(c, c^*)$ among functions $c(\mathbf{x}, t)$, $c^*(\mathbf{x}, t)$ which have continuous first derivatives in time and continuous second derivatives in space, and satisfy $c = c_s$, $c^* = c_s^*$ on S_1, $c = c_0$ at $t = 0$, $c^* = c_0$ at $t = t_f$:

$$I(c, c^*) = \int_0^{t_f} \int \left[\mathscr{D} \nabla c \cdot \nabla c^* + \frac{1}{2}\left(c^* \frac{\partial c}{\partial t} - c \frac{\partial c^*}{\partial t} \right) + \tfrac{1}{2}\mathbf{u} \cdot (c^* \nabla c - c \nabla c^*) \right] dV\, dt + \int_0^{t_f} \int_{S_2} \left[(\beta + \tfrac{1}{2}\mathbf{u} \cdot \mathbf{n}) c c^* - (c + c^*) g \right] ds\, dt + \tfrac{1}{2} \int \left[c_0 (c^* - c) \right]_0^{t_f} dV. \tag{9.103}$$

The Euler equations are

$$\frac{\partial c}{\partial t} + \mathbf{u} \cdot \nabla c - \nabla \cdot (\mathscr{D} \nabla c) = 0, \tag{9.104a}$$

$$-\frac{\partial c^*}{\partial t} - \mathbf{u} \cdot \nabla c^* - \nabla \cdot (\mathscr{D} \nabla c^*) = 0, \tag{9.104b}$$

and the natural boundary conditions are

$$\left.\begin{aligned} \mathscr{D}\mathbf{n} \cdot \nabla c + \beta c &= g \\ \mathscr{D}\mathbf{n} \cdot \nabla c^* + \beta c^* + \mathbf{u} \cdot \mathbf{n} c^* &= g \end{aligned}\right\} \text{ on } S_2. \tag{9.105}$$

The functional apparently has no physical significance and is only stationary so that some of the advantages of variational principles are lost. If calculations are performed using trial functions

$$c(\mathbf{x}, t) = \sum_{i=1}^{N} A_i(t) c_i(\mathbf{x}), \tag{9.106a}$$

$$c^*(\mathbf{x}, t) = \sum_{i=1}^{N} B_i(t) c_i^*(\mathbf{x}), \tag{9.106b}$$

where $c_i(\mathbf{x})$ is known and the A_i and B_i functions are to be determined, the results are the same as if one applied MWR to Eqs. (9.104) and (9.105) using weighting functions c_j^*, c_j, respectively. When the same trial functions are appropriate for both c and c^*, that is, $c_i = c_i^*$, the variational method is equivalent to the Galerkin method (Schmit, 1956). This is certainly the case for heat conduction problems. Consequently, even though variational principles are obtained using the adjoint equations, we do not obtain new computational techniques. The variational method is more powerful than MWR only if the functional is used to monitor errors of successive approximations or if the functional itself is a quantity of interest. This latter situation is seldom the case in heat and mass transfer problems.

9.4 Applications to Heat Transfer

Variational principles for heat transfer problems are discussed in several sections above. Steady-state heat conduction is treated in Section 7.2, unsteady-state heat conduction in Section 9.3 [see Eq. (9.103)], and entry-length problems, treated by separation of variables to reduce the problem to an eigenvalue problem, are discussed in Section 3.3 and 7.5. Here we examine variational principles for unsteady-state problems using convolution integrals and Laplace transforms. Both approaches are limited to linear problems and give identical results. We conclude with an application to radiative heat transfer.

Gurtin (1964) used convolution integrals to reduce an initial-value problem to an integral equation, for which he formulated a variational principle. The convolution of two functions u and v is defined as

$$\int_v u^*v\, dV = \int_v \int_0^t u(t-\tau, \mathbf{x})v(\tau, \mathbf{x})\, d\tau\, dV. \tag{9.107}$$

The convolution satisfies a commutative relation $u^*v = v^*u$, as may be seen by transforming the variables in Eq. (9.107), $\eta = t - \tau$:

$$\int_0^t u(t-\tau, \mathbf{x})v(\tau, \mathbf{x})\, d\tau = \int_0^t v(t-\eta, \mathbf{x})u(\eta, \mathbf{x})\, d\eta. \tag{9.108}$$

We also define

$$\nabla u^*\, \nabla v = \int_0^t \nabla u(t-\tau, \mathbf{x}) \cdot \nabla v(\tau, \mathbf{x})\, d\tau. \tag{9.109}$$

Gurtin's variational principle is for the problem

$$\partial u/\partial t = \kappa\, \nabla^2 u \quad \text{in } V, \tag{9.110a}$$

$$u = u_0(\mathbf{x}) \quad \text{at } t = 0, \tag{9.110b}$$

$$u = u_s(\mathbf{x}) \quad \text{on } S. \tag{9.110c}$$

Make the functional Φ stationary for each t and all functions u which have continuous second derivatives in spatial coordinates and continuous first derivatives in time and satisfy $u = u_s$, a given function of $\mathbf{x}$, on the boundary S:

$$\Phi(u; t) = \int_v [u^* u + \kappa^* \nabla u^* \nabla u - 2u_0{}^* u] \, dV. \tag{9.111}$$

The first variation gives

$$\delta\Phi = 2\int_v \delta u^* [u - \kappa^* \nabla^2 u - u_0] \, dV. \tag{9.112}$$

The term in brackets is the Euler equation. If Eq. (9.110a) is integrated with respect to time we obtain

$$u - u_0 = \int_0^t \kappa \nabla^2 u(\tau, \mathbf{x}) \, d\tau = \kappa^* \nabla^2 u, \tag{9.113}$$

which is the same as the Euler equation in (9.112). Thus the variational principle (9.111) applies to (9.110).

Consider applications of this principle. Assume a trial function of the form

$$u^N(\mathbf{x}, t) = u_s(\mathbf{x}) + \sum_{i=1}^{N} A_i(t) u_i(\mathbf{x}). \tag{9.114}$$

Then $\delta u = \partial u^N / \partial A_i$ and (9.112) is

$$\int_0^t \int_v u_i(\mathbf{x}) \left[u(\tau) - \int_0^\tau \kappa \nabla^2 u(\mathbf{x}, s) \, ds - u_0(\mathbf{x}) \right] dV \, d\tau = 0. \tag{9.115}$$

Differentiate twice with respect to t to get

$$\int_v u_i(\mathbf{x}) \left[\frac{\partial u}{\partial t} - \kappa \nabla^2 u \right] dV = 0 \tag{9.116}$$

as the equations governing the approximate solution derived from the variational principle. These equations, however, are also derived by the Galerkin method so that the variational method applied to (9.111) using the trial function (9.114) is equivalent to a Galerkin method. If more general trial functions are used $u^N = u^N(A_i, \mathbf{x}, t)$ additional terms appear in (9.116) and the Galerkin method is not equivalent. Rafalski and Zyszkowski (1969) apply the same procedure (9.114) for a coupled system of equations, the transient energy equation and constitutive relation. Wilson and Nickell (1966) use the variational principle (9.111) to apply the finite element method to transient heat conduction.

Next consider the Laplace transform method of solving (9.110). Taking the Laplace transform [see Eq. (3.66)] gives

$$s\bar{u} - u_0 = \kappa \nabla^2 \bar{u}. \tag{9.117}$$

Divide by s, take the inverse Laplace transform, and use the fact that multiplication in the Laplace transform domain is equivalent to convolution in the time domain (Churchill, 1958, p. 37):

$$\mathscr{L}^{-1}[\bar{u}\bar{v}] = u^*v = \int_0^t u(t-\tau)v(\tau)\,d\tau,$$
$$u - u_0 = \mathscr{L}^{-1}\left[\frac{\kappa}{s}\nabla^2\bar{u}\right] = \kappa^*\,\nabla^2 u. \tag{9.118}$$

This is just (9.113) so the ideas are very similar. O'Toole (1967) applied this idea to time-dependent transport processes. He pointed out that the functional (9.119) provides a means for comparing trial functions, for various s. Similarly the functional (9.111) serves the same purpose, except as a function of t. The variational principle for (9.117) is the following.

Make the variational integral Φ stationary among functions $\bar{u}(s, \mathbf{x})$ which have continuous second spatial derivatives, are continuous functions of s, and satisfy $\bar{u} = u_s/s$ on S.

$$\Phi[\bar{u}; s] = \int_v \left[\bar{u}^2 - 2\frac{u_0}{s}\bar{u} + \frac{\kappa}{s}\nabla\bar{u}\cdot\nabla\bar{u}\right] dV. \tag{9.119}$$

The Euler equations are easily seen to be Eq. (9.117), and the first variation is equivalent to Eq. (9.112). Similar ideas were applied to entry-length problems by Savkar (1970).

We conclude that a variational principle for the Laplace transform of an initial-value problem is equivalent to the variational principle for the convoluted problem. Applications of either method are equivalent to the Galerkin method, provided the trial function is of the form (9.114).

Variational principles also govern heat transfer by radiation. These problems are usually formulated as integral equations, which have not been discussed thus far in this book. All the approximate methods are equally applicable, of course, and variational methods are applicable to a problem of the type

$$z(x) = f(x) + \int_a^b K(x, y)z(y)\,dy \tag{9.120}$$

provided the kernel K is symmetric in y and x. This corresponds to the self-adjointness property for differential equations. Examples of radiation problems are given in the literature by Sparrow (1960), for two finite parallel plates, Sparrow and Haji–Sheikh (1965), and Usiskin and Siegel (1960) for a cylindrical enclosure with specified wall heat flux. We present a simple example treated by Sparrow (1960).

Consider two plates of width L and infinite length placed a distance h apart. Each plate is maintained at the same temperature T and the system loses energy by radiation through the sides, which are exposed to black bodies at zero temperature. Then the radiation flux is only to the surroundings and none is returned. The governing equation is an integral equation with a symmetric kernel:

$$B(x) = \varepsilon\sigma T^4 + \frac{\rho h^2}{2}\int_{-L/2}^{L/2} B(y)\,\frac{dy}{[(y-x)^2 + h^2]^{3/2}}. \qquad (9.121)$$

Here ε is the emissivity, σ is Stefan's constant, $\rho = 1 - \varepsilon$ is the ref.ec'.vity, y is the coordinate perpendicular to the surfaces, and x is the coordinate parallel to the surfaces. The center of the coordinate system is at the midpoint between the surfaces. $B(x)$ represents the total radiation flux leaving area dA due to emission and reflection. In dimensionless form this is, with $\gamma = h/L$, the aspect ratio,

$$\beta(x) = 1 + \frac{\rho\gamma^2}{2}\int_{-1/2}^{1/2} \beta(y)\,\frac{dy}{[(y-x)^2 + \gamma^2]^{3/2}}. \qquad (9.122)$$

The variational principle is the following.

Make stationary the variational integral I among all continuous functions β:

$$I = \frac{\rho\gamma^2}{2}\int_{-1/2}^{1/2}\int_{-1/2}^{1/2} \frac{\beta(x)\beta(y)\,dx\,dy}{[(y-x)^2 + \gamma^2]^{3/2}} - \int_{-1/2}^{1/2} \beta^2(x)\,dx + 2\int_{-1/2}^{1/2} \beta(x)\,dx.$$

(9.123)

Sparrow does calculations using a trial function which is symmetric in x and y, since the problem obviously has that symmetry. A simple power series is used:

$$\beta(x) = \sum_{i=1}^{N} c_i x^{2i-2}. \qquad (9.124)$$

For the third approximation some of the integrals arising in Eq. (9.123) must be computed numerically. The solution depends on the reflectivity and the aspect ratio. Shown in Table 9.2 is the heat transfer to the surroundings per unit width:

$$\frac{Q/L}{\varepsilon\sigma T^4} = \left[1 - \varepsilon\int_{-1/2}^{1/2} \beta\,dx\right]\Big/\rho. \qquad (9.125)$$

TABLE 9.2

RADIANT HEAT TRANSFER; VALUES OF $Q/L\varepsilon\sigma T^4$

ε	$\gamma=1$	$\gamma=0.5$		$\gamma=0.1$	
	$N=2$	$N=2$	$N=3$	$N=2$	$N=3$
0.1	0.9338	0.8576	0.8576	0.4434	0.4423
0.3	0.8246	0.6687	0.6690	0.2294	0.2277
0.5	0.7382	0.5490	0.5455	0.1607	0.1598
0.7	0.6686	0.4667	0.4662	0.1257	0.1239
0.9	0.6070	0.4104	0.4031	0.1048	0.1059

Comparison of the successive approximations suggests that the variational approximation is very adequate.

9.5 Applications to Mass Transfer

The transport equation (1.5) applies equally as well to heat and mass transfer. Variational principles applicable to heat transfer are obviously applicable to mass transfer. For that reason we discuss here only situations which are unique to mass transfer. One complication is that a system may include many different chemical species, thus necessitating the treatment of systems of equations. In addition the reaction rate term is important. Both these complications will be discussed.

If one considers isothermal, isobaric, multicomponent system with a first-order chemical reaction, the equations are (Toor, 1964; Fitts, 1962)

$$\frac{\partial c_i}{\partial t} + \mathbf{u} \cdot \nabla c_i - \sum_{j=1}^{N} \nabla \cdot (D_{ij} \nabla c_j) - \sum_{j=1}^{N} K_{ij}\, c_j = 0. \tag{9.126}$$

Focus on the multicomponent aspect and assume steady-state and no motion. Then introduce a matrix notation to simplify the equations

$$\nabla \cdot (\bar{\bar{D}}\, \nabla \bar{c}) = \bar{\bar{K}} \bar{c}. \tag{9.127}$$

The equations adjoint to these are found by Eq. (9.47) extended to systems of equations:

$$\nabla \cdot (\bar{\bar{D}}^T\, \nabla \bar{c}) = \bar{\bar{K}}^T \bar{c}. \tag{9.128}$$

Clearly the system is self-adjoint and a variational principle exists only if $D_{ij} = D_{ji}$ and $K_{ij} = K_{ji}$. In that case the variational integral is a simple extension of those used previously:

$$I(\bar{c}) = \sum_{i,\, j=1}^{N} \int [D_{ij}\, \nabla c_i \cdot \nabla c_j + K_{ij}\, c_i\, c_j]\, dV. \tag{9.129}$$

Consider next the material balance for tubular reactors with axial diffusion, Eqs. (5.87) and (5.88),

$$\frac{1}{\text{Pe}}\frac{d^2c}{dz^2} - \frac{dc}{dz} - R(c) = 0,$$
$$c - \frac{1}{\text{Pe}}\frac{dc}{dz} = 1 \text{ at } z = 0, \qquad \frac{dc}{dz} = 0 \quad \text{at} \quad z = 1. \tag{9.130}$$

The variational principle for this problem was first derived by Pakes and Storey (1967). We know from (9.86) that the first derivative can be eliminated by an integrating factor. The chemical reaction term always has a symmetric Fréchet derivative [see Eq. (9.98)] so that term presents no problem. The variational integral can be derived by analogy from (9.88), (9.89), and (9.98).

Make stationary the integral I among all functions c with continuous second derivatives:

$$I = \int_0^1 \left[\frac{1}{\text{Pe}} e^{-\text{Pe}z}\left(\frac{dc}{dz}\right)^2 + 2e^{-\text{Pe}z}\int_0^c R(c')\,dc'\right] dx + [c(0) - 1]^2. \tag{9.131}$$

The Euler equation and natural boundary conditions are

$$-\frac{1}{\text{Pe}}\frac{d}{dz}\left(e^{-\text{Pe}z}\frac{dc}{dz}\right) + e^{-\text{Pe}z}R(c) = 0,$$
$$-\frac{1}{\text{Pe}}c' + c = 1 \quad \text{at} \quad z = 0, \tag{9.132}$$
$$e^{-\text{Pe}}\frac{1}{\text{Pe}}c' = 0 \quad \text{at} \quad z = 1.$$

Pakes and Storey (1967) present calculations for two special cases. First they treat the linear equation for $\text{Pe} = 1$, $R = 2c$:

$$c'' - c' - 2c = 0 \tag{9.133}$$

and take trial functions of the form

$$c = A + B\exp a_1 x + a_2 \exp a_3 x. \tag{9.134}$$

They require the boundary conditions to be satisfied and make the integral stationary among the remaining parameters. Even though the problem is linear, the parameters enter in a nonlinear fashion so that the set of algebraic equations derived from the variational principle is nonlinear and has multiple solutions. This was indeed found to be the case and two such solutions, obtained with different initial guesses, are shown in Table 9.3. Note that the

TABLE 9.3

FIRST-ORDER REACTION IN TUBULAR REACTOR[a]

	Case 1	Case 2
a_1	2.00000	−0.40867
a_2	0.50626	0.74374
a_3	−1.00010	0.46378
I	−0.78315	−0.78313
Values of $c(z)$		
$z = 0.0$	0.51895	0.51959
0.2	0.43337	0.43337
0.4	0.36749	0.36662
0.6	0.31977	0.31905
0.8	0.28998	0.29049
1.0	0.27945	0.28095

[a] Pakes and Storey (1967). Used by permission of the copyright owner, the Institution of Chemical Engineers.

solutions are very close to each other even though the parameters are different. These results reflect the insensitivity of exponential approximations.

The second example treated by Pakes and Storey has a second-order reaction and Pe = 1:

$$c'' - c' - 2c^2 = 0. \tag{9.135}$$

The same trial function was used and the results are shown in Table 9.4. Note that the third approximation is close to the second approximation so that higher approximations are not considered. The third approximation is also very close to the numerical solution, and in fact the error in the third approximation is much less than the difference between the second and third approximation. These results can be compared to the results in Table 5.10 for the orthogonal collocation method. The orthogonal collocation method is much simpler to apply since the integral (9.131) need not be calculated.

Consider the same problem except with another reaction, $A \rightleftarrows B \rightarrow C$, where the reaction rates are given by $R_A(A, B)$, $R_B(A, B)$. The mass balance on the reactor then takes the form

$$\begin{aligned} D_A \frac{d^2A}{dx^2} - u\frac{dA}{dx} - R_A(A, B) &= 0, \\ D_B \frac{d^2B}{dx^2} - u\frac{dB}{dx} - R_B(A, B) &= 0. \end{aligned} \tag{9.136}$$

TABLE 9.4

SECOND-ORDER REACTION IN TUBULAR REACTOR[a]

	Second approximation	Third approximation	Finite difference solution
a_1	0.17271	0.20826	—
a_2	0.00569	−0.06296	—
a_3	—	0.03390	—
I	−0.54553	−0.54554	−0.54554
Values of $c(x)$			
0.0	0.63754	0.63678	0.63678
0.4	0.52255	0.52335	0.52336
0.8	0.46610	0.46547	0.46548
1.0	0.45915	0.45759	0.45759

[a] Pakes and Storey (1967). Used by permission of the copyright owner, the Institution of Chemical Engineers.

We examine whether a variational principle exists for these equations. We know how to treat the first derivative, so concentrate on the reaction rate terms. From Eq. (9.38b) the Fréchet derivative is symmetric and a variational principle exists if

$$\frac{\partial R_A}{\partial B} = \frac{\partial R_B}{\partial A}. \tag{9.137}$$

This clearly places restrictions on the reaction rate expressions for which variational principles exist. For example if

$$\begin{aligned} R_A &= -k_1 A + k_{-1} B, \\ R_B &= +k_1 A - k_{-1} B - k_2 B, \end{aligned} \tag{9.138}$$

then Eq. (9.137) is satisfied only if $k_{-1} = k_1$, and this clearly would be a special case. Consequently a variational principle exists only rarely for the coupled set of equations (9.136).

Non-isothermal diffusion and reaction in a catalyst is governed by Eq. (5.55). We consider a first-order irreversible reaction with an Arrhenius temperature dependence and general boundary conditions at the surface of the catalyst:

$$\nabla^2 y = \phi^2 y \exp[\gamma(z-1)/z], \tag{9.139a}$$

$$\nabla^2 z = -\beta\phi^2 y \exp[\gamma(z-1)/z], \tag{9.139b}$$

$$\left.\begin{aligned} -\mathbf{n}\cdot\nabla y &= \frac{\text{Sh}}{2}\frac{\mathscr{D}_g}{\mathscr{D}}(y-1) \\ -\mathbf{n}\cdot\nabla z &= \frac{\text{Nu}}{2}\frac{k_g}{k}(z-1) \end{aligned}\right\} r = 1. \qquad \begin{aligned}(9.139\text{c})\\(9.139\text{d})\end{aligned}$$

Here $\text{Sh} = k_g' D/\mathscr{D}_g$, $\text{Nu} = hD/k_g$, where k_g' and h are the mass and heat transfer coefficients, D is the particle diameter, $\mathscr{D}_g$ and k_g are the diffusivity and thermal conductivity of the gas, and $\mathscr{D}$ and k are the effective diffusivity and thermal conductivity of the catalyst.

First examine the possibility of a variational principle for the system of equations. Equation (9.38b) requires

$$\phi^2 y \frac{\partial}{\partial z} \exp[\gamma(z-1)/z] = -\beta\phi^2 \exp[\gamma(z-1)/z]. \tag{9.140}$$

This is not satisfied, so that no variational principle exists for the system of equations.

The equations can be combined by multiplying (9.139a) by β and adding it to (9.139b) to get $\nabla^2(z + \beta y) = 0$. Integrate once to obtain $\nabla(z + \beta y) = K_1$. At $r = 0$ the value of $\nabla(z + \beta y)$ is zero, so that $K_1 = 0$. Integrate once again to obtain

$$z + \beta y = K_2. \tag{9.141}$$

Next multiply (9.139c) by β and add it to (9.139d):

$$-\mathbf{n}\cdot\nabla(z + \beta y) = \frac{\text{Nu}}{2}\frac{k_g}{k}(z(1) - 1)) + \frac{\beta\,\text{Sh}}{2}\frac{\mathscr{D}_g}{\mathscr{D}}(y(1) - 1)) = 0. \tag{9.142}$$

Thus

$$z(1) = 1 + \beta\,\delta(1 - y(1)), \qquad \delta = \frac{\text{Sh}\,\mathscr{D}_g}{\mathscr{D}}\frac{k}{\text{Nu}k_g}. \tag{9.143}$$

Combination of (9.141)–(9.143) gives $z(r)$ as a function of $y(r)$. Thus (9.139a,c) can be solved for $z(r)$ and a variational principle exists, Eq. (9.98).

It is of interest to form a variational principle for the system of equations (9.139), and their "adjoint" equations, as discussed in Section 9.2. Define

$$\left.\begin{aligned} N(T, c) &= \alpha'\,\nabla^2 T + \beta' R(c, T) = 0 \\ M(T, c) &= \alpha\,\nabla^2 c + \beta R(c, T) = 0 \end{aligned}\right\} \text{in } V, \qquad \begin{aligned}(9.144\text{a})\\(9.144\text{b})\end{aligned}$$

$$\left.\begin{aligned} -\mathbf{n}\cdot\nabla T &= \text{Nu}(T - T_s) \\ -\mathbf{n}\cdot\nabla c &= \text{Sh}(c - c_s) \end{aligned}\right\} \text{on } S. \qquad \begin{aligned}(9.144\text{c})\\(9.144\text{d})\end{aligned}$$

The variational principle is then a generalization of (9.75) to systems of equations. Consider the variational integral

$$I(T, c, T^*, c^*) = \int_v [T^* N(T, c) + c^* M(T, c)]\, dV. \tag{9.145}$$

Taking the first variation gives

$$\delta I = \int_v [\delta T^* N(T, c) + \delta c^* M(T, c) + T^* N_T' \delta T + c^* M_T' \delta T + T^* N_c' \delta c + c^* M_c' \delta c] \, dV \tag{9.146}$$
$$= \int_v [\delta T^* N + \delta c^* M + \delta T(\tilde{N}_T' T^* + \tilde{M}_T' c^*) + \delta c(\tilde{N}_c' T^* + \tilde{M}_c' c^*] \, dV.$$

This suggests we define the adjoint equations as

$$\left.\begin{aligned} N^* &= \alpha' \nabla^2 T^* + \beta' \left(\frac{\partial R}{\partial T}\right)_{T, c} T^* + \beta \left(\frac{\partial R}{\partial T}\right)_{T, c} c^* \qquad (9.147\text{a}) \\ M^* &= \alpha \nabla^2 c^* + \beta' \left(\frac{\partial R}{\partial c}\right)_{T, c} T^* + \beta \left(\frac{\partial R}{\partial c}\right)_{T, c} c^* \qquad (9.147\text{b}) \end{aligned}\right\} \text{in } V,$$

$$\left.\begin{aligned} -\mathbf{n} \cdot \nabla T^* &= \text{Nu}(T^* - T_s) \qquad (9.147\text{c}) \\ -\mathbf{n} \cdot \nabla c^* &= \text{Sh}(c^* - c_s) \qquad (9.147\text{d}) \end{aligned}\right\} \text{on } S.$$

To obtain the correct boundary conditions from (9.145) take T^* times (9.144c) and c^* times (9.144d), integrate over the surface and add to (9.145). Integration by parts gives the following variational integral:

$$I(T, c, T^*, c^*) = \int_v [-\alpha' \nabla T^* \cdot \nabla T - \alpha \nabla c^* \cdot \nabla c + T^* \beta' R + c^* \beta R] \, dV - \int_s [\alpha' \, \text{Nu}(T^* - T_s)(T - T_s) + \alpha \, \text{Sh}(c^* - c_s)(c - c_s)] \, dS. \tag{9.148}$$

Make the integral I stationary among functions T, c, T^*, c^*, which have continuous second derivatives, and T, T^* or c, c^* take prescribed values on any portion of the boundary which has an infinite Nusselt or Sherwood numbers.

It is easily verified that the Euler equations are (9.144a,b) and (9.147a,b) and the natural boundary conditions are (9.144c,d) and (9.147c,d). This is an example of the procedure to form a variational principle for *any* set of nonlinear equations.

Finally we consider the question of whether the solution to (9.139) is unique. Let us rewrite (9.139) to be in terms of the temperature function, z, rather than the concentration, y, and write the equations allowing a more general dependence upon concentration and temperature:

$$\begin{aligned} \nabla^2 z + \phi^2 f(z) &= 0 \quad \text{in } V, \\ z &= 1 \quad \text{on } S. \end{aligned} \tag{9.149}$$

Luss and Amundson (1967) show how to use enclosure theorems for eigenvalue problems (see Section 7.6) to give sufficient conditions for uniqueness.

Suppose there are two solutions, z_1 and z_2. The difference $v = z_1 - z_2$ must satisfy

$$\begin{aligned} \nabla^2 v + \phi^2 A(x) v &= 0 \quad \text{in } V, \\ v &= 0 \quad \text{on } S, \\ A(x) &\equiv \frac{f(z_2(x)) - f(z_1(x))}{z_2(x) - z_1(x)} = f'(z^*(x)). \end{aligned} \tag{9.150}$$

The first expression for $A(x)$ comes from rearrangement of the equations which z_1 and z_2 satisfy (subtract them to get an equation for v) and the second expression for $A(x)$ follows from the mean-value theorem for some $z^*(x)$.

Consider the comparison equation:

$$\begin{aligned} \nabla^2 v + \lambda A(x) v &= 0 \quad \text{in } V, \\ v &= 0 \quad \text{on } S. \end{aligned} \tag{9.151}$$

If (9.151) has only negative eigenvalues or if the smallest positive eigenvalue is larger than ϕ^2 then the only solution of (9.150) is the null solution and (9.149) is unique. This follows from the maximum principle (see Theorem 11.30 for a related version of the maximum principle). From Application 4 in Section 7.6 we know that if $A(x)$ is increased everywhere the eigenvalues of (9.151) decrease. Thus define the problem

$$\begin{aligned} \nabla^2 v + \lambda^* B v &= 0 \quad \text{in V}, \\ v &= 0 \quad \text{on } S, \\ B &= \begin{cases} \text{Max } f'(z) \geq A(x), \\ 1 \leq z \leq 1 + \beta. \end{cases} \end{aligned} \tag{9.152}$$

Combining all conditions gives as a sufficient condition for uniqueness

$$\lambda \geq \lambda^* \geq \phi^2 \tag{9.153}$$

The eigenvalues λ^* are known for simple geometries, giving the sufficient condition for uniqueness

$$\phi^2 \leq \mu/B, \tag{9.154a}$$

$$\lambda^* B = \mu, \tag{9.154b}$$

$$\phi^2 = k_0 R^2/\mathscr{D}, \tag{9.154c}$$

$$\mu = \begin{cases} \pi^2/4 \quad \text{slab}, \quad R \text{ is half-thickness}, \\ (2.405)^2 \quad \text{cylinder}, \\ \pi^2 \quad \text{sphere}. \end{cases} \tag{9.154d}$$

Luss and Amundson also point out that among catalyst particles of the same volume V, the eigenvalues of (9.152) are lowest for the sphere. Thus if R in (9.154) is regarded as the radius of a sphere of equivalent volume, the result holds for catalyst particles of arbitrary configuration. Luss and Amundson also treat several simultaneous reactions. Comparison with numerical solutions to (9.149) shows that the criterion (9.154a) is within a factor of two of the exact point of bifurcation, where multiple solutions occur. Thus (9.154a) gives a convenient, and reasonably accurate, criterion for uniqueness.

9.6 Upper Bound for Heat Transport by Turbulent Convection

We discuss here the problem of heat transport between two flat, infinite plates whose temperature is fixed. In Chapter 6 we treated the onset of convection when the temperature difference across the layer is large enough. Here we consider even larger temperature differences and attempt to provide bounds for the heat flux across the fluid layer. The bounds are provided by a variational principle of a very special kind. The Euler equations are not the governing equations but rather are averages of the governing equations. Furthermore, the maximum principle must be solved exactly, which is a difficult task. The solution to the maximum principle exhibits features that closely resemble actual measurements of heat flux by turbulent convection and this makes the results of great interest. The detailed results are found in Howard (1963), Busse (1969) and, for a similar problem, Nickerson (1969). Here we only outline the results.

The basic equations are the Navier–Stokes equations and energy equation using the Boussinesq approximation. The temperature T^* is divided into two parts, $T(z)$ which is the horizontal average, and the remaining fluctuating part θ. An overbar denotes a horizontal average and brackets denote the average over the layer:

$$\bar{f} = \int f(x, y, z)\, dx\, dy \Big/ \int dx\, dy, \qquad \langle f \rangle = \int_{-1/2}^{+1/2} \bar{f}(z)\, dz. \tag{9.155}$$

The equations are then, in dimensionless form,

$$\nabla^2 \mathbf{v} + \mathbf{k}\theta - \nabla p = \frac{1}{\mathrm{Pr}}\left(\mathbf{v}\cdot\nabla\mathbf{v} + \frac{\partial \mathbf{v}}{\partial t}\right), \tag{9.156a}$$

$$\nabla \cdot \mathbf{v} = 0, \tag{9.156b}$$

$$\nabla^2\theta - w\frac{\partial T}{\partial z} = \mathbf{v}\cdot\nabla\theta - \overline{\mathbf{v}\cdot\nabla\theta} + \frac{\partial\theta}{\partial t}, \tag{9.156c}$$

$$\frac{\partial^2 T}{\partial z^2} = \frac{\partial}{\partial z}\overline{w\theta} + \frac{\partial T}{\partial t}, \qquad w = \mathbf{k}\cdot\mathbf{v}. \tag{9.156d}$$

The last equation is derived by taking the horizontal average of the energy equation. We further assume that the horizontal averages are time independent. This permits an integration of (9.156d) to obtain

$$\frac{dT}{dz} = -R - \langle w\theta\rangle + \overline{w\theta}. \tag{9.157}$$

The constant of integration, $R + \langle w\theta\rangle$ is the dimensionless heat flux across the layer and is determined by the boundary conditions.

We derive from (9.156) two integral equalities. Multiply (9.156a) by $\mathbf{v}$ and (9.156c) by θ and integrate over the layer. Apply the boundary conditions, $\mathbf{v} = 0$, $T^*(-\frac{1}{2}) = T_0$, and $T^*(\frac{1}{2}) = T_1$, and use the fact that the fluid is incompressible. This gives two equations, called "power integrals" by Howard:

$$\langle|\nabla \times \mathbf{v}|^2\rangle = \langle w\theta\rangle, \tag{9.158a}$$

$$\langle|\nabla\theta|^2\rangle = R\langle w\theta\rangle - \langle(\overline{w\theta} - \langle w\theta\rangle)^2\rangle. \tag{9.158b}$$

Consider the following variational problem. Maximize the quantity $\langle w\theta\rangle$ for a given R among all functions $\mathbf{v}$ and θ which vanish on the boundary, have zero horizontal averages, satisfy the continuity equation and Eq. (9.158). Note first that R is given so that the principle corresponds to maximizing the dimensionless heat flux across the layer, $N = R + \langle w\theta\rangle$. There is a most important difference between this maximum principle and those treated earlier. The functions $\mathbf{v}$ and θ which solve the variational principle may not be solutions to (9.156), since they are not the Euler equations for the principle [see (9.161)]. The variational solution provides an upper bound for the exact heat flux, given by the solution to (9.156), because it provides the maximum heat flux from a class of functions which includes the solution to (9.156) but includes other functions as well. Consequently, the variational principle gives a heat flux which is at least as large as the exact solution and perhaps larger. Furthermore, for the principle to be of use in deriving rigorous upper bounds we must actually find the maximum. Approximations will not suffice since they may give heat fluxes below the exact solution.

The actual problem solved by Howard and Busse is inverse to that stated, which will be given.

Given $\mu > 0$, find the minimum $R(\mu)$ of the functional

$$R(\mathbf{v}, \theta, \mu) \equiv [\langle|\nabla \times \mathbf{v}|^2\rangle\langle|\nabla\theta|^2\rangle + \mu\langle(\overline{w\theta} - \langle w\theta\rangle)^2\rangle]/\langle w\theta\rangle^2. \tag{9.159}$$

among all fields $\mathbf{v}, \theta$ that vanish at $z = +\frac{1}{2}$, satisfy the equation of continuity, and have zero horizontal averages.

The functional is homogeneous of degree zero with respect to $\mathbf{v}$ and θ so that the amplitude of both quantities is left undetermined. We therefore impose the conditions,

$$\langle w\theta \rangle = \mu, \qquad \langle |\nabla \times \mathbf{v}|^2 \rangle = \mu. \tag{9.160}$$

This ensures that (9.158a) is satisfied, and (9.158b) is contained in the functional, (9.159).

The first step in the solution of the variational principle is to remove one of the constraints—the continuity equation. This enlarges the class of comparison functions, and hence increases the maximum heat flux. The bound is therefore worse, but it is easier to find. Howard (1963) shows that the Euler equation can be reduced to

$$\begin{gathered} \mu^{-1}\langle f'^2 \rangle f'' + (\mathscr{H} + 1)\langle f^2 \rangle f - f^3 = 0, \\ \mathscr{H}(f) = [\langle f^4 \rangle - \langle f^2 \rangle^2 + \mu^{-1}\langle f'^2 \rangle^2]/\langle f^2 \rangle^2, \end{gathered} \tag{9.161}$$

where f is a function on the interval (0, 1). Howard solves this equation to provide the bound denoted by H in Fig. 9.1.

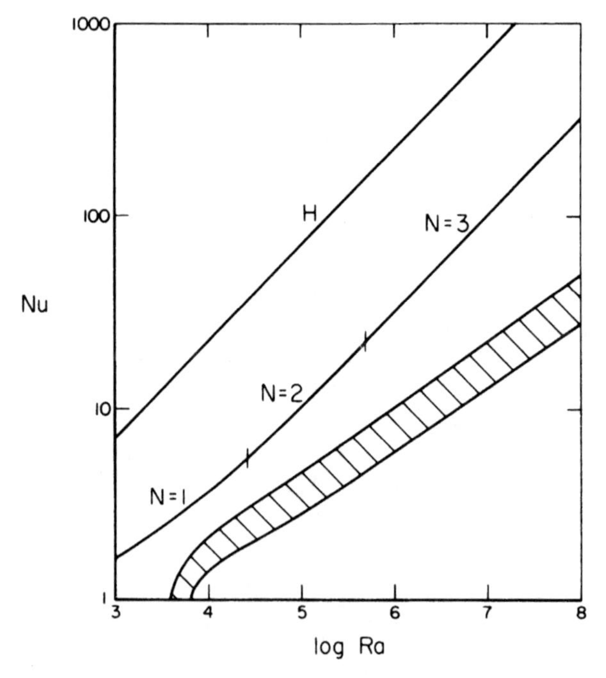

Fig. 9.1. Upper bound for heat transport by turbulent convection (Busse, 1969; reprinted from the *Journal of Fluid Mechanics* with permission of the copyright owner, Cambridge University Press).

We next consider the variational principle (9.159). With the continuity equation as a constraint the class of functions is smaller and the maximum is expected to be lower (or at least the same). Howard derives the Euler equations,

$$\begin{gathered}\langle|\nabla \times \mathbf{v}|^2\rangle \nabla^2\theta + [(R+\mu)\langle w\theta\rangle - \mu\overline{w\theta}]w = 0,\\ \langle|\nabla\theta|^2\rangle \nabla^2\mathbf{v} + [(R+\mu)\langle w\theta\rangle - \mu\overline{w\theta}]\theta\mathbf{k} - \mu\langle w\theta\rangle^2 \nabla p = 0.\end{gathered} \tag{9.162}$$

Busse (1969) shows that the velocity field can be represented

$$\mathbf{v} = \nabla \times (\nabla \times \mathbf{k})v. \tag{9.163}$$

The structure of the equations (9.162) suggests that the solutions can be written in terms of functions with a periodic structure in the xy plane. Busse writes

$$v = \sum_{n=1}^{N} \phi_n(x, y)\frac{w_n(z)}{\alpha_n^2}, \qquad \theta = \sum_{n=1}^{N} \phi_n(x, y)\theta_n(z), \tag{9.164}$$

where ϕ_n satisfies

$$\left(\frac{\partial^2}{\partial x^2} + \frac{\partial^2}{\partial x^2}\right)\phi_n(x, y) = -\alpha_n^2\phi_n(x, y), \qquad \langle\phi_n^2\rangle = 1. \tag{9.165}$$

Howard (1963) earlier had solved the equations using $N = 1$, assuming that this gave the maximum flux. This assumption was later proved false, since for some range of R the heat flux for $N = 2$ or 3 is above that for $N = 1$. Busse (1969) used boundary-layer methods to solve the variational problem in the limit when μ tends to infinity. The solution is then not strictly a maximum for finite μ although this is assumed. The problem without the multiple boundary layer structure has not been solved.

Additional constraints could be imposed as well. In that case the maximum nature of the principle would have to be verified. The maximum, for a fixed R, would decrease or stay the same compared to solutions with fewer constraints. Consequently, we would expect that as more constraints are imposed the class of admissible functions would eventually be restricted to the solution to (9.156). There is no assurance that this is true, however, and the more complicated variational problems with more constraints might be insoluble. Consequently, this method of getting upper bounds does not lead to a convergent process in a practical sense. The solutions, however, do contain features which are observed in experiments.

The solution derived by Busse (1969) is plotted in Fig. 9.1. Notice that as the Rayleigh number is increased the maximum heat transport curve corresponds first to $N = 1$, then to $N = 2$ and 3. (The curves for $N > 3$ are very close to those for $N = 3$.) This leads to kinks in the curve of heat transport versus Rayleigh number. The same phenomenon is observed experimentally (Malkus, 1954; Willis and Deardorff, 1967). Apparently as the

Rayleigh number increases, more modes of convection are activated successively and the heat transport increases with discontinuities in the slope of the curve. Other features of the experiments are not so well modeled by the bounding solution. As R approaches infinity the bounding solution predicts that the Nusselt number should depend on $R^{1/2}$ rather than on $R^{1/3}$ as observed. It is possible that the experiments have not been done for large enough Rayleigh number to include the region where the asymptotic analysis is valid. In spite of these quantitative deficiencies, the variational principle gives useful qualitative information.

The energy method has also been applied to study stability of natural convection between flat plates and in spherical shells. As discussed in Section 8.7 the best bounds ensuring stability are derived from a variational principle. We do not give details of the method here, but note that the details are similar to those in Section 8.7. The fluid layer problem has been treated by Joseph (1965, 1966), who shows that for Couette flow heated from below the flow is stable, provided

$$\mathrm{Re}^2 + R < 1708, \tag{9.166}$$

where $\mathrm{Re} = dV/\nu$, and the upper and lower plates have velocities $\pm V$. In particular, for the problem with no imposed motion on the surfaces ($\mathrm{Re} = 0$) subcritical, nonlinear instabilities cannot occur. Joseph and Shir (1966) study the same problem with heat generation, while Joseph and Carmi (1966) study spherical shells and Joseph (1970) studies buoyancy-driven convection with both heat and mass transport.

EXERCISES

9.1. Verify Eqs. (9.A), (9.C), and (9.D) of Table 9.1 as is done for (9.B) in (9.23)–(9.26).

9.2. Determine the conditions on $a_i(x)$ so that the operator

$$L(u) = \sum_{i=0}^{4} a_i(x) \frac{d^i u}{dx^i}$$

is self-adjoint.

Answer: $a_4''' - a_2' + a_1 = 0, \quad 2a_4' = a_3.$

9.3. Consider the following problem,

$$\sum_{i=0}^{3} a_i \frac{d^i u}{dx^i} = f(x), \qquad 0 < x < 1, \quad u(0) = u(1) = u'(0) = 0,$$

where a_i are known constants and $f(x)$ is known. Find the adjoint problem (differential equation and boundary conditions). Formulate a variational principle for the original problem and its adjoint.

Answer: $L^* v = \sum_{i=0}^{3} (-1)^i a_i \, d^i v/dx^i$, $v(0) = v(1) = v'(1) = 0$.

9.4. Apply the variational principle (9.103) using the trial functions (9.106). Then apply the Galerkin method to (9.104a) using (9.106a). Compare the equations governing $A_i(t)$.

9.5. Determine if a variational principle exists for (5.94).
Answer: Only if $\mathrm{Pe}_H = \mathrm{Pe}_M$ and $\partial R/\partial T = \beta\, \partial R/\partial c$.

REFERENCES

Ames, W. F., and de la Cuesta, S. H. (1963). Techniques for the Determination of Heat Transfer in Fluids and Nonhomogeneous Bodies, *Ind. Eng. Chem. Fund.* **2**, 21–26.

Busse, F. H. (1969). On Howard's Upper Bound for Heat Transport by Turbulent Convection, *J. Fluid Mech.* **37**, 457–477.

Churchill, R. V. (1958). "Operational Mathematics," McGraw-Hill. New York.

Courant, R., and Hilbert, D. (1953). "Methods of Mathematical Physics," Vol. I. Wiley (Interscience), New York.

Fitts, D. D. (1962). "Nonequilibrium Thermodynamics." McGraw-Hill, New York.

Gurtin, M. E. (1964). Variational Principles for Linear Initial-Value Problems, *Quart. Appl. Math.* **22**, 252–256.

Howard, L. N. (1963). Heat Transport by Turbulent Convection, *J. Fluid. Mech.* **17**, 405–433.

Joseph, D. D. (1965). On the Stability of the Boussinesq Equations, *Arch. Rat. Mech. Anal.* **20**, 59–71.

Joseph, D. D. (1966). Nonlinear Stability of the Boussinesq Equations, *Arch. Rat. Mech. Anal.* **22**, 163–184.

Joseph, D. D. (1970). Global Stability of the Conduction-Diffusion Solution, *Arch. Rat. Mech. Anal.* **36**, 285–292.

Joseph, D. D., and Carmi, S. (1966). Subcritical Convective Instability, Part 2. Spherical Shells, *J. Fluid Mech.* **26**, 769–777.

Joseph, D. D., and Shir, C. C. (1966). Subcritical Convective Instability, Part 1. Fluid Layers *J. Fluid Mech.* **26**, 753–768.

Krajewski, B. (1964). A Variational Approach to the Three-Dimensional Theory of Convective Heat Transfer, *Arch. Mech. Stos.* **16**, 533–536.

Lanczos, C. (1961). "Linear Differential Operators." Van Nostrand-Reinhold, Princeton, New Jersey.

Lewins, J. (1965). "Importance: The Adjoint Function." Pergamon Press, Oxford.

Li, J. C. M. (1964). Thermokinetic Analysis of Heat Conduction, *Int. J. Heat Mass Transfer* **7**, 1335–1339.

Luss, D., and Amundson, N. R. (1967). Uniqueness of the Steady State Solutions for Chemical Reaction Occurring in a Catalyst Particle or in a Tubular Reactor with Axial Diffusion, *Chem. Eng. Sci.* **22**, 253–266.

Malkus, W. V. R. (1954). Discrete Transitions in Turbulent Convection, *Proc. Roy. Soc. (London)* **A225**, 185–195.

Mikhlin, S. G. (1964). "Variational Methods in Mathematical Physics." Macmillan, New York.

Morse, P. M., and Feshback, H. (1953). "Methods of Theoretical Physics." Vol I. McGraw-Hill, New York.

Nichols, R. A., and Bankoff, S. G. (1965). Adjoint Variational Principle for Convective Diffusion, *Int. J. Heat Mass Transfer* **8**, 329–335.

Nickerson, E. C. (1969). Upper Bounds on the Torque in Cylindrical Couette Flow, *J. Fluid Mech.* **38**, 807–815.

O'Toole, J. T. (1967). Variational Principles for Time-Dependent Transport Processes, *Chem. Eng. Sci.* **22**, 313–318.

Pakes, H. W., and Storey, C. (1967). Solution of the Equations for a Tubular Reactor with Axial Diffusion by a Variational Technique, *Trans. Inst. Chem. Eng.* **45**, CE96–98, 108.

Rafalski, P. and Zyszkowski, W. (1969). On the Variational Principles for the Heat Conduction Problem, *AIAA J.* **7**, 606–609.

Roussopoulos, P. (1953). Méthodes Variationnelles en Théorie des Collisions. *C. R. Acad. Sci. Paris* **236**, 1858–1860.

Savkar, S. D. (1970). On a Variational Formulation of a Class of Thermal Entrance Problems, *Int. J. Heat Mass Transfer* **13**, 1187–1197.

Schmit, L. A. (1956). Application of the Variational Method, the Galerkin Technique, and Normal Coordinates in a Transient Temperature Distribution Problem, Wright Air Develop Cent. Tech. Rep. 56–287, AD–97326.

Slattery, J. C. (1964). A Widely Applicable Type of Variational Integral. I, *Chem. Eng. Sci.* **19**, 801–806.

Southwell, R. V. (1946–1956). "Relaxation Methods in Theoretical Physics," Vols. I, **II**, Oxford Univ. Press (Clarendon) London and New York.

Sparrow, E. M. (1960). Application of Variational Methods to Radiation Heat-Transfer Calculations, *J. Heat Transfer, Trans. ASME, Ser. C*, **82**, 375–380.

Sparrow, E. M., and Haji-Sheikh, A. (1965). A Generalized Variational Method for Calculating Radiant Interchange between Surfaces, *J. Heat Transfer, Trans. ASME, Ser. C* **87**, 103–109.

Tonti, E. (1969). Variational Formulation of Nonlinear Differential Equations. I, II, *Bull. Acad. Roy. Belg.* (*Classe des Sci.*) (5) **55**, 137–165, 262–278.

Toor, H. L. (1964) Solution of the Linearized Equations of Multicomponent Mass Transfer: I, *AIChE J.* **10**, 448–465.

Usiskin, C. M., and Siegel, R. (1960). Thermal Radiation from a Cylindrical Enclosure with Specified Wall Heat Flux, *J. Heat Transfer, Trans ASME, Ser. C* **82**, 369–374.

Vainberg, M. M., (1964). "Variational Methods for the Study of Nonlinear Operators." Holden-Day, San Francisco, California.

Wang, C. F. (1970). A Variational Principle of Steady-State Transport Processess, *AIAA J.* **8**, 335–338.

Washizu, K. (1955). Application of the Variational Method to Transient Heat Conduction Problems, Aeroelastic and Structures Res. Lab., Mass. Inst. Tech., Tech. Rep. 25–17, AD59951.

Willis, G. E., and Deardorff, J. W. (1967). Confirmation and Renumbering of the Discrete Heat Flux Transitions of Malkus, *Phys. Fluids* **10**, 1861–1866.

Wilson, E. L., and Nickell, R. E. (1966). Application of the Finite Element Method to Heat Conduction Analysis, *Nucl. Eng. Design* **4**, 276–286.

Chapter

10

On the Search for Variational Principles

10.1 Introduction

The variational principles presented in the previous two chapters show that there are problems in fluid mechanics and heat transfer for which variational principles exist. In special cases the variational integral is a minimum or maximum. Often the variational integral is only stationary, and in the general case no variational principle exists at all, except after inclusion of the adjoint problem. The goal of describing processes in terms of maximum and minimum principles is an alluring one which has occupied scientists for centuries. The first such principles were Fermat's principle of least time in optics and Maupertuis' principle of least action in mechanics (see Yourgrau and Mandelstam, 1968). In 1744 the mathematician Euler expressed the hope:

> As the construction of the universe is the most perfect possible, being the handiwork of an all-wise Maker, nothing can be met with in the world in which some maximal or minimal property is not displayed. There is, consequently, no doubt but that all the effects of the world can be derived by the method of maxima and minima from their final causes as well as from their efficient ones.

It is clear from Chapters 8 and 9 that this hope is not to be realized. The theorem of minimum rate of entropy production has sometimes been advanced to serve for this general minimum principle and we examine its validity below. As Kaplan (1969) remarks "...this is one expression of the desire for unity and simplicity which motivates all scholars." While the goal is admirable, we need to be explicit about what has been achieved.

We use the term variational principle only for the classical formalism described in Section 7.1. To find a variational principle of the nonlinear or linear problem

$$N(u) - f = 0, \tag{10.1}$$

we use the "adjoint" equation using Fréchet differentials, as in (9.75). The variational integral is†

$$I(u, v) = \int [vN(u) - vf - ug] \, dV, \tag{10.2}$$

and the first variation is

$$\delta I = \int \{\delta v[N(u) - f] + \delta u[\widetilde{N_u'} v - g]\} \, dV = 0. \tag{10.3}$$

Thus we can define an "adjoint variational principle" for the problem (10.1) and its adjoint

$$\widetilde{N_u'} v - g \equiv N^*(u, v) - g = 0. \tag{10.4}$$

In contrast we can write a variational equation

$$\overline{\delta J} = \int [N(u) - f] \, \delta u \, dV, \tag{10.5}$$

where the bar over δJ indicates that J may not exist, that is, Eq. (10.5) is not necessarily the variation of any functional. Equation (10.5) is called a "quasi-variational principle." It is a variational equation and a variational integral may not exist. Another approach is to write (10.2) as

$$K(u, u^0) = \int [N(u^0) - f] u \, dV. \tag{10.6}$$

Variations are made with respect to u while keeping u^0 fixed:

$$\delta_u I = \int [N(u^0) - f] \delta u \, dV. \tag{10.7}$$

† In time-dependent problems integration over time is included as well.

The notation $\delta_u I$ refers to variations only in u. After the variation we set $u = u^0$ and recover Eq. (10.1) as the "Euler equation." This type of principle is called a "restricted variational principle." It differs from an adjoint variational principle because, in place of the Euler equation for the adjoint (10.4), we use $u = u^0$.

Clearly the adjoint variational principle makes the functional (10.2) stationary. In the quasi-variational principle there is no variational integral to be made stationary. For the restricted variational principle the functional is not stationary unless

$$N^*(u, u) - g = 0, \tag{10.8}$$

which results from setting $v = u = u^0$ in (10.4).

The variational approaches of Rosen, Glansdorff and Prigogine, and Biot are examined in the light of these definitions after the entropy balance equation is introduced.

10.2 Entropy Production

Begin with the Cauchy momentum equation (1.13), the continuity equation of Eq. (1.15) and a total energy balance:

$$\rho \frac{d\mathbf{u}}{dt} = \rho \hat{\mathbf{F}} - \nabla p + \nabla \cdot \mathbf{T}, \tag{10.9a}$$

$$\frac{d\rho}{dt} = -\rho \nabla \cdot \mathbf{u} \tag{10.9b}$$

$$\rho \frac{d(\frac{1}{2}u^2 + U)}{dt} = -\nabla \cdot \mathbf{q} + \rho \mathbf{u} \cdot \hat{\mathbf{F}} + \nabla \cdot (\mathbf{u} \cdot \mathbf{T}) - \nabla \cdot (\mathbf{u}p). \tag{10.9c}$$

We leave completely free the choice of constitutive relations and equation of state necessary to completely specify a particular material. Implicit in the equations is the assumption of local equilibrium, that is, the internal energy function can be defined at each position and time even though the system is not in thermodynamic equilibrium, and the same equation of state is valid for equilibrium situations as for nonequilibrium situations.

The kinetic energy equation is derived by taking the scalar product of $\mathbf{u}$ into the momentum equation. This is then subtracted from the energy balance to give the internal energy equation:

$$\rho \frac{dU}{dt} = -q_{i,i} - pu_{j,j} + u_{i,j} T_{ij}. \tag{10.10}$$

The entropy balance is derived from the First Law of Thermodynamics in the form

$$dU = T\,dS - p\,dV. \tag{10.11}$$

By definition $V = 1/\rho$ and we assume that (10.11) is valid in a convected coordinate frame:

$$\frac{dU}{dt} = T\frac{dS}{dt} + \frac{p}{\rho^2}\frac{d\rho}{dt}. \tag{10.12}$$

Substitution of (10.12) into (10.10), and use of the continuity equation (10.9b) gives

$$\rho T\frac{dS}{dt} = -q_{i,i} + u_{i,j}T_{ij}. \tag{10.13}$$

The last term in the equation is the viscous dissipation and is denoted by Φ. The entropy balance can be rearranged to give the form of a balance equation:

$$\rho\frac{dS}{dt} = -\nabla\cdot\left(\frac{\mathbf{q}}{T}\right) + \mathbf{q}\cdot\nabla\left(\frac{1}{T}\right) + \frac{\Phi}{T}. \tag{10.14}$$

This is in the general form of a balance equation,

$$\rho\frac{dS}{dt} = \nabla\cdot\mathbf{j} + \sigma. \tag{10.15}$$

When this equation is integrated over the volume dV, and when Reynold's transport theorem (Aris, 1962) and the divergence theorem are applied, then the equation takes the form†

$$\frac{d}{dt}\int_v \rho S\,dV = \int_s \mathbf{n}\cdot\mathbf{j}\,dS + \int_v \sigma\,dV. \tag{10.16}$$

The terms can be identified as the rate of change of entropy, the rate of transfer of entropy between the surroundings and the system, and the rate of entropy production. Thus we can call the last two terms in (10.14) the rate of entropy production per unit volume,

$$\sigma = \mathbf{q}\cdot\nabla(1/T) + \Phi/T. \tag{10.17}$$

For an incompressible, Newtonian fluid in an isothermal situation when inertial terms are negligible and body forces are conservative, the variational principle (8.102) means that the viscous dissipation, Φ, is minimized. Since

† The symbol S denotes both entropy and surface area. The usage is clear from the context.

the dissipation differs by only a constant factor from the rate of entropy production, we can also interpret this physical situation as one which minimizes the rate of entropy production. Possible extensions to more general situations are discussed in Section 10.4.

The Dirichlet integral is minimized for the steady-state temperature distribution (see Section 7.2) under assumptions of no motion, Fourier's law, and constant thermal conductivity:

$$D(T) = \tfrac{1}{2} \int k \, \nabla T \cdot \nabla T \, dV. \tag{10.18}$$

This integral differs slightly from the rate of entropy production for the same situation

$$\sigma = \mathbf{q} \cdot \nabla\left(\frac{1}{T}\right) = -\frac{(\mathbf{q} \cdot \nabla T)}{T^2} = \frac{k}{T^2} \nabla T \cdot \nabla T. \tag{10.19}$$

Here T represents the absolute temperature, and for small temperature changes near room temperature the factor T^{-2} is approximately constant. Then the governing variational principle (10.18) can be interpreted as minimizing the rate of entropy production.

10.3 Heat Transfer

We consider here variational principles for heat transfer which have been motivated by the principle of minimum rate of entropy production. The variational principle for steady-state heat conduction (Section 7.2) simplifies under assumptions of no heat generation, thermal conductivity a function of position only, and the temperature is specified on the boundary S.

Minimize $\Phi(T)$ among all functions satisfying $T = T_1$ on S which are continuous and have continuous second derivatives:

$$\Phi(T) = \int \tfrac{1}{2} k(\mathbf{x}) \nabla T \cdot \nabla T \, dV. \tag{10.20}$$

The Euler equation is $\nabla \cdot (k \nabla T) = 0$. In applications the energy equation is thus satisfied approximately, whereas the constitutive relation $\mathbf{q} = -k\nabla T$ is satisfied exactly. This situation is reversed for the reciprocal variational principle.

Maximize $\Psi(\mathbf{q})$ among functions $\mathbf{q}$ which have continuous first derivatives and which satisfy $\nabla \cdot \mathbf{q} = 0$ in V,

$$\Psi(\mathbf{q}, T) = \int_v \left(-\frac{1}{2k} \mathbf{q} \cdot \mathbf{q}\right) dV - \int_s \mathbf{n} \cdot \mathbf{q} T_1 \, dS. \tag{10.21}$$

The restriction is handled using a Lagrange multiplier and adding $\lambda\nabla \cdot \mathbf{q}$ to the functional. The Euler equations and natural boundary conditions are then

$$\begin{aligned} \mathbf{q}/k + \nabla T &= 0 \quad \text{in } V, \\ T - T_1 &= 0 \quad \text{on } S, \end{aligned} \tag{10.22}$$

where the Lagrange multiplier has been interpreted as the temperature.

Onsager (1931) presented similar variational principles based on the rate of entropy production. Define the three integrals

$$\begin{gathered} \dot{S}(\mathbf{q}) = -\int [(\nabla \cdot \mathbf{q})/T]\, dV, \qquad \dot{S}^*(q_n) = \int (\mathbf{n} \cdot \mathbf{q}/T)\, dS, \\ \Phi(\mathbf{q}) = \int (\mathbf{q} \cdot \mathbf{R} \cdot \mathbf{q}/2T)\, dV, \end{gathered} \tag{10.23}$$

which can be identified as the entropy accumulation [see (10.13)], the entropy outflow [see (10.16)] and a dissipation function which is one-half the rate of entropy generation [related to (10.17)]. $\mathbf{R}$ is assumed to be symmetric and positive definite. The first principle is the following.

Maximize the functional $A(\mathbf{q}, T)$ among vector functions $\mathbf{q}$ which have continuous first derivatives and satisfy $\nabla \cdot \mathbf{q} = 0$ in V. The temperature is a prescribed function in V which satisfies the boundary condition $T = f$ on S:

$$\begin{aligned} A(\mathbf{q}, T) &= \dot{S}^*(q_n) - \Phi(\mathbf{q}), \\ A &= \int [(\nabla \cdot \mathbf{q})/T]\, dV + \int [\mathbf{q} \cdot \nabla(1/T) - \mathbf{q} \cdot \mathbf{R} \cdot \mathbf{q}/2T]\, dV. \end{aligned} \tag{10.24}$$

The first variation is

$$\delta A = \int [(\nabla \cdot \delta\mathbf{q})/T]\, dV + \int \delta\mathbf{q} \cdot [\nabla(1/T) - \mathbf{R} \cdot \mathbf{q}/T]\, dV. \tag{10.25}$$

Since the equation $\nabla \cdot \mathbf{q} = 0$ is satisfied by the variation, too, the Euler equation becomes

$$\nabla(1/T) = \mathbf{R} \cdot \mathbf{q}/T. \tag{10.26}$$

For an isotropic media with $R_{ij} = \delta_{ij}/kT$ this reduces to Fourier's law,

$$-k\, \nabla T = \mathbf{q}. \tag{10.27}$$

Consequently the Euler equation is the constitutive relation. The second variation is negative definite and hence satisfies the necessary condition for a maximum principle:

$$\delta^2 A = -\int (\delta\mathbf{q} \cdot \mathbf{R} \cdot \delta\mathbf{q}/T)\, dV \le 0. \tag{10.28}$$

For the exact solution $\dot{S}^* = 2\Phi$ so that the function A becomes just Φ and the principle corresponds to maximizing the rate of entropy production.

Onsager also gives a principle which is reciprocal to this, except for different boundary conditions.

Minimize $\Phi(\mathbf{q})$ among vector functions $\mathbf{q}$ which have continuous first derivatives and satisfy

$$\nabla \cdot \mathbf{q} = 0 \quad \text{in } V, \qquad \mathbf{n} \cdot \mathbf{q} = f \quad \text{on } S. \tag{10.29}$$

The first variation and Euler equations are

$$\delta \int_v [\mathbf{q} \cdot \mathbf{R} \cdot \mathbf{q}/2T + \lambda \nabla \cdot \mathbf{q}]\, dV = 0, \tag{10.30a}$$

$$\delta \mathbf{q}: \mathbf{R} \cdot \mathbf{q}/T - \nabla \lambda = 0, \tag{10.30b}$$

$$\delta \lambda: \nabla \cdot \mathbf{q} = 0. \tag{10.30c}$$

Identifying the Lagrange multiplier $\lambda = 1/T$ gives Eq. (10.26). The second variation satisfies the necessary condition for a minimum principle:

$$\delta^2 \Phi = \int [\delta \mathbf{q} \cdot \mathbf{R} \cdot \delta \mathbf{q}/T]\, dV \geq 0. \tag{10.31}$$

Thus the rate of entropy production is being minimized.

Comparing the principle based on entropy production (10.29) to the variational principle (10.21) [or (7.59) for the same boundary conditions] reveals an important difference. The principle (10.29) requires that the temperature be a known function of position, whereas this is not required in (10.21) or (7.59). Consider Eqs. (10.29) and (10.30) for the case when $\mathbf{R} = \mathbf{U}/kT$. Then the solution to the variational principle satisfies

$$\nabla \cdot (kT^2\, \nabla \lambda) = 0. \tag{10.32}$$

This is valid when T is any arbitrary, known function of position, and the variational principle makes $\mathbf{q}$ satisfy (10.30b) and λ satisfy (10.32). Only if T is the exact solution to $\nabla \cdot (k\, \nabla T) = 0$ can we make the identification $\lambda = 1/T$ to have Eqs. (10.30b), (10.30c), and (10.32) correspond to $\nabla \cdot (k\, \nabla T) = 0$. Thus we can interpret the variational principle (10.29): when the temperature is known, the heat flux which corresponds to the minimum rate of entropy production is the heat flux satisfying Eqs. (10.30b) and (10.30c).

Another form of the theorem of minimum rate of entropy production is the following.

Minimize the functional $I(T)$ among all functions T which have continuous second derivatives and satisfy $T = f$ on S:

$$I(T) = \int L\nabla(1/T) \cdot \nabla(1/T)\, dV = \int (L/T^4)\, \nabla T \cdot \nabla T\, dV. \tag{10.33}$$

If L is independent of temperature this yields the Euler equation

$$\nabla \cdot [L\nabla(1/T)] = -\nabla \cdot [(L/T^2)\nabla T] = 0. \tag{10.34}$$

Clearly this reduces to the heat conduction equation only when $L/T^2 = k$. When kT^2 is not constant the theorem of minimum rate of entropy production does not correspond to steady-state heat conduction. When $1/T^2$ is approximately constant the rate of entropy production for the actual temperature distribution is not much different from the actual minimum.

Rosen (1953) manipulated Onsager's results in order to apply them to the unsteady-state heat transfer. The equations are now

$$\rho\, C_v \frac{\partial T}{\partial t} = -\nabla \cdot \mathbf{q}, \qquad \mathbf{q} = -k\, \nabla T. \tag{10.35}$$

The generalization to (10.29) is the following.

Minimize $\Phi(\mathbf{q})$ among vector functions $\mathbf{q}$ which have continuous first derivatives and satisfy $-\nabla \cdot \mathbf{q} = \rho C_v\, \partial T/\partial t$ in V and $\mathbf{n} \cdot \mathbf{q} = f$ on S:

$$\Phi(\mathbf{q}) = \int (\mathbf{q} \cdot \mathbf{q}/2k)\, dV. \tag{10.36}$$

The Euler equation is derived using Lagrange multipliers:

$$\mathbf{q} = k\, \nabla\lambda = -\, k\, \nabla T. \tag{10.37}$$

Notice that again T (or rather $\partial T/\partial t$) must be known throughout the region and the functional is also a function of time. Rosen also presented another type of variational principle.

Make the functional $I(T, \partial T/\partial t)$ "stationary" to variations in T with continuous second derivatives while keeping $\partial T/\partial t$ fixed and requiring $T = f$ on S:

$$I\left(T, \frac{\partial T}{\partial t}\right) = \int \left[\rho C_v T \frac{\partial T}{\partial t} + \frac{k}{2} \nabla T \cdot \nabla T\right] dV. \tag{10.38}$$

Rosen calls this a restricted variational principle because the time derivative is fixed while the temperature is varied. Rosen's treatment is clearly outside the framework of the calculus of variations. To see this let us calculate the variation by substituting $T(x, t) = T^*(x, t) + \varepsilon\eta(x)$ into the integral of Eq. (10.38). The T^* refers to the "stationary" solution (stationary with respect to the variational principle, not time) and η is restricted to be a function of x because the time derivative cannot be varied. We get

$$\left.\frac{d\Phi}{d\varepsilon}\right|_{\varepsilon=0} = \int \eta \left[\rho C_v \frac{\partial T^*}{\partial t} - \nabla \cdot (k\, \nabla T^*)\right] dV = 0. \tag{10.39}$$

The next step in calculus of variations is to invoke the fundamental lemma. Here, however, we run into a difficulty. The term in brackets is a function of time and the $\eta(x)$ is a function only of x. Pick a time t and suppose the term in brackets is nonzero for some x. Since η is arbitrary it can be chosen such that the integral is positive, which violates the condition that the result be zero. Thus the term in brackets must be zero at that time. Next assume the term in brackets is not zero for some other time. Now η is not arbitrary because it is not a function of time and the same η must be used for all time. Thus it is not possible to conclude that the term in brackets is zero. Consequently in this type of restricted variational principle the integral is not stationary. When the restricted variational principle is used for computations we might expect that the results are equivalent to those derived by the Galerkin method, since this is also the case for problems with true variational principles. This has been shown by Finlayson and Scriven (1965, 1966, 1967), and the reader can easily verify it by taking $T = T(a_i(t), \mathbf{x}, t)$ and applying the two methods. In the Galerkin method the weighting function is $\partial T/\partial a_i$ [in place of η in (10.39)].

Gyarmati (1969, 1970) employs similar variational principles. In Gyarmati (1969) he takes the variation of

$$\Phi = \int [\mathbf{J} \cdot \mathbf{X} - \tfrac{1}{2}\mathbf{X} \cdot \mathbf{L} \cdot \mathbf{X} - \tfrac{1}{2}\mathbf{J} \cdot \mathbf{R} \cdot \mathbf{J}]\, dV \tag{10.40}$$

with respect to $\mathbf{X}$ and $\mathbf{J}$ (the forces and fluxes) subject to the balance equation

$$\rho a' + \nabla \cdot \mathbf{J} = \mathbf{J} \cdot \mathbf{X}. \tag{10.41}$$

The time derivative a', is held constant in the variation,† just as $\partial T/\partial t$ was held constant in (10.36). Thus this principle is a restricted variational principle, too.

Another extension of Onsager's approach is the local potential method (Glansdorff and Prigogine, 1964; Prigogine and Glansdorff, 1965; see also the symposium volume, Donnelly *et al.*, 1966). We outline their approach for the following nonlinear problem:

$$\begin{aligned} &\rho C_v \frac{\partial T}{\partial t} - \nabla \cdot (k \nabla T) = 0 \quad \text{in } V, \\ &k\mathbf{n} \cdot \nabla T + h(T - T_s) = 0 \quad \text{on } S, \\ &T = g(\mathbf{x}) \quad \text{in } V, \qquad t = 0. \end{aligned} \tag{10.42}$$

† This is explicitly pointed out in Gyarmati (1970, p. 165). Furthermore, if the constraint (10.41) or (10.35) is incorporated into the functional by means of Lagrange multipliers and variations taken *without* keeping a' or $\partial T/\partial t$ fixed, incorrect results are obtained; e.g., $\partial(\lambda\rho)/\partial t = 0$. The λ must be $\lambda(x, t)$ and arbitrary, however, to recover the constraint (10.35), (10.41) for all time, as we found in (10.39).

The initial temperature field g and the external temperature T_s are given. The density ρ, heat capacity C_v, thermal conductivity k, and heat transfer coefficient h are all functions of temperature. We obtain the local potential by multiplying Eqs. (10.42) by a variation δT and integrating over the volume and surface, respectively. Combination of the equations gives

$$\int \tfrac{1}{2}k\,\delta(\nabla T)^2\,dV + \int (\delta T)\rho C_v \frac{\partial T}{\partial t}\,dV + \int \tfrac{1}{2}h\,\delta(T - T_s)^2\,dS = 0. \quad (10.43)$$

Clearly there is no functional and this is a quasi-variational principle. The quasi-variational principle can be converted into a restricted variational principle by introducing an alias $T_0(x, t)$ for temperature and setting

$$\begin{gathered} k = k_0 \equiv k(T_0), \qquad h = h_0, \\ \frac{\partial T}{\partial t} = \frac{\partial T_0}{\partial t}, \\ \rho = \rho_0, \qquad C_v = C_0, \end{gathered} \quad (10.44)$$

In any variation T_0 is held fixed so that the variations in (10.43) can be taken outside the integral sign, and the result integrated with respect to t to give the local potential:

$$\begin{aligned} \Phi(T, T_0) \equiv \int_0^t \int_v \left[\frac{1}{2} k_0(\nabla T)^2 + \rho_0 C_0 T \frac{\partial T_0}{\partial t}\right] dV\,dt \\ + \int_0^t \int_s \frac{1}{2} h_0(T - T_s)^2\,dS\,dt. \end{aligned} \quad (10.45)$$

The restricted variational principle is the following.

Make the local potential $\Phi(T, T_0)$ "stationary" with respect to variations in T while holding T_0 constant. After the variation set $T_0 = T$.

We derive the Euler equations by considering $\phi(T + \varepsilon Z, T_0 + \varepsilon_0 Z_0)$. Since the functional depends on two variables the first variation is

$$\delta\Phi = \varepsilon \left.\frac{\partial \Phi}{\partial \varepsilon}\right|_{\varepsilon = \varepsilon_0 = 0} + \varepsilon_0 \left.\frac{\partial \Phi}{\partial \varepsilon_0}\right|_{\varepsilon = \varepsilon_0 = 0}. \quad (10.46)$$

The Euler equations and natural boundary conditions are, after the subsidiary conditions $T = T_0$, $Z = Z_0$ are invoked,

$$\rho C_v \frac{\partial T}{\partial t} - \nabla \cdot (k\,\nabla T) = 0, \quad (10.47a)$$

$$k\mathbf{n} \cdot \nabla T + h(T - T_s) = 0, \quad (10.47b)$$

$$\frac{1}{2}\frac{dk}{dT}(\nabla T)^2 + \frac{d(\rho C_v)}{dT} T \frac{\partial T}{\partial t} - \frac{\partial(\rho C_v T)}{\partial t} = 0, \tag{10.47c}$$

$$\frac{1}{2}\frac{dh}{dT}(T - T_s)^2 = 0 \quad \text{on } S, \qquad \rho C_v \, \delta T = 0 \quad \text{at } t = t. \tag{10.47d}$$

Clearly these cannot all be satisfied and the variational integral is therefore not stationary when only Eqs. (10.47a) and (10.47b) are satisfied. Computations using this "principle" are also equivalent to the Galerkin method (Finlayson and Scriven, 1967).

Next, assume we know the solution to Eqs. (10.47a) and (10.47b), which we call T^*. Then set $T_0 = T^*$ in the local potential (10.45), which is then a functional of T alone. The variation of Φ is easily found:

$$\phi(\varepsilon) = \Phi(T + \varepsilon\eta, T^*), \qquad \delta\Phi = \varepsilon \left.\frac{d\phi}{d\varepsilon}\right|_{\varepsilon=0}. \tag{10.48}$$

The Euler equation and natural boundary condition are

$$\begin{gathered} \rho^* C^* \frac{\partial T}{\partial t} - \nabla \cdot (k^* \nabla T) = 0, \\ k^* \mathbf{n} \cdot \nabla T + h^*(T - T_s) = 0. \end{gathered} \tag{10.49}$$

The solution to Eqs. (10.49) is the solution to Eqs. (10.47a) and (10.47b), too. Thus we obtain the correct Euler equations. Furthermore, the functional $\Phi(T, T^*)$ is a minimum for the T which solves Eqs. (10.49), since

$$\delta^2\Phi \equiv \frac{\varepsilon^2}{2}\phi''(0) = \frac{\varepsilon^2}{2}\int_0^t \left\{\int_V k^*\nabla\eta \cdot \nabla\eta \, dV + \int_S h^*\eta^2 \, dS\right\} dt \geq 0. \tag{10.50}$$

We have the somewhat puzzling situation that if the exact solution is known we can find a variational principle for it, but if it is not known the functional (10.45) is not even stationary. The local potential was interpreted as an evolutionary criterion by Prigogine (1966). It is clear that *if we take* $T_0 = T^*$, *where* T^* *is the known exact solution to* (10.47a) and (10.47b), *then the local potential* $\Phi(T, T^*)$ *is a minimum when* T *satisfies* (10.47a) *and* (10.47b).† *If, however,* T_0 *is not known, and the variational principle* (10.45) *is applied, with* $T = T_0$ *after the variation, then the functional is not stationary, or a minimum*. This has been proved by Finlayson and Scriven (1967), and follows from Eqs. (10.47a)–(10.47d). Despite this, advocates of the local potential

† This means that during the evolution towards a steady state the local potential $\Phi(T, T^*)$ is above its value achieved at steady state, since the transient solution to Eqs. (10.47a,b) is an admissible function and the minimum of $\Phi(T, T^*)$ occurs for the steady state. This is *not* true of the rate of entropy production. See Li (1962) for an example.

method continue to say that Φ in (10.45) is minimized with respect to the variational parameters. This is not true if the principle is applied as described in (10.45).

There is still a third approach for the same problem—the Lagrangian thermodynamics advanced by Biot (1957, 1959, 1962, 1967, 1970). This approach is a quasi-variational principle—there are no variational integrals and the principle is stated in terms of a variational equation. Consider a heat transfer problem with convection:

$$\begin{gathered} \rho C_v \left(\frac{\partial T}{\partial t} + \mathbf{u} \cdot \nabla T \right) = \nabla \cdot (k \nabla T) \quad \text{in } V, \\ k\mathbf{n} \cdot \nabla T + h(T - T_s) = 0 \quad \text{on } S, \\ T = g(\mathbf{x}) \quad \text{in } V, \qquad t = 0. \end{gathered} \tag{10.51}$$

For convenience here we take the physical properties to be functions of position but not of temperature. Several auxiliary functions are introduced: a total heat, h, and a heat flow vector, $\mathbf{H}$, defined (not uniquely) by

$$h \equiv \int_0^T \rho C_v \, dT, \tag{10.52a}$$

$$h = -\nabla \cdot \mathbf{H} \quad \text{in } V, \tag{10.52b}$$

$$\mathbf{n} \cdot \frac{\partial \mathbf{H}}{\partial t} = k\mathbf{n} \cdot \nabla T \quad \text{on } S. \tag{10.52c}$$

The thermal potential, V, and variational invariants, $\overline{\delta D}$ and $\overline{\delta D_s}$, are introduced (note that D and D_s are not defined):

$$\begin{gathered} V = \int_v \int_0^h T \, dh \, dV, \qquad \delta V = \int_v T \, \delta h \, dV = -\int T \nabla \cdot \delta \mathbf{H} \, dV, \\ \overline{\delta D} = \int \frac{1}{k} \left(\frac{\partial \mathbf{H}}{\partial t} - h\mathbf{u} \right) \cdot \delta \mathbf{H} \, dV, \qquad \overline{\delta D_s} = \int \frac{1}{h} \mathbf{n} \cdot \frac{\partial \mathbf{H}}{\partial t} \mathbf{n} \cdot \delta \mathbf{H} \, dS. \end{gathered} \tag{10.53}$$

The "variational principle" is then expressed as a quasi-variational principle:

$$\delta V + \overline{\delta D} + \overline{\delta D_s} = -\int_s T_s \, \mathbf{n} \cdot \delta \mathbf{H} \, dS. \tag{10.54}$$

By means of the divergence theorem, (10.54) can be written to show the Euler equation and natural boundary condition:

$$\int_v \delta \mathbf{H} \cdot \left[\nabla T + \frac{1}{k} \left(\frac{\partial \mathbf{H}}{\partial t} - h\mathbf{u} \right) \right] dV + \int_s \mathbf{n} \cdot \delta \mathbf{H} \left[T_s - T + \frac{1}{h} \mathbf{n} \cdot \frac{\partial \mathbf{H}}{\partial t} \right] dS = 0. \tag{10.55}$$

The divergence of k times the first term in brackets gives the original equations (10.51).

This quasi-variational principle does reproduce the original equations, but it is unduly complicated. Note that Eqs. (10.52b) and (10.52c) must be solved exactly. Since it is usually easy to make good guesses for trial functions for the temperature, we must then solve a partial differential equation exactly before proceeding to apply the quasi-variational principle. For two- or three-dimensional problems this is a serious disadvantage. Since no variational integrals are made stationary, that common advantage of variational principles is lost. The Galerkin method has been shown to give the same approximate solution, if formulated in terms of the same variables, $\mathbf{H}$ and h, so that the quasi-variational principle gains no advantage in applications (Finlayson and Scriven, 1965, 1967). The Galerkin method applied in terms of temperature alone, without the introduction of "total heat" and "heat flow vectors," is much simpler and certainly would be preferred in computations. The examples solved by Biot do provide an illustration of approximate solutions useful in a variety of one-dimensional problems (see Biot, 1970; Hiraoka and Tanaka, 1968).

Biot attempts to interpret his quasi-variational principle in terms of minimum rate of entropy production. As shown elsewhere (Finlayson and Scriven, 1967) the interpretation applies only to the approximate solution, not necessarily to the exact solution, so that the physical meaning of the principle is somewhat vacuous.

In summary: steady-state heat conduction with constant thermal conductivity is governed by the Dirichlet principle. This is approximately the same as the minimum rate of entropy production. If the thermal conductivity varies as $k = \text{constant}/T^2$ then the theorem of minimum rate of entropy production governs the problem. The entropy production theorem is not valid for other $k(T)$ or when convection or transient effects are included. Computations based on the principles of Rosen, Gyarmati, Glansdorff and Prigogine, or Biot are equivalent to those based on the Galerkin method.

10.4 Fluid Mechanics

The concept of a minimum rate of entropy production has also been applied to fluid mechanics. Onsager felt the principle governed all fluid flow. He reports (Onsager, 1969) "... for all the complications of hydrodynamics a 'principle of least dissipation' derived by Helmholtz assured the symmetry." The results of Chapter 8 indicate that this is an oversimplification. Millikan (1929) was unsuccessful in deriving a variational principle for the full, steady-state Navier–Stokes equations, non-Newtonian fluids often do not have minimum principles, and the inertial terms are not included in any of the variational principles involving dissipation. We see below that both inertial and dissipation terms are included in variational principles only by means of restricted variational principles, in which case the variational integral is not stationary, or by including the adjoint equation as in Chapters 8 and 9.

In heat transfer problems we faced a similar problem. We could formulate a variational principle for the steady-state part of the equation but not the time derivative. In a restricted variational principle the time derivative was held fixed during variation. In fluid flow we can do the same thing—formulate a variational principle for the viscous part of the equation and hold the inertial terms fixed during the variation. Restricted variational principles similar to this were presented by Miche (1949), (and criticized by Gerber, 1950), and Herivel (1954a,b). We discuss here the approach taken by Schechter (1962), which is based on the Local Potential, and contrast it with other variational principles for non-Newtonian fluids.

We consider the steady-state momentum balance for an incompressible, Reiner–Rivlin fluid in a volume V, and take boundary conditions as velocity specified on the boundary S:

$$\begin{aligned} \rho u_j u_{i,j} &= \rho \hat{\mathrm{F}}_i - p_{,i} + T_{ji,j}, \\ T_{ij} &= G_1(\mathrm{II}, \mathrm{III}) d_{ij} + G_2(\mathrm{II}, \mathrm{III}) d_{ik} d_{kj}. \end{aligned} \tag{10.56}$$

(See Section 8.4 for the notation.) When the inertial terms are negligible we have the variational principle of Eq. (8.102). The restricted variational principle given by Schechter (1962) includes the inertial terms and permits more general functions G_1 and G_2.

Make "stationary" the integral $I(\mathbf{u}, \mathbf{u}^0)$ among functions $\mathbf{u}$ which satisfy the boundary conditions, $\mathbf{u} = \mathbf{f}$ on S and are solenoidal. After the variation set $\mathbf{u} = \mathbf{u}^0$:

$$I(\mathbf{u}, \mathbf{u}^0) = \int_v [-\rho u_i^0 u_j^0 d_{ij} + \tfrac{1}{2} G_1^0 d_{ij} d_{ij} + \tfrac{1}{3} G_2^0 d_{ik} d_{jk} d_{ij} - \rho \hat{\mathrm{F}}_i u_i] \, dV. \tag{10.57}$$

The partial first variation is

$$\delta_u I = \int_v \{ [\rho u_i^0 u_j^0 - G_1^0 d_{ij} - G_2^0 d_{ik} d_{kj}]_{,j} \delta u_i - [\lambda_{,i} + \rho \hat{\mathrm{F}}_i] \delta u_i \} \, dV = 0. \tag{10.58}$$

After setting $u_i^0 = u_i$ we obtain (10.56) as the Euler equation. The first variation with respect to u^0 does not vanish, however, so that the integral is not stationary.

It is perhaps worthwhile to summarize the situations when a theorem of minimum rate of entropy production is valid. Beginning with the full momentum equations, if time-dependent terms or inertial terms are included then the theorem is not true. If these terms are absent, but the fluid is non-Newtonian, the theorem is true only if the forces are conservative, $S_t = 0$ (the velocity is specified on the boundary), and $d_{ij} T_{ij}$ in (10.13) is proportional to Γ in (8.102). As revealed by Eqs. (8.131) and (8.132) this is true for the power law fluid, or a special case of it, the Newtonian fluid. In more general cases the variational principle which reproduces the equations of motion does not involve an integral which is proportional to the rate of entropy production, so that the theorem of minimum rate of entropy production is not valid. The restricted variational principles and quasi-variational principles are equivalent to a Galerkin method. An adjoint variational principle, the Galerkin method, or another criterion of MWR, is then preferred due to its simplicity and generality.

EXERCISES

10.1. Consider heat transfer in a slab with constant thermal conductivity:

$$\frac{d^2T}{dx^2} = 0, \qquad T(0) = 1, \qquad T(1) = 1 + a, \quad a \geq 0.$$

The solution is $T = 1 + ax$. The variational principle governing this problem is Eq. (7.48). Minimize

$$I(T) = \int_0^1 \left(\frac{dT}{dx}\right)^2 dx$$

among continuous $T(x)$ with continuous second derivatives and which satisfy $T(0) = 1$, $T(1) = 1 + a$.

The local potential for this problem is

$$\Phi(T, T_0) = \int_0^1 T_0^2(x) \left[\frac{dT^{-1}}{dx}\right]^2 dx,$$

and $\Phi(T, T)$ is the rate of entropy production. Find the Euler equation for the local potential (vary T^{-1} but not T_0; after variation set $T = T_0$) and solve it subject to the boundary conditions on T. This T then "minimizes" $\Phi(T, T_0)$ among all T satisfying the boundary conditions. Calculate the value of Φ for this T (and $T_0 = T$); call it Φ_1.

Next find the minimum of $\Phi(T, T)$ among functions satisfying the boundary conditions; call it Φ_2. Prove $\Phi_2 \leq \Phi_1$. Thus the lowest value of the rate of entropy production does not occur for the solution to the steady-state heat conduction equation. The principle of minimum rate of entropy production does not hold.

Answer: $T = 1 + ax$ "minimizes" $\Phi(T, T_0)$ if T_0 is held fixed during the variation and $T = T_0$ after the variation; $\Phi_1 = \Phi(1 + ax, 1 + ax) = a^2/(1 + a)$; $\ln T = x \ln (1 + a)$ minimizes $\Phi(T, T)$ and $\Phi_2 = [\ln (1 + a)]^2$.

10.2. Consider the Exercise 3.15, to be solved for small time with a trial solution of the form of Eq. (3.20), with $c = (1 - \eta)^2$.

(a) First apply Lagrangian thermodynamics. Show that with this c, Eqs. (10.52) gives

$$H = \tfrac{1}{3}q - x + x^2/q - x^3/(3q^2).$$

Why must $H = 0$ at $x = q$? Next calculate δH and $\partial H/\partial t$, and substitute into Eq. (10.54). The variation δH is taken with respect to q:

$$\delta q q q' \int_0^1 (\tfrac{1}{3} - \eta^2 + \tfrac{2}{3}\eta^3)^2 \, d\eta - 2\,\delta q \int_0^1 (1 - \eta)^2(\eta^2 - \eta) \, d\eta = \delta q/3.$$

Solve for $q(t)$.

(b) Next apply the Galerkin method to Eq. (10.55), using the same trial functions for c and H.

(c) Finally, apply the Galerkin method as done in Chapter 3, Eq. (3.23). Which Galerkin method is easier to apply, b or c?

Answer: (a) $q^2 = 11.3t$, (b) $q^2 = 11.3t$, must be the same as in (a), (c) $q^2 = 13.3t$.

REFERENCES

Aris, R. (1962). "Vectors, Tensors, and the Basic Equations of Fluid Mechanics." Prentice-Hall, Englewood Cliffs, New Jersey.

Biot, M. A. (1957). New Methods in Heat Flow Analysis with Applications to Flight Structures, *J. Aero. Sci.* **24**, 857–873.

Biot, M. A. (1959). Further Developments of New Methods in Heat-Flow Analysis, *J. Aerosp. Sci.* **26**, 367–381.

Biot, M. A. (1962). Lagrangian Thermodynamics of Heat Transfer in Systems Including Fluid Motion, *J. Aerosp. Sci.* **29**, 568–577.

Biot, M. A. (1967). Complementary Forms of the Variational Principle for Heat Conduction and Convection, *J. Franklin Inst.* **283**, 372–378.

Biot, M. A. (1970). "Variational Principles in Heat Transfer." Oxford Univ. Press (Clarendon), London and New York.

Donnelly, R. J., Herman, R., and Prigogine, I. (eds.) (1966). "Non-Equilibrium Thermodynamics, Variational Techniques, and Stability." Univ. of Chicago, Chicago, Illinois.

Finlayson, B. A., and Scriven, L. E. (1965) The Method of Weighted Residuals and Its Relation to Certain Variational Principles for the Analysis of Transport Processes, *Chem. Eng. Sci.* **20**, 395–404.

Finlayson, B. A., and Scriven, L. E. (1966). Galerkin's Method and the Local Potential, *in* "Non-Equilibrium Thermodynamics, Variational Techniques and Stability," (R. J. Donnelly, R. Herman, and I. Prigogine, eds.) pp. 291–294. Univ. of Chicago, Chicago, Illinois.

Finlayson, B. A., and Scriven, L. E. (1967). On the Search for Variational Principles, *Int. J. Heat Mass Transfer* **10**, 799–821.

Gerber, R. (1950). Observations sur un Travail Récent de M. Miche, *J. Math. Pures Appl.* [9] **32**, 79–84.

Glansdorff, P., and Prigogine, I. (1964). On a General Evolution Criterion in Macroscopic Physics, *Physica* **30**, 351–374.

Gyarmati, I. (1969). On the "Governing Principle of Dissipative Processes" and its Extension to Non-Linear Problems, *Ann. Phys.* **23**, 353–378.

Gyarmati, I. (1970). "Non-Equilibrium Thermodynamics." Springer-Verlag, Berlin.

Herivel, J. W. (1954a). A General Variational Principle for Dissipative Systems, *Proc. Royal Irish Acad.* **56A**, 37–44.

Herivel, J. W. (1954b). A General Variational Principle for Dissipative Systems. II, *Proc. Royal Irish Acad.* **56A**, 67–75.

Hiraoka, M., and Tanaka, K. (1968). A Variational Principle for Transport Phenomena, Part 2, Applications to Several Problems (in English), *Memoirs Fac. Eng., Kyoto Univ.* **30**, 397–418.

Kaplan, S. (1969). Variational Methods in Nuclear Engineering, *Advan. Nucl. Sci. Tech.* **5**, 185–221.

Li, J. C. M. (1962). Carathéodory's Principle and the Thermokinetic Potential in Irreversible Thermodynamics, *J. Phys. Chem.* **66**, 1414–1420.

Miche, R. (1949). Sur la Réduction à un Principe Variationnel du Mouvement non Lent des Fluides Visqueux, *J. Math. Pures Appl.* **28**, 151–179.

Millikan, C. B. (1929). On the Steady Motion of Viscous Incompressible Fluids; with Particular Reference to a Variation Principle, *Phil. Mag.* [7] **7**, 641–662.

Onsager, L. (1931). Reciprocal Relations in Irreversible Processes. I, *Phys. Rev.* **37**, 405–426.

Onsager, L. (1969). The Motion of Ions: Principles and Concepts, *Science* **166**, 1359–1364.

Prigogine, I. (1966). Evolution Criteria, Variational Properties, and Fluctuations *in* "Non-Equilibrium Thermodynamics, Variational Techniques, and Stability," (R. J. Donnelly, *et al.*, eds.), pp. 3–16, Univ. of Chicago, Chicago, Illinois.

Prigogine, I., and Glansdorff, P. (1965). Variational Properties and Fluctuation Theory, *Physica* **31**, 1242–1256.

Rosen, P. (1953). On Variational Principles for Irreversible Processes, *J. Chem. Phys.* **21**, 1220–1221.

Schechter, R. S. (1962). On a Variational Principle for the Reiner–Rivlin Fluid, *Chem. Eng. Sci.* **17**, 803–806.

Yourgrau, W., and Mandelstam, S. (1968). "Variational Principles in Dynamics and Quantum Theory," 3rd ed. Pitman, London.

Chapter

11

Convergence and Error Bounds

We have postponed the question of convergence and error bounds because those subjects require special notation and mathematical background. In the previous chapters we considered only numerical convergence of the successive approximations. This is sufficient for many purposes, but more detailed mathematical justification is often available. We present a summary of results to acquaint the reader with the scope of information available. The theorems are stated in as much detail as space permits, and the mathematically inclined reader is referred to the original references for complete details.

11.1 Definitions

These definitions serve merely to identify the terminology, which is presented in more detail in standard references on analysis (cf. Courant and Hilbert, 1953, Mikhlin, 1964, Collatz, 1966a).

We take $x = (x_1, \ldots, x_n)$ as a point in n-dimensional Euclidian space E_n, V is a bounded domain in E_n, S is the boundary of V, $\overline{V} = V \cup S$, $Q = V \times (0, T)$, where T is a positive number, $0 \leq t \leq T$.

The class of functions which is k-times continuously differentiable is denoted by $C^k(V)$. The class of functions that are square integrable over Q is denoted by $\mathscr{L}_2(Q)$, with the scalar product and norm

$$(u, v) = \int_Q uv \, dx \, dt, \qquad \|u\| = (u, u)^{1/2}. \tag{11.1}$$

A sequence of functions ϕ_k is said to be orthonormal if $(\phi_k, \phi_m) = \delta_{km}$. $W^1(V)$ is the Hilbert space of functions $u(x)$ such that u and $\partial u/\partial x_i \in \mathscr{L}_2(V)$, with scalar product

$$(u, u)_1 = \int_v uv \, dx + \int_v \sum_{i=1}^n \frac{\partial u}{\partial x_i} \frac{\partial v}{\partial x_i} \, dx. \tag{11.2}$$

The space $W^{1,1}(Q)$ is the Hilbert space of functions $u(x, t)$ for which u, $\partial u/\partial x_i$, and $\partial u/\partial t \in \mathscr{L}_2(Q)$ with the scalar product

$$(u, v)_{1,1} = \int_Q uv \, dx \, dt + \int_Q \left(\sum_{i=1}^n \frac{\partial u}{\partial x_i} \frac{\partial v}{\partial x_i} + \frac{\partial u}{\partial t} \frac{\partial v}{\partial t} \right) dx \, dt. \tag{11.3}$$

A subscript zero, $W_0(V)$, means that the functions vanish on S, and for $W_0(Q)$ the functions vanish on $S \times (0, T)$.

Consider the linear operator L with a field of definition D_L, that is, Lu is defined for $u \in D_L$. The inner product is

$$(u, v)_L = (u, Lv) = \int_v uLv \, dx. \tag{11.4}$$

The operator is symmetric if for elements u and v in D_L

$$(u, Lv) = (v, Lu). \tag{11.5}$$

The operator is positive definite if for any function in D_L, not identically zero,

$$(u, Lu) \geq 0, \tag{11.6}$$

and is positive bounded below if for any $u \in D_L$

$$(u, Lu) \geq \gamma(u, u), \qquad \gamma > 0. \tag{11.7}$$

Take the functional $F(u)$ and a sequence of functions $\{u_n\}$. If

$$\lim_{n \to \infty} F(u_n) = d, \qquad \text{where} \quad d = \inf F(u), \tag{11.8}$$

then the functions $\{u_n\}$ constitute a minimizing sequence for the functional F. The infimum (or inf) is the greatest lower bound. Even though the functional converges to d, the minimizing sequence may not converge to a function in the space, as is illustrated in Section 11.2. There may be no function in the space which gives the minimum value.

If we wish to solve

$$Lu - f = 0, \tag{11.9}$$

the solution is called a classical solution if Eq. (11.9) is obeyed everywhere in V. A generalized solution is one which obeys

$$(\phi, Lu - f) = 0 \tag{11.10}$$

for all possible ϕ in a given class.

There are several types of convergence. Uniform convergence: given $\varepsilon > 0$ we can find an n such that

$$|u(x) - u_n(x)| < \varepsilon. \tag{11.11}$$

Convergence in the mean implies

$$\|u - u_n\| < \varepsilon. \tag{11.12}$$

The energy is defined

$$|u| \equiv (u, Lu) \tag{11.13}$$

and consequently involves derivatives of u. Convergence in energy requires

$$|u - u_n| < \varepsilon. \tag{11.14}$$

With convergence in the mean the converging sequence may not approach the limit function at every point in the domain, but the regions in which the converging sequence differs from the limit function become vanishingly small as $n \to \infty$. Convergence in energy permits similar things to happen to derivatives of the sequence. The corresponding error bounds are pointwise error (11.11), mean square error (11.12) and energy error (11.14) bounds. A sequence u_n converges weakly to an element u of a space if

$$\lim_{n \to \infty} (u_n, \phi) = (u, \phi) \tag{11.15}$$

holds for all ϕ in the space. The Galerkin method sometimes yields sequences which converge weakly to a generalized solution.

A set of functions is said to be linearly independent if the only solution to

$$\sum_{i=0}^{n} C_i f_i = 0 \tag{11.16}$$

is $C_i = 0$. If a set of functions is linearly dependent, then one function, say f_n, can be written in terms of the other functions. Inclusion of f_n in a series of approximating functions then represents needless duplication.

The most important property (with respect to convergence) of a set of trial functions is that they form a complete set of functions. A set of functions is complete in a space if any function in the space can be expanded in terms of the set of functions:

$$\left\| u - \sum_{i=1}^{\infty} a_i \phi_i \right\| < \varepsilon. \tag{11.17}$$

A set of functions complete for one space need not be complete for another space, so that it is necessary to specify the space, or class of functions. For example in Eq. (11.4) the class of functions is $u \in D_L$ or those functions for which Lu is defined. The following theorem (Mikhlin, 1964, p. 66) holds:

THEOREM 11.1. If an orthonormal set of functions $\{\phi_i\}$ is complete in the sense of convergence in the mean, with respect to some class of functions, then the Fourier series of any function u of the given class

$$u = \sum_{i=1}^{\infty} (u, \phi_i)\phi_i \tag{11.18}$$

converges in the mean to this function.

A system of functions is said to be complete in energy if, for n sufficiently large,

$$\left| u - \sum_{i=1}^{n} C_i \phi_i \right| < \varepsilon. \tag{11.19}$$

One method to prove a set of functions is complete is to show that the only function orthogonal to each member of the set is the null function (Courant and Hilbert, 1953, p. 110). We do not prove completeness here, but summarize some known results. Mikhlin and Smolitskiy (1967, p. 237) state the following theorem:

THEOREM 11.2. Let $\phi_n \in D_L$, where L is a positive definite operator and suppose that the sequence $\{L\phi_n\}$ is complete in the given Hilbert space H. The sequence $\{\phi_n\}$ is complete in the energy space H_L.

Completeness in energy is required of trial functions, and Theorem 11.2 means that trial functions must be capable of representing functions ϕ as well as derivatives $L\phi$.

If L and N are positive definite operators and the spaces H_L and H_N contain the same members, any system that is complete in H_L is complete in H_N and vice versa. This is useful for proving completeness. Consider the problem

$$\begin{gathered} Lu = -(p(x)u')' + q(x)u = f(x), \\ u(0) = u(1) = 0, \\ p(x) \geq \gamma > 0, \qquad q(x) \geq 0. \end{gathered} \tag{11.20}$$

If we define

$$Nu = -u'' \qquad u(0) = u(1) = 0, \tag{11.21}$$

then the spaces H_L and H_N consist of the same elements: functions for which $u(0) = u(1) = 0$ and

$$\int_0^1 [u'(x)]^2 \, dx < \infty. \tag{11.22}$$

Functions which are complete for Eq. (11.21) are then complete for Eq. (11.20). Examples are (Mikhlin and Smolitskiy, 1967, p. 238; Collatz, 1966a, p. 70)

$$\phi_n = x^n (1-x), \qquad \phi_n = \sin n\pi x. \tag{11.23}$$

Weierstrass's theorem (Courant and Hilbert, 1953, p. 65) says that any continuous function on $a \le x \le b$ can be approximated uniformly by polynomials. The derivatives can be approximated as well, which is a special case of Theorem 11.5. Collatz (1966a, p. 74) shows also that once we have an orthonormal system of functions it is possible to generate new systems by means of a weight function $p(x)$ which is positive and continuous on (a, b) and lies between positive bounds m and M. Similarly the polynomials used in orthogonal collocation are complete (see Theorem 11.19).

For the problem

$$\begin{gathered} Lu = \sum_{i=0}^{2m} (-1)^i \frac{d^i}{dx^i}\left(p_i(x)\frac{d^i u}{dx^i}\right) = f(x), \\ u = u' = \cdots = u^{(m-1)} = 0 \qquad \text{at} \quad x = 0, 1, \end{gathered} \tag{11.24}$$

the following system is complete in H_L (assuming $p_{2m} > 0$ and the p_i are such that Lu is positive definite):

$$\phi_n = x^{n+2m}(1-x)^{2m}. \tag{11.25}$$

For the two-dimensional problem

$$\begin{gathered} Lu \equiv -\sum_{i,j=1}^{2} \frac{\partial}{\partial x_i}\left(A_{ij}\frac{\partial u}{\partial x_j}\right) + Cu = f(x) \qquad \text{in } V, \\ u = 0 \qquad \text{on } S, \\ \sum_{i,j=1}^{2} A_{ij}\xi_i\xi_j \ge \mu(\xi_1{}^2 + \xi_2{}^2), \end{gathered} \tag{11.26}$$

any system which is complete in the energy of the operator $-\nabla^2$ is also complete in the energy of the operator L. Two such systems for rectangular and circular geometries are

$$\begin{gathered} \phi_{i,j} = \sin(i\pi x/a)\sin(j\pi y/b), \\ \phi_{i,j} = C_{i,j}J_j(\gamma_{j,i}r)\cos j\theta, \end{gathered} \tag{11.27}$$

where $\gamma_{j,i}$ is the nth positive root of the Bessel function $J_j(x)$ and $C_{i,j}$ is chosen such that $\|\phi_{i,j}\| = 1$. Babusko *et al.* (1966) discuss optimal approximations to give guidance to the choice of trial functions. They find, for example, that both sets of trial functions (11.23) are optimal. The calculations using polynomials are unstable for large $n(>7)$ because the functions are "almost" linearly dependent. This result suggests that for high-order approximations the polynomials must be orthogonal, as in the orthogonal collocation method.

Eigenvalue problems provide a convenient source of complete sets of functions, according to the theorem by Courant and Hilbert (1953, p. 424).

THEOREM 11.3. Consider the variational problem

$$\lambda = \min_{u \in \mathscr{F}} \frac{D(u)}{H(u)}, \tag{11.28}$$

where $H(u)$ is positive definite. Then the corresponding eigenfunctions, ϕ_i, are complete in the space of functions for which $H(u) = \|u\|$ is defined.

Many of the theorems which follow are for nonlinear problems. An operator is linear if, for any elements u, v in its field of definition and constants a, b,

$$L(au + bv) = aLu + bLv. \tag{11.29}$$

An equation is semilinear if the nonlinearities occur only in the function, but not its derivatives, for example,

$$Lu = f(x, u). \tag{11.30}$$

For these problems we need the concept of a Lipschitz constant K,

$$|f(x, u) - f(x, v)| < K|u - v|. \tag{11.31}$$

11.2 Boundary-Value Problems

The power of theorems on convergence and error bounds depends upon the type of problem: the number of dimensions, the existence of a variational principle, and the effect of boundary conditions. Courant (1943) pointed out that convergence is improved as the order of the highest derivative increases and convergence becomes worse as the number of independent variables increases. The theorems below follow this pattern: pointwise convergence is established for ordinary differential equations and two-dimensional problems, whereas convergence in the mean holds for three-dimensional cases. Error bounds are available for problems with variational principles whereas few are available for nonlinear or non-self-adjoint problems.

We first consider convergence of a minimizing sequence, that is, one which makes the variational integral converge, $I(u_n) \to I(u)$. It is tempting to say that the minimizing sequence converges to the solution, $u_n \to u$. This is correct for certain second-order ordinary differential equations, but is not necessarily true for partial differential equations, as is illustrated by the following example from Kantorovich and Krylov (1958, p. 337).

Minimize $I(u)$ among all continuous functions $u(r, \theta)$ which have piece-wise continuous first derivatives and which satisfy $u = 0$ on $r = 1$:

$$I(u) = \int_0^1 \int_0^{2\pi} [u_r^2 + (u_\theta/r)^2] r \, dr \, d\theta. \tag{11.32}$$

$I(u)$ represents the Dirichlet integral, and the obvious solution is $u = 0$, giving $I(u) = 0$. Consider the following sequence of functions:

$$u_n = \begin{cases} 1, & 0 \le r \le a_n^2, \\ \ln r/\ln a_n - 1, & a_n^2 \le r \le a_n, \\ 0, & a_n \le r \le 1. \end{cases} \tag{11.33}$$

This presents a surface which is one at the center and falls rapidly to zero as r increases. For this function the integral (11.32) converges to zero as a_n approaches zero. The function itself does not converge to the solution, since $u_n(0, \theta) = 1$ for all n. Thus convergence of the functional does not necessarily imply uniform convergence of the function itself. In this case, however, convergence in the mean obtains since $\int (u_n - u)^2 r \, dr \, d\theta \to 0$ as $a_n \to 0$.

Three-Dimensional Heat Conduction; Rayleigh–Ritz Method

Steady-state heat conduction with a constant thermal conductivity is governed by the equation

$$u_{xx} + u_{yy} + u_{zz} = f(x, y, z). \tag{11.34}$$

Mikhlin (1964, Chapter 3) proves Theorem (11.4).

THEOREM 11.4. Consider the equation $L(u) = f$ and assume that there is at least one solution with finite energy and that the operator is positive bounded below. The variational integral is $F(u) = (Lu, u) - 2(u, f)$. Then

(1) There can be only one solution.

(2) Any minimizing sequence for the functional converges in energy and in the mean to the solution.

(3) When the Rayleigh–Ritz method is applied, using coordinate functions which are complete in energy and linearly independent, the approximate solution constitutes a minimizing sequence for the functional and hence converges in the mean to the solution.

To use this theorem it is only necessary to prove that the operator in question is positive bounded below and that the coordinate functions are complete. Mikhlin (1964, Section 22) does this for Eq. (11.34) under a variety of boundary conditions. It is positive bounded below for the set of functions which are continuous together with their first and second derivatives in the domain $\bar{V}$ and satisfy the boundary conditions

$$u = 0, \tag{11.35a}$$

or

$$\partial u/\partial n + \sigma u = 0 \qquad \text{on } S, \tag{11.35b}$$

where σ is a nonnegative function, not identically zero. For the Neumann problem, with boundary condition

$$\partial u/\partial n = 0, \tag{11.36}$$

Eq. (11.34) is positive bounded below for the set of functions satisfying the same continuity requirements, and boundary condition (11.36) and $\int u\, dS = 0$. For the Neumann problem to be unique it is also necessary that $\int f\, dS = 0$.

To prove the completeness of a system of functions we can use the following theorem, which is a generalization to two dimensions of Weierstrass's theorem in one dimension (Kantorovich and Krylov, 1958, p. 275).

THEOREM 11.5. If the function $v(x, y)$ is continuous in a closed bounded region V, together with its partial derivatives, v_x and v_y, then for $\varepsilon > 0$ one can find a polynomial $P(x, y)$ such that in the region V the following inequalities will be satisfied:

$$|v(x, y) - P(x, y)| < \varepsilon, \qquad |v_x - P_x| < \varepsilon, \qquad |v_y - P_y| < \varepsilon. \tag{11.37}$$

Kantorovich and Krylov (1958, p. 276) construct such a polynomial for the boundary condition of the first kind. Assume there is a function $w(x, y)$ which is continuous and has continuous, bounded derivatives, w_x and w_y, and satisfies

$$w(x, y) > 0 \quad \text{in } V, \qquad w(x, y) = 0 \quad \text{on } S. \tag{11.38}$$

The following system of functions is then complete:

$$\phi_0 = w, \quad \phi_1 = wx, \quad \phi_2 = wy, \quad \phi_3 = wxy, \ldots. \tag{11.39}$$

Kantorovich and Krylov then describe various ways to construct the function w. Oftentimes the bounding contour is of the form $f(x, y) = 0$ and it may be possible to take $w = \pm f$. For convex polygons with n boundaries $a_i x + b_i y + c_i = 0$, a possibility is

$$w = \prod_{i=1}^{n} (a_i x + b_i y + c_i). \tag{11.40}$$

More difficult cases are also discussed by Kantorovich and Krylov. For boundary conditions of the third kind the following polynomial is complete:

$$u_n = a_1 + a_2 x + a_3 y + a_4 xy + a_5 x^2 + \cdots. \tag{11.41}$$

The constants can be chosen to satisfy Eq. (11.35b) or the boundary condition can be satisfied as a natural boundary condition. Schultz (1969c) proves convergence and estimates the error (in energy) when the approximating functions are spline functions. Theorems (11.4) and (11.5) imply convergence of the Galerkin and Rayleigh–Ritz methods applied to the linearized form of steady-state heat conduction [Eqs. (2.48), (2.50), and (7.47)]. They also imply convergence of the same methods applied to laminar flow through ducts [Eqs. (4.1), (7.62), and (7.63)].

Error bounds are less available. Kantorovich and Krylov (1958, p. 342) prove a theorem on error bounds for the following equation:

$$(a\, u_x)_x + (b\, u_y)_y - cu + f(x, y), \qquad a, b > 0, \quad c \geq 0, \tag{11.42}$$

which has the corresponding variational integral

$$I(u) = \int_s [a\, u_x{}^2 + b\, u_y{}^2 + cu^2 - 2fu]\, dx\, dy. \tag{11.43}$$

Under many assumptions they prove

$$|u_n(x, y) - u(x, y)| \leq c[\varepsilon_n \ln (n/\varepsilon_n)]^{1/2}, \qquad \varepsilon_n = I(u_n) - I(u), \tag{11.44}$$

which means that uniform convergence obtains. To calculate the error bound it is necessary to know a lower bound for $I(u)$, and this can be obtained by means of the reciprocal variational principle. Generally the kind of additional assumption which results in uniform convergence rather than convergence in the mean is that the derivatives of the approximate solute do not increase too rapidly by comparison to ε_n, for example

$$\int_\alpha^\beta (u_n)_y{}^2\, dy \leq K_n, \qquad \lim_{n\to\infty} \varepsilon_n \ln K_n = 0. \tag{11.45}$$

The interested reader is referred to Kantorovich and Krylov (1958) for additional details. In the proofs of these results the existence of a variational integral plays an indispensible role. Yasinsky and Kaplan (1968) have applied reciprocal variational principles to estimate the error in approximate solutions. The variational principle gives $I(u_n)$ and the reciprocal variational principle gives a lower bound for $I(u)$.

Transport Equation, Galerkin Method

The steady-state transport equation

$$-\sum_{i,j=1}^{3} \frac{\partial}{\partial x_i}\left(A_{ij} \frac{\partial u}{\partial x_j}\right) + \sum_{i=1}^{3} B_i \frac{\partial u}{\partial x_i} + Cu = f, \tag{11.46}$$

$$A_{ij}(x), \quad B_i(x), \quad C(x), \quad f(x),$$

is non-self-adjoint unless $B_i = 0$. There is no variational principle so that we consider convergence of the Galerkin method. Assume there exists a constant μ_0 such that for any point in $\bar{V}$ and real numbers t_i,

$$\sum_{i,j=1}^{3} A_{ij} t_i t_j \geq \mu_0 \sum_{i=1}^{3} t_i^2. \qquad (11.47)$$

The coefficients A_{ij} and their first derivatives are continuous in V and the coefficients B_i and C are continuous in $\bar{V}$. We assume the problem has a unique solution. Mikhlin (1964, Section 24) proves that the operator

$$L_0 u = -\sum_{i,j=1}^{3} \frac{\partial}{\partial x_i}\left(A_{ij} \frac{\partial u}{\partial x_j}\right) \qquad (11.48)$$

is positive bounded below for the set of functions which vanishes on S. The convergence proofs are done in terms of the operator L_0 and the remaining terms in the differential equation are bounded in terms of this operator and its inverse. Mikhlin (1964, p. 480) proves the following theorem.

THEOREM 11.6. Assume the problem (11.46) is unique for the boundary conditions

$$u = 0 \quad \text{on } S_1, \qquad \partial u/\partial n = 0 \quad \text{on } S_2, \qquad \partial u/\partial n + \sigma u = 0 \quad \text{on } S_3. \qquad (11.49)$$

Apply the Galerkin method using trial functions which are complete in energy of L_0, are linearly independent, and satisfy $u = 0$ on S_1. Then the approximate solution u_n converges in energy of L_0 to the solution.

Since the operator L_0 is positive bounded below, convergence in energy also implies convergence in the mean. A similar theorem was proved earlier by Keldysh (1942).

These theorems are not directly relevant to the heat and mass transfer problems in Sections 2.2, 2.5 because the theorems do not allow infinite domains. Theorem 11.6 would be applicable to heat and mass transfer problems with known convection in a finite domain. When (11.46) is nonlinear because the coefficients A_{ij}, B_i, and C depend on u, convergence of the Galerkin method follows from theorems stated below.

Mean-square error bounds for the Galerkin method (as well as other schemes of MWR) can be derived in terms of the mean-square residual (see Section 11.6). Pointwise error bounds can sometimes be constructed using the maximum principle (see Section 11.5).

Sturm–Liouville Equation

Consider the ordinary differential equation of the Sturm–Liouville type:

$$\frac{d}{dx}\left(p(x)\frac{dy}{dx}\right) - q(x)y = f, \qquad y(0) = y(1) = 0, \tag{11.50}$$
$$p > 0, \quad q \geq 0, \quad 0 \leq x \leq 1.$$

If the boundary conditions are instead $u(0) = u_0$, $u(1) = u_1$, then the problem can be transformed into the form of Eq. (11.50),

$$\begin{aligned} u &= y + xu_1 + (1 - x)u_0, \\ y(0) &= y(1) = 0, \end{aligned} \tag{11.51}$$

The problem (11.50) is self-adjoint and the variational integral is

$$I(y) = \int_0^1 [p(y')^2 + qy^2 + 2uf]\, dx. \tag{11.52}$$

Kantorovich and Krylov (1958, p. 241) prove that a solution exists and is unique. Then they prove (p. 262) the following theorem.

THEOREM 11.7. If the Rayleigh–Ritz method is applied to Eq. (11.52) using a set of trial functions which is complete, the functions and their first derivatives are continuous in [0, 1], then the variational integral and the function converge to the solution,

$$I(y_n) \to I(y), \qquad y_n \to y. \tag{11.53}$$

As trial functions we can use sine functions or polynomials (11.23).

Error bounds have been established by Kantorovich and Krylov (1958, p. 327). Using the sine functions and the Rayleigh–Ritz method for the case $p = 1$ they show that the error is

$$\begin{aligned} |y - y_n| \leq \frac{\sqrt{2}\max q^{1/2}}{\pi^{3/2}(n+1)^{3/2}} \Bigg\{ 1 + \frac{\max q}{4[6\pi(n+1)]^{1/2}} \left(1 + \frac{4\sqrt{2}}{n^{3/2}\pi^2}\right) \\ \times \left(1 - \frac{\max q}{\pi^2(n+1)^2}\right)^{-1} \Bigg\} \left[\int_0^1 \frac{f^2}{q}\, dx\right]^{1/2}. \end{aligned} \tag{11.54}$$

For the special case of $q = f = 1$ the values of the error are tabulated in Table 11.1. When compared to the actual error, the error bound is quite conservative.

TABLE 11.1

ERROR BOUNDS FOR STURM–LIOUVILLE PROBLEM. VALUES SHOWN ARE FOR $|y(\frac{1}{2}) - y_n(\frac{1}{2})|$

n	1	3	5	9	99
Actual error	0.004	0.00077	0.00026	0.000058	—
Error bound	0.094	0.033	0.018	0.008	0.00025

The form of the equation so far is suitable for planar geometry, but not cylindrical or spherical geometry, since then the coefficients become unbounded at $x = 0$. For spherical geometry the transformation $y = v/x$ will reduce the equation

$$\frac{1}{x^2}\frac{d}{dx}\left(x^2\frac{dy}{dx}\right) - qy = f \tag{11.55}$$

to the form $v'' + qv = xf$ for which the theorems are applicable. For cylindrical geometry there is apparently no similar transformation which results in an equation of the form of (11.50) with bounded coefficients and bounded domain. When spline functions are used, error bounds are provided by Birkhoff *et al.* (1968), who use piecewise Hermite polynomials in each interval, (x_{i-1}, x_i).

Non-Self-Adjoint Ordinary Differential Equations

Non-self-adjoint ordinary differential equations must be solved with the Galerkin method (or another MWR) and convergence proofs are known (Mikhlin, 1964). Consider the problem,

$$(-1)^m u^{(2m)} - Ku = f(x), \tag{11.56a}$$

$$u = u' = \cdots = u^{(m-1)} = 0 \quad \text{at} \quad x = 0, 1. \tag{11.56b}$$

Here, Ku is a linear differential operator of order $2m - 1$. We use coordinate functions which are from the field of definition of $L_0 u = (-1)^m u^{(2m)}$ and which are complete in a space H_0 with inner product $(L_0 u, v)$. The proof of convergence depends essentially on showing that the operator L_0 is positive-bounded-below and using the consequences of that fact. Mikhlin (1964, p. 477) proves the following theorem.

THEOREM 11.8. The Galerkin method applied to Eqs. (11.56) gives a convergent sequence in H_0 provided:

(1) the problem has a unique solution:

(2) the coordinate functions are in the field of definition of L_0 and satisfy the boundary conditions (11.56b).

For sufficiently large n the following inequalities hold:

$$|u_n^{(k)}(x) - u^{(k)}(x)| < \varepsilon, \qquad k < m,$$

$$\int_0^1 [u_n^{(m)}(x) - u^{(m)}(x)]^2 \, dx < \varepsilon. \tag{11.57}$$

Thus pointwise convergence holds for the function and its first $m - 1$ derivatives, whereas the mth derivative converges in the mean. Trial functions can be taken as polynomials, such as

$$u_n = x^m(1 - x)^m \sum_{i=1}^{n} a_i x^{i-1}. \tag{11.58}$$

Szalek (1970) provides bounds for the functional arising in non-self-adjoint nuclear engineering problems.

For semilinear problems we have (Schultz, 1969a, b) the following theorem.

THEOREM 11.9. The problem

$$\begin{gathered} -D^2u + p(x)Du + q(x)u = f(x, u(x)), \\ u(0) = u(1) = 0, \end{gathered} \tag{11.59}$$

where

$$p(x) \in C^1[0, 1], \qquad q(x) \in C[0, 1],$$

$$\left| \frac{f(x, u) - f(x, v)}{u - v} \right| \le M(c) \qquad \text{for} \quad |u|, |v| \le c, \tag{11.60}$$

has a unique generalized solution. The trial functions must form a complete set of functions in W_0^1 on [0, 1]. The Galerkin solution and its first derivatives converge in the mean to the solution and its first derivative.

Error bounds are given for the mean-square error. Ciarlet *et al.* (1967) study the problem

$$Lu = f(x, u(x)), \tag{11.61}$$

where L is an nth order, self-adjoint ordinary differential operator. The problem thus has a variational principle, which is used to calculate mean-square error bounds when using piecewise cubic Hermite polynomials. Shampine (1968) uses reciprocal variational principles to derive bounds for the same problem. Leipholz (1965) studies a similar problem and proves uniform convergence of the Galerkin method when the trial functions are taken as the eigenfunctions to the adjoint problem $Lu = -\lambda u$. He considers the case when the trial functions do not satisfy the natural boundary conditions as well (Leipholz, 1967a, b).

Theorems 11.7–11.9 specify such simple boundary conditions that they are not applicable to many engineering problems. For example, the reactor with axial diffusion [Eqs. (5.87), (5.88)] has boundary conditions of the third kind.

Visik (1961a, 1962) uses the Galerkin method to prove the existence of weak solutions to the following nonlinear, elliptic, system of equations:

$$\sum_{|\alpha|} (-1)^{|\alpha|} D_\alpha A_\alpha(x, D_\alpha u) = h(x), \tag{11.62}$$

where

$$u = (u_1, \dots, u_N), \qquad A_\alpha = (A_\alpha{}^1, \dots, A_\alpha{}^N), \dots .$$

The detailed statement of the theorem is similar to that given for initial-value problems in Section 11.3. Since the Galerkin method is used to prove the existence of solutions, the convergence is guaranteed.

Finn and Smith (1967) construct a generalized solution to Oseen's equation using the Galerkin method:

$$\nabla^2 \mathbf{u} - \mathbf{u}_0 \cdot \nabla \mathbf{u} - \nabla p = 0, \qquad \nabla \cdot \mathbf{u} = 0. \tag{11.63}$$

The trial functions must form a complete system of functions, be infinitely differentiable and solenoidal. Dubinskii (1967b) and Heywood (1970) prove the Galerkin method converges for the steady-state Navier–Stokes equation.

Least Squares Method

Consider the linear equation

$$Lu = f \quad \text{in } V, \qquad u = 0 \quad \text{on } S \tag{11.64}$$

and its solution by the least-squares method. We minimize the functional

$$J(u) = \int (Lu - f)^2 \, dV. \tag{11.65}$$

If the approximation solution is represented in the form

$$u_n = \sum_{i=1}^{n} a_i \phi_i, \tag{11.66}$$

we obtain the equations governing the approximate solution:

$$\sum_{i=1}^{n} a_i (L\phi_i, L\phi_k) = (f, L\phi_k). \tag{11.67}$$

Mikhlin (1964, Chapter 10) proves the following theorem.

THEOREM 11.10. When the homogeneous equation $Lu = 0$ has only a trivial solution (i.e., the inverse operator exists) approximate solutions can be constructed by the least-squares method and they are unique. The method gives a sequence of approximate solutions which converge in the mean to the exact solution if

(1) the sequence of coordinate functions is L-complete,
(2) Eq. (11.64) is soluble,
(3) there exists a constant K such that for any u in the field of definition of the operator L the following inequality holds:

$$\|u\| \leq K \|Lu\|. \tag{11.68}$$

The term L-complete means that for any $\varepsilon > 0$ it is possible to find a positive integer n and constants $\{a_k\}$ such that

$$\|Lu - Lu_n\| < \varepsilon. \tag{11.69}$$

For a second-order operator it must be possible to approximate second derivatives in the mean, whereas in the Rayleigh–Ritz method it was only necessary to approximate first derivatives. For bounded operators, however, ordinary completeness implies L-completeness:

$$\|Lu - Lu_n\| \leq \|L(u - u_n)\| \leq \|L\| \|u - u_n\| < \varepsilon. \tag{11.70}$$

If the sequence is complete, then $\|u - u_n\|$ can be made arbitrarily small and if the operator is bounded $\|L\|$ is bounded, giving L-completeness.

For the two-dimensional heat conduction equation the results can be more specific. Mikhlin (1964, p. 503) considers the problem

$$u_{xx} + u_{yy} = f(x, y) \tag{11.71}$$

and proves the inequality,

$$[u(x, y) - u_n(x, y)]^2 \leq \int_v G^2(x, y; \xi, \eta)\, d\xi\, d\eta \int_v (f - \nabla^2 u_n)^2\, dV, \tag{11.72}$$

where G is the Green's function. The first integral is bounded and the second integral is the quantity minimized by the least-squares method. Thus Eq. (11.72) provides uniform convergence of the approximate solution as well as pointwise error bounds. Furthermore, the error bounds are improved when more terms are included in the approximate solution. Despite this advantage of the least-squares method, the convergence is slower than the Rayleigh–Ritz method. Consider the two methods applied with the same trial functions, and let u_n denote the solution by least squares, and v_n the solution by Rayleigh–

Ritz. Since the Rayleigh–Ritz method minimizes the variational integral (11.43),

$$I(u_n) \geq I(v_n). \tag{11.73}$$

This implies, however, that

$$|u_n - u_0| \geq |v_n - u_0|, \tag{11.74}$$

which means that the Rayleigh–Ritz method converges faster, since it has the smallest energy norm.

Nitsche (1969) considers the Sturm–Liouville boundary-value problem (11.50) and compares the convergence of two methods. The Rayleigh–Ritz method is at least as good as the least-squares method when measured in terms of the asymptotic behavior of the error norms, and the convergence of the Rayleigh–Ritz method is better when using trigonometric polynomials. Similar error bounds for nonlinear problems are treated in Section 11.6.

Collocation Method

Kadner (1960) applies the collocation method to the following non-self-adjoint ordinary differential equation.

$$\begin{aligned} L(y) &\equiv \sum_{\nu=0}^{n} f_\nu(x) y^{(\nu)}(x) = r(x), \\ U_\mu(y) &\equiv \sum_{\kappa=0}^{n-1} [\alpha_{\mu\kappa}\, y^{(\kappa)}(a) + \beta_{\mu\kappa}\, y^{(\kappa)}(b)] = 0, \qquad \mu = 1, \ldots, n, \end{aligned} \tag{11.75}$$

under the assumptions that $f_\nu(x)$ are real and n-times continuously differentiable in $[a, b]$, the forcing function $r(x)$ is real and continuous and not identically zero, the homogeneous problem ($r \equiv 0$) has only the trivial solution, and the boundary conditions have real coefficients and are linearly independent. The proofs of convergence and error bounds require use of the adjoint operator to convert the problem to self-adjoint form of order twice as high:

$$\begin{aligned} L_A(y) &\equiv \sum_{\nu=0}^{n} (-1)^\nu [f_\nu(x) y(x)]^{(\nu)} = 0, \\ V_\mu(y) &\equiv \sum_{\kappa=0}^{n-1} [\bar{\alpha}_{\mu\kappa}\, y^{(\kappa)}(a) + \bar{\beta}_{\mu\kappa}\, y^{(\kappa)}(b)] = 0, \qquad \mu = 1, \ldots, n, \\ M(y) &\equiv L_A\, L(y) = \sum_{\nu=0}^{n} (-1)^\nu \left[f_\nu \sum_{\rho=0}^{n} f_\rho\, y^{(\rho)} \right]^{(\nu)} = L_A\, r, \\ U_\mu(y) &= \sum_{\kappa=0}^{n-1} [\alpha_{\mu\kappa}\, y^{(\kappa)}(a) + \beta_{\mu\kappa}\, y^{(\kappa)}(b)] = 0, \\ V_\mu(Ly) &= \sum_{\kappa=0}^{n-1} \left[\bar{\alpha}_{\mu\kappa} \sum_{\nu=0}^{n} (f_\nu\, y^{(\nu)})^{(\kappa)} \Big|_{x=a} + \bar{\beta}_{\mu\kappa} \sum_{\nu=0}^{n} (f_\nu\, y^{(\nu)})^{(\kappa)} \Big|_{x=b} \right] = 0. \end{aligned} \tag{11.76}$$

The self-adjoint problem has a symmetric Green's function denoted by $G(x, s)$. Kadner proves the following theorem.

THEOREM 11.11. The solution is expanded in a polynomial satisfying the boundary conditions

$$y_n = \sum_{j=1}^{n} c_j u_j(x), \tag{11.77}$$

the u_j are complete in $\mathscr{L}_2(a, b)$, and the collocation points are arbitrary, except that the corresponding quadrature weights must be positive. Then

$$|y^{(\sigma)} - y_n^{(\sigma)}|^2 \le \int_a^b [G^{(\sigma,0)}(x, s)]^2\, ds \int_a^b \varepsilon_k^2(s)\, ds, \qquad \sigma = 1, \ldots, n-1,$$

$$\int_a^b |y^{(n)} - y_n^{(n)}|^2\, dx \le \int_a^b \varepsilon_k^2(s)\, ds \left\{\left[\int_a^b \int_a^b (G^{(n,0)}(x, s))^2\, ds\, dx\right]^{1/2} + \frac{1}{|f_n(\xi)|}\right\}^2, \tag{11.78}$$

where ε_k is the residual. This theorem gives error bounds and where $a < \xi < b$, uniform convergence of derivatives up to order $n - 1$ and convergence in the mean of the nth derivative.

Vainikko (1965) proves similar results for ordinary differential equations without converting the problem to a higher-order equation. He considers

$$u^{(m)} + \sum_{j=0}^{m-1} e_j(x) u^{(j)} = f(x),$$
$$\sum_{j=0}^{m-1} [\alpha_{ij} u^{(j)}(a) + \beta_{ij} u^{(j)}(b)] = 0. \tag{11.79}$$

The approximate solution is expressed in polynomials $\phi_k(s)$ of order $m + k$ which satisfy the boundary conditions. The collocation points are taken as the roots to the $(n + 1)$st polynomial of a system of orthogonal polynomials on $[a, b]$ with respect to a weighting function $\rho(x)$ which satisfies

$$\rho(x) > 0, \qquad \int_a^b \frac{dx}{\rho(x)} < \infty. \tag{11.80}$$

THEOREM 11.12. Suppose $e_j(x), f(x)$ are continuous on $[a, b]$ and suppose Eq. (11.79) has a unique solution. Then for n sufficiently large the approximate solution is unique and a sequence of approximate solutions $u_n(s)$ together with derivatives up to $m - 1$ inclusive converge uniformly to $u(s)$ and its derivatives, while the mth derivative converges in the mean with respect to $\rho(s)$:

$$\left[\int_a^b p(x)\,|u_n^{(m)}(x) - u^{(m)}(x)|^2\,dx\right]^{1/2} \le CE_n(u^{(m)}), \tag{11.81}$$
$$\max_{a\le x\le b} |u_n^{(j)}(x) - u^{(j)}(x)| \le CE_n(u^{(m)}), \qquad j = 0, \ldots, m-1,$$

where C is a constant not depending on n or $f(a)$ and $E_n(u^{(m)})$ is the best uniform approximation to $u^{(m)}(s)$ by a polynomial of degree not exceeding n.

To use this theorem to obtain error bounds it is necessary to estimate $E_n(u^{(m)})$, which can sometimes be done using the Jackson Theorems (see Cheney, 1966, p. 139). Vainikko continues to prove that the collocation method is stable if and only if the sequence $\phi_k^{(m)}(x)$ is strongly minimal. The condition of strongly minimal means that matrix

$$A_{ik} = \int_a^b \rho(s)\phi_i^{(m)}(s)\overline{\phi_k^{(m)}(s)}\,ds, \qquad i, k = 1, \ldots, n \tag{11.82}$$

is positive bounded below. Stability of the computations means that errors in the solution grow no more rapidly than the norm of the error of the coefficients. He notes that instability can occur only through errors in the matrix, whereas in the Rayleigh–Ritz method instability can occur through errors in the matrix or the forcing term.

Karpilovskaya (1963) proves convergence for the partial differential equation on the square region v: $0 < s, t < \pi$,

$$u_{xx} + u_{yy} + \lambda q(x, y) = f(x, y). \tag{11.83}$$

He expands the solution in a series of sines and cosines

$$u_{mn} = \sum_{k=1}^{m} \sum_{l=1}^{n} a_{kl} \sin kx \sin ly. \tag{11.84}$$

THEOREM 11.13. If $q(x, y)$ and $f(x, y)$ are continuous in $\overline{V}$, λ is not an eigenvalue of Eq. (11.83), and the collocation points are $\big(\pi(2k-1)/(2m+1), \pi(2l-1)/(2n+1)\big)$, $k = 1, \ldots, m$; $l = 1, \ldots, n$; the approximate solution and its first derivatives converge uniformly to the exact solution and its first derivatives.

The order of convergence is also established. Karpilovskaya also considers the expansion

$$u_n(x, y) = \sum_{k=1}^{n} a_k(x) \sin ky \tag{11.85}$$

and proves that the collocation method, applied at the points $y = \pi(2k-1)/(2n+1)$ gives an approximate solution which converges uniformly to the solution.

In summary, the general conditions needed to prove convergence are that the system of expansion functions is complete. Other conditions depend on the problem. Many types of linear, ordinary differential equations can be handled using Green's functions to convert the problem to an integral equation. For partial differential equations, convergence in energy is valid if the operator is positive definite, convergence in the mean is valid for positive-bounded-below operators. For non-self-adjoint operators it is sometimes possible to separate the problem into a self-adjoint, positive-bounded-below part and bound the remaining non-self-adjoint part. The error bounds are specially derived in each case and usually make use of some special property of the equations, such as a variational integral, a Green's function etc. Convergence theorems do not appear to be available for boundary layer problems (Sections 4.2 and 4.3), problems in infinite domains (Sections 2.2 and 2.5), boundary methods (Section 2.3), or for non-Newtonian fluids (Section 8.4).

11.3 Initial-Value Problems

Convergence proofs for the Galerkin method have been proved for a variety of linear and nonlinear initial-value problems, including the transport equation and the Navier–Stokes equation.

Galerkin Method, Linear Equations

Green (1953) considered the problem,

$$\begin{aligned} L(u) = u_{xx} - u_t - g(x, t)u &= f(x, t), \\ u(x, 0) = u(0, t) = u(\pi, t) &= 0. \end{aligned} \tag{11.86}$$

The functions f and g are continuous in Q, together with their first partial derivatives. In addition $f(0, 0) = f(\pi, 0) = 0$. The approximate solution is written in the form

$$u_n(x, t) = \sum_{k=1}^{n} C_{nk}(t) \sin kx \tag{11.87}$$

and determined as the solution to the set of ordinary differential equations:

$$\int [L(u_n) - f] \sin jx \, dx = 0, \qquad j = 1, 2, \ldots, n, \tag{11.88}$$

$$C_{nk}(0) = 0.$$

Green proves the problem is unique and then proves the following theorem.

THEOREM 11.14. The sequence u_n converges uniformly to the solution u and the first derivatives u_{nt} and u_{nx} converge uniformly to u_t and u_x, respectively. The function u_{nxx} converges in the mean to u_{xx}. The mean-square error satisfies (11.89) where $F_n^*(t) \to 0$ as $n \to \infty$:

$$\int (u_n - u)^2 \, dx \leq t^2 F_n^{*2}(t),$$
$$F_n^*(t) \equiv \max_{\tau \leq t} \left(\int [f(x, \tau) - Lu_n(x, \tau)]^2 \, dx \right)^{1/2}. \tag{11.89}$$

The error bound is thus given in terms of the mean-square residual. Note that when g = constant the method is just an expansion in the eigenfunctions of the spatial operator, or a Fourier series method.

A more general transport equation is treated by Il'in *et al.* (1962):

$$L(u) = \sum_{i,j=1}^{n} \frac{\partial}{\partial x_i} \left(A_{ij}(x, t) \frac{\partial u}{\partial x_j} \right)$$
$$+ \sum_{i=1}^{n} B_i(x, t) \frac{\partial u}{\partial x_i} + C(x, t)u - \frac{\partial u}{\partial t} = F(x, t), \tag{11.90}$$
$$u(x, 0) = \phi(x), \qquad u\,|_s = 0.$$

The functions B_i, C, and F are continuous in Q, and A_{ij} is continuous and has continuous first derivatives. The initial function $\phi(x) \in W_0^{\,1}(V)$, and the coefficient A_{ij} satisfies (11.47). The boundary of V is in the class A^2 which means it can be expressed in the form $x_1 = g(x_2, \ldots, x_n)$ in the neighborhood of any of its points and $g \in C^2$. Il'in *et al.* first prove the following theorem.

THEOREM 11.15. If the boundary of V belongs to the class A^2, then there exists an orthogonal system of functions $X_n(x)$ in $\mathscr{L}_2(V)$ with the following properties:

(1) The $X_n(x)$ are twice differentiable with continuous derivatives in $\bar{V}$ and are zero on the boundary S of V;

(2) For every function $\phi \in W_0^{\,1}(V)$ and $\varepsilon > 0$ there is a linear combination $\phi_N = \sum_{k=1}^{N} C_k X_k$ such that

$$\|\phi - \phi_N\|_1 < \varepsilon; \tag{11.91}$$

(3) For the function $v(x, t)$ in $C^\infty(Q)$ that is zero in a neighborhood of the boundary of Q and for any $\varepsilon > 0$ there is a linear combination $u_N(x, t) = \sum_{k=1}^{N} C_k(t) X_k(x)$ so that

$$\int_0^T \|u - u_N\|_1 \, dt < \varepsilon. \tag{11.92}$$

The approximate solution to Eq. (11.90) is found in the form

$$u_N(x, t) = \sum_{k=1}^{N} C_{Nk}(t) X_k(x), \tag{11.93}$$

with the functions $C_{Nk}(t)$ determined by the ordinary differential equations

$$\int_v [L(u_N) - F] X_k(x)\, dx = 0, \qquad k = 1, \ldots, N. \tag{11.94}$$

The initial conditions are determined by $C_{Nk}(0) = (\phi, X_k)$ and $\phi(x) = |_s 0$ is required. Il'in *et al.* prove the following theorem.

THEOREM 11.16. The approximate solution $u_N(x, t)$ converges in the mean to a function $u(x, t) \in W^{1,1}(Q)$ which is a generalized solution to (11.90). The first spacial and time derivatives converge weakly to the corresponding generalized derivatives.

A suitable choice of expansion functions are the eigenfunctions for the Laplacian operator for the domain in question. For boundary conditions of the third kind, convergence follows from Theorem 11.18 for nonlinear problems. Anderssen (1969) proves for the adjoint variational method applied to initial-value problems [hence also for the Galerkin method for Eqs. (9.104)] that the necessary and sufficient condition for stability of the process is that the system of functions be strongly minimal Eq. (11.82).

Visik (1963) considers the linear problem for $u(x, t)$,

$$A \frac{\partial^2 u}{\partial t^2} + B \frac{\partial u}{\partial t} + Cu = h, \tag{11.95}$$

where A, B, C are spatial operators and proves that the Galerkin method converges. Veliev (1964) deduces the stability of the Galerkin calculations in the following circumstances:

(1) $A = 0$, $B = I$.
(2) $A = I$, $B = \alpha I$, $\alpha > 0$.
(3) $A = I$, $B = 0$.
(4) A, B, C positive definite, self-adjoint.
(5) $A = 0$, B, C, positive definite, self-adjoint.

The trial functions must be linearly independent, complete in a space H_0 (which depends on the problem), and normal in the space H_0, that is strongly minimal with a bounded matrix (11.82).

Galerkin Method, Nonlinear Equations

Many existence theorems are proved using the Galerkin method: Visik and Ladyženskaya (1958), Dubinskii (1967a), Ladyženskaya *et al.* (1968), and Douglas and DuPont (1970). A simplified statement of Theorem 6.7 in Ladyženskaya *et al.* (1968) is the following theorem.

THEOREM 11.17. The problem

$$u_t - \frac{\partial}{\partial x_i} a_i(x, t, u, u_x) + a(x, t, u, u_x) = 0, \tag{11.96}$$

$$u\big|_{S_T} = 0, \qquad u(x, 0) = g(x),$$

for any $g(x) \in \mathscr{L}_2(V)$ has at least one generalized solution u if the following conditions are satisfied:

(1) The functions a_i and a satisfy the inequalities

$$\begin{aligned} |a_i(x, t, u, p)| &\le \phi_1(x, t) + c|u|^{q^*/2} + c|p|, \\ |a(x, t, u, p)| &\le \phi_2(x, t) + c|u|^{q^*/q'} + c|p|^{m^*/q'}, \end{aligned} \tag{11.97}$$

where $q^* < q = 2(n + 2)/n$, $q' = q/(q - 1)$, $m^* < 2$.

(2) For any function $u(x)$ from $W_0{}^1(V)$

$$\int_v [a_i(x, t, u, u_x)u_{x_i} + a(x, t, u, u_x)u]\, dx \ge \nu \int_v |u_x|^2\, dx - c(t) \int_v (1 + u^2)\, dx, \tag{11.98}$$

$$\nu > 0, \qquad \int_0^T c(t)\, dt \le c.$$

(3) A monotonicity condition is valid

$$\int_v [a_i(x, t, v, v_x) - a_i(x, t, v, u_x)](v_{x_i} - u_{x_i})\, dx \ge \int_v \nu(|v_x|, |u_x|)|v_x - u_x|\, dx, \tag{11.99}$$

where $\nu(t_1, t_2)$ is a continuous positive function for $t_1 \ge 0$ and $t_2 \ge 0$, while u and v are arbitrary elements of $W_0{}^1(V)$.

(4) Galerkin's method is applied using a complete system of functions $\{\psi_k(x)\}$ in $W_0{}^1(V)$ such that $(\psi_k, \psi_l) = \delta_{kl}$ and $\max_v[|\psi_k|, |\psi_{kx}|] = C_k < \infty$. The approximate solution determined from the equations

$$(u_t, \psi_k) + (a_i, \psi_{kx_i}) + (a, \psi_k) = 0, \qquad C_k{}^N(0) = (g, \psi_k), \tag{11.100}$$

converges weakly in V and uniformly in $(0, T)$ to the solution to (11.96). First derivatives $u_{x_i}^N$ converge weakly to u_{x_i}.

The theorem can be extended to systems of equations and boundary conditions of the third kind (Ladyženskaya *et al.* 1968; Finlayson, 1971). In this case $u = \big(u_1(x, t), \ldots, u_M(x, t)\big)$ and $|u|^2 = \sum_{i=1}^{M} u_i^2$.

THEOREM 11.18. The problem

$$\frac{\partial u_j(x, t)}{\partial t} - \alpha_j \nabla^2 u_j(x, t) + R_j(u, x, t) = 0,$$

$$[\mathbf{n} \cdot \nabla u_j + \delta_j(s, t) u_j]_{s_T} = \psi_j(s, t), \qquad s \in S, \tag{11.101}$$

$$u_j|_{t=0} = \psi_{j0}(x), \qquad j = 1, \ldots, M,$$

has at least one generalized solution if $\psi_{j0} \in \mathscr{L}_2(V)$,

$$|R(u, x, t)| \le \phi(x, t) + c|u|^{q^*/q'}, \qquad \phi \in L^{q'}(Q_T), \tag{11.102}$$

and for any functions $u \in W^1(V)$,

$$\sum_{j=1}^{M} u_j R_j \ge -c(t)(1 + |u|^2), \qquad \int_0^T c(t)\, dt \le c. \tag{11.103}$$

The Galerkin method is applied using a complete system of functions $\{\psi_k(x)\}$ in $W^1(V)$ such that $(\psi_k, \psi_l) = \delta_{kl}$ and $\max_v[|\psi_k|, |\psi_{kx}|] = C_k < \infty$. The approximate solution determined from the equations

$$\int_{Q_T} [-u_j \phi_t + \alpha_j \nabla u_j \cdot \nabla \phi + R_j \phi]\, dx\, dt$$

$$+ \int_v u_j \phi\, dx - \int_v \psi_{j0} \phi(x, 0)\, dx + \alpha_j \int_{S_T} (\delta_j u_j - \psi_j)\phi\, ds\, dt = 0 \tag{11.104}$$

for any smooth $\phi \in W^{1,1}(Q_T)$, converges weakly in V and uniformly in $(0, T)$ to the solution to Eqs. (11.101). First derivatives $u_{x_i}^N$ converge weakly to u_{x_i}.

This theorem applies to several problems arising in chemically reacting systems, Eqs. (5.53 and 5.98), with the following reaction rate expressions (see Finlayson, 1971):

$$R = cK(T), \qquad R = cK(T)/[1 + |c|K(T)],$$

$$K(T) = \exp[\gamma(1 - (1/|T|))], \tag{11.105}$$

$$R = (1 - c)K(T) - \kappa c \exp[\gamma - (\gamma_2/|T|)].$$

For the orthogonal collocation method we need the following theorem.

THEOREM 11.19. The orthogonal collocation method applied to (11.101) converges if the Galerkin method converges.

As discussed in Section 5.6 the orthogonal collocation method can be regarded as a discrete form of Galerkin's method, which causes a quadrature error in evaluating integrals involving the nonlinear function $R_j(u, x, t)$. Yet as the number of terms in the approximation solution is increased, this quadrature error goes to zero, so that the collocation and Galerkin solutions become identical. Thus if the Galerkin method converges, the orthogonal collocation method converges, too. The Jacobi polynomials used in the orthogonal collocation method are complete in $W^1(0, 1)$ (Courant and Hilbert, 1953, Chapter 2).

These theorems also apply to steady-state problems, assuming that a steady state exists. Thus the Galerkin and orthogonal collocation method converge for the nonlinear heat conduction problems [Eqs. (2.3) and (2.48)]. Since the Rayleigh–Ritz method is equivalent to a Galerkin method, the variational methods applied to (7.47) converge. If the heat transfer coefficient depends on temperature, the theorems must be generalized. Theorems 11.18 and 11.19 imply convergence of the Galerkin and collocation methods applied to the steady-state chemical reaction problems [Eqs. (5.58) and (9.139)]. For transient heat and mass transfer and entry-length problems convergence also follows (Sections 3.2–3.4, 5.2, and 7.5).

We illustrate the application of the theorems by considering the nonlinear transport equation, in which all physical properties are temperature dependent and the velocity is a known, bounded function of position and time:

$$\rho C \frac{\partial T}{\partial t} - \nabla \cdot (k \nabla T) + \rho C \mathbf{u} \cdot \nabla T = f(x, t, T) \quad \text{in } V,$$
$$T(\mathbf{x}, 0) = T_0(\mathbf{x}), \qquad T = f(\mathbf{x}_s) \quad \text{on } S. \tag{11.106}$$

The theorem requires homogeneous boundary conditions. Suppose there exists a function $T_s(x)$ which satisfies $T_s = f$ on S, $T_s \in \mathscr{L}_2(V)$. Consider the new function $u = T - T_s$. We also must divide the equation by ρC, which is assumed positive; the positivity is a consequence of the second law of thermodynamics. After rearrangement, the problem is

$$Nu \equiv u_t - \nabla \cdot (\alpha \nabla u) - \gamma \nabla u \cdot \nabla u - \left(\alpha \frac{d \ln \rho C}{dT} + \frac{1}{\rho C}\frac{dk}{dT}\right) \nabla u \cdot \nabla T_s$$
$$+ \mathbf{u} \cdot \nabla u - g(x, t, u) = 0,$$
$$g \equiv \alpha \nabla^2 T_s + \frac{1}{\rho C}\frac{dk}{dT} \nabla T_s \cdot \nabla T_s + \frac{1}{\rho C} f - \mathbf{u} \cdot \nabla T_s, \tag{11.107}$$
$$\alpha = k/\rho C, \qquad \gamma = \alpha\, d \ln \rho C/dT,$$
$$u = 0 \quad \text{on } S, \qquad u(\mathbf{x}, 0) = \phi(\mathbf{x}) = T_0 - T_s.$$

In the notation of the theorem we define

$$a_i \equiv \alpha\, u_{x_i},$$

$$a \equiv -\gamma\, \nabla u \cdot \nabla u - \left(\alpha \frac{d \ln \rho C}{dT} + \frac{1}{\rho C} \frac{dk}{dT}\right) \nabla u \cdot \nabla T_s + \mathbf{u} \cdot \nabla u - g. \tag{11.108}$$

The theorem is satisfied if the following conditions hold,

$$\begin{aligned} &0 < \alpha_1 \le \alpha \le \alpha_2, && |\gamma| \le \gamma_2, \\ &|g| \le c_1 + c_2 |u|^2, && 1 - u\, d(\ln \rho C)/dT \ge c_3 > 0, \\ &-gu \ge -c_4(1 + u^2), && (1/\rho C) dk/dT \le c_5, \end{aligned} \tag{11.109}$$

where α_1, α_2, γ_2, c_i are constants. Basically this puts restrictions on the rate of growth of the function g and the temperature dependence of the physical parameters. The satisfaction of these inequalities is aided by use of the maximum principle, discussed in Section 11.5. If we can find values of z and Z in Eq. (11.140) the temperature can be bounded *a priori* and the physical parameters can be easily bounded. Thus we have the following theorem.

THEOREM 11.20. A generalized solution to the problem (11.106) exists provided the conditions (11.109) are satisfied. Galerkin's method converges in the mean when the trial solution is taken in the form

$$u = \sum_{i=1}^{N} c_i(t) \psi_i(x), \tag{11.110}$$

where $\{\psi_i\}$ satisfy the conditions of Theorem 11.17 and the functions of time $c_i(t)$, are determined from Eqs. (11.100), (11.107), and (11.108).

Quasilinear Parabolic Systems

Quasilinear parabolic systems are also treated by Visik (1961b). The equation in Q is

$$\frac{\partial u}{\partial t} + \sum_{|\alpha|, |\gamma| \le m} (-1)^{|\alpha|} D_\alpha A_\alpha(x, t, D_\gamma u) \equiv \frac{\partial u}{\partial t} + L(u) = h(x, t), \tag{11.111}$$

where

$$\begin{aligned} &h = (h^1, \dots, h^M), && D_\alpha = \partial^{|\alpha|}/\partial x_{\alpha_1}, \cdots \partial x_{\alpha_n}, \\ &\alpha = (\alpha_1, \dots, \alpha_n), && |\alpha| = \alpha_1 + \cdots + \alpha_n . \end{aligned}$$

The initial and boundary conditions are

$$u(x, 0) = \psi(x), \qquad u = \phi(x, t), \dots, D_w u = \phi_w(x, t) \tag{11.112}$$

for $x \in S \times (0, T)$, $|w| \leq m - 1$. The main condition on the nonlinear operator is the semiboundedness of $L(u)$:

$$A(w; v, v) \equiv \sum_{|\alpha|, |\beta| \leq m} (A_{\alpha\beta}(x, t, D_{\gamma w}) D_\beta v, D_\alpha v)$$
$$\geq c^2 \left(\sum_{j, |\alpha| = m} |D_\alpha w^j|^{\delta_j} D_\alpha v^j, D_\alpha v^j \right) - K(v, v), \qquad (11.113)$$

where $A_{\alpha\beta} = \partial A_\alpha / \partial D_\beta u$, w, and v are arbitrary sufficiently smooth functions and v satisfies the homogeneous boundary conditions (11.112) with $\phi_j = 0$. The constants c^2 and K do not depend on w, v, or t.

By a solution we mean a function $u(x, t)$ satisfying (11.112) and

$$\int_0^T \left[\left(\frac{\partial u}{\partial t}, v \right) + \sum_{|\alpha|, |\gamma| \leq m} (A_\alpha(x, t, D_\gamma u), D_\alpha v) \right] dt = \int_0^T (h, v)\, dt \qquad (11.114)$$

for any function v satisfying the homogeneous boundary conditions. Visik proves the generalized solution is unique.

THEOREM 11.21. The Galerkin method with $u_N = f + \sum C_{N_i} X_i(x)$ converges weakly to the generalized solution. The first time derivative and spatial derivatives $|\alpha| \leq m$ converge weakly to the corresponding derivatives of the generalized solution. The complete system $\{X_i(x)\}$ is capable of approximating in the mean derivatives up to order m. The function f satisfies the nonhomogeneous boundary conditions.

Navier-Stokes Equations

For the Navier–Stokes equations, Serrin (1963) gives an excellent review of the existence theorems, which have been proved using the Galerkin method. Taking $\rho = 1$, $\mu = 1$, the equations are:

$$\begin{aligned} \mathbf{u}_t + \mathbf{u} \cdot \nabla \mathbf{u} &= -\nabla p + \nabla^2 \mathbf{u}, \\ \nabla \cdot \mathbf{u} &= 0, \\ \mathbf{u}(\mathbf{x}, 0) &= \mathbf{u}_0(\mathbf{x}), \qquad \mathbf{x} \in V, \\ \mathbf{u}(\mathbf{x}, t) &= 0, \qquad \mathbf{x} \in S, \quad t \in (0, T). \end{aligned} \qquad (11.115)$$

Hopf (1950, 1951) was the first to prove that a generalized solution existed. Kiselev and Ladyženskaya (1963) proved the same result under weaker assumptions about the initial data ($\mathbf{u}_0$) and showed, in addition, that the first spatial and time derivatives are uniformly bounded until a time T, which is infinite for a two-dimensional problem or a three-dimensional problem with small enough initial data. In addition it has been shown when the weak solution has a certain regularity, it is unique and a classical solution exists almost everywhere. Since these results are proved using Galerkin's method they imply weak convergence of the Galerkin method.

11.4 Eigenvalue Problems

Consider the eigenvalue problem,

$$Lu - \lambda u = 0 \quad \text{in } V, \qquad u = 0 \quad \text{on } S. \tag{11.116}$$

Define the inner products $A(u, v) = (u, v)_L$, $B(u, v) = (u, v)$, for elements u and v in $\mathscr{F}$ which vanish on S and are twice continuously differentiable. We assume that the operator is self-adjoint, such that the bilinear form A is symmetric, $A(u, v) = A(v, u)$, and that the quotient $A(u, u)/B(u, u)$ is bounded below by a positive constant for all functions u in $\mathscr{F}$. The quadratic function $B(u, u)$ is called completely continuous with respect to $A(u, u)$ if every uniformly bounded sequence $\{u_n\}$ contains a subsequence $\{u_n'\}$ such that $B(u_n' - u_m', u_n' - u_m') \to 0$ as $n, m \to \infty$. Weinberger (1962) proves the following theorems.

THEOREM 11.22. If the positive definite quadratic functional $B(u, u)$ is completely continuous with respect to $A(u, u)$ then a minimizing vector exists such that the Rayleigh quotient (11.117) assumes its minimum value for some vector u in the space $\mathscr{F}$:

$$\lambda = A(u, u)/B(u, u). \tag{11.117}$$

THEOREM 11.23. Consider the square region $D = \{0 \leq x, y \leq 1\}$ and the space $\mathscr{F}$:

$$A(u, u) \equiv \int (\nabla u \cdot \nabla u + u^2)\, dV, \qquad B(u, u) = \int u^2\, dV. \tag{11.118}$$

Then $B(u, u)$ is completely continuous with respect to $A(u, u)$.

These theorems mean that there exists a minimizing function; otherwise the eigenvalue defined by (11.117) may not actually be attained for a function u in $\mathscr{F}$. Convergence of the Rayleigh–Ritz method also holds.

THEOREM 11.24. Let $X_i(x)$ be a system of linearly independent functions which are complete in energy. Then the Rayleigh–Ritz approximate eigenvalues $\lambda_j^{(n)}$ converge to the true eigenvalues λ_j as n approaches infinity.

A corresponding theorem on eigenfunctions is more difficult to establish. Indeed it sometimes happens (for the Galerkin method) approximate eigenfunctions do not converge to the true eigenfunctions (see Mikhlin, 1964, p. 476).

Consider the eigenvalue problem

$$-\sum_{i,j=1}^{3} \frac{\partial}{\partial x_i}\left(A_{ij}\frac{\partial u}{\partial x_j}\right) + \sum_{i=1}^{3} B_i \frac{\partial u}{\partial x_i} + \lambda C(x)u = 0 \quad \text{in } V, \tag{11.119}$$

$$u = 0 \quad \text{on } S,$$

where A_{ij} satisfies (11.47). Mikhlin (1964, p. 480) proves the following theorem.

THEOREM 11.25. If the Galerkin method is applied to Eqs. (11.119) using a system of functions which is complete in energy and linearly independent, then the approximate eigenvalues converge to the true eigenvalue.

This proves convergence of the Galerkin method in Section 3.1 when applied to (11.119).

Ordinary Differential Equations

The following general theorem holds.

THEOREM 11.26. Consider the problem (11.120) where f is continuous, $\partial f/\partial y$ and $\partial f/\partial y'$ are bounded,

$$\lambda = \min \frac{I(y)}{J(y)}, \qquad I(y) = \int_a^b f(x, y, y')\, dx, \qquad J(y) = \int_a^b y^2\, dx. \tag{11.120}$$

Let $X_i(x)$ be a system of functions satisfying the boundary conditions such that for any $\varepsilon > 0$ it is possible to find an n and constants c_i such that $v_n = \sum_{i=1}^{n} c_i X_i$:

$$|y - v_n| < \varepsilon, \qquad |y' - v_n'| < \varepsilon. \tag{11.121}$$

Then

$$\lim_{n\to\infty} \frac{I(v_n)}{J(v_n)} = \lambda. \tag{11.122}$$

Thus the eigenvalues converge. For the Sturm–Liouville system a stronger result is valid (Gelfand and Fomin, 1963, p. 198).

THEOREM 11.27. For the problem

$$\begin{gathered}(p(x)y')' - \lambda q(x)y = 0, \qquad y(0) = y(1) = 0,\\ p(x) > 0, \qquad q(x) \geq 0, \qquad 0 \leq x \leq 1,\end{gathered} \tag{11.123}$$

apply the Rayleigh–Ritz method using the trial functions $X_i(x) = \sin i\pi x$. Then the approximate eigenvalues converge to the true value and the approximate eigenfunctions converge uniformly to the exact eigenfunctions.

Error bounds for the case $p = 1$ are available (Kantorovich and Krylov, 1958, p. 336):

$$\left|\frac{\lambda_j - \lambda_j^{(n)}}{\lambda_j}\right| < \frac{\lambda_j^{(n)}}{\pi^4(n+1)^4}\left[|\lambda_j^{(m)}| \max q^2 + \max\left|\frac{q'^2}{q}\right|\right]. \qquad (11.124)$$

Mikhlin (1964, p. 477) also proves convergence of the Galerkin method for the eigenvalue problem given by (11.56) with $f = 0$ under the same conditions as in Theorem 11.8. Birkhoff *et al.* (1966) and Johnson (1969) prove convergence and provide error bounds when the trial functions are piecewise cubic functions. Leipholz (1967) proves convergence of the Galerkin method when the trial functions satisfy only the essential boundary conditions and certain boundary terms are added to the residual to account for the natural boundary conditions.

Stability Problems

Convergence proofs exist for several of the stability problems treated in Chapter 6. DiPrima and Sani (1965) prove the following theorem.

THEOREM 11.28. Apply the Galerkin method to the Taylor–Dean problem (11.125):

$$(D^2 - a^2)^2 u = f(x)v, \qquad (D^2 - a^2)v = -a^2 T g(x) u,$$
$$u = Du = v = 0 \quad \text{at} \quad x = 0, 1, \qquad (11.125)$$

for continuous functions f and g and trial functions $u_n = (1-x)^2 x^{n+1}$, $v_n = (1-x)x^n$ or the Reid and Harris functions [Eqs. (6.8) and (6.9)]. The approximate eigenvalues converge to the exact eigenvalues. The eigenfunctions converge in the Hilbert space with the norm

$$\|u, v\| = \int_0^1 [|(D^2 - a^2)u|^2 + |Dv|^2 + a^2|v|^2]\, dx. \qquad (11.126)$$

Sani (1968) proves convergence when u is expanded in a complete set of functions in $\mathscr{L}_2(0, 1)$ (such as u_n given above), the second equation is solved exactly for v_n, and the first equation is solved approximately by the Galerkin or variational methods. Petryshyn (1965, 1968) provides a general theorem which is useful for proving convergence of the Galerkin method.

The completeness of the trial function is most important. Petrov (1940) claims to prove convergence of the Galerkin method applied to plane Couette and plane Poiseuille flow Eq. (6.180). The trial functions are $f_s = \phi_s'' - \alpha^2\phi_s$, where ϕ_s is an eigenfunction of

$$\phi^{\text{IV}} - 2\alpha^2\phi'' + \alpha^4\phi + \mu(\phi'' - \alpha^2\phi) = 0,$$
$$\phi = \phi' = 0 \quad \text{at} \quad x = 0, 1. \qquad (11.127)$$

Gallagher (1969) recently showed that Petrov's results were in error because the expansion functions are not complete in $\mathscr{L}_2(0, 1)$ for all values of α.

For nonlinear stability problems Eckhaus (1965) and DiPrima (1967) have applied a method which can be related to MWR. The eigenfunctions to the linear stability problem are used as expansion functions with time-dependent coefficients in the nonlinear equations. The nonlinear equations are then made orthogonal to the adjoint eigenfunction, leaving a set of ordinary differential equations in the expansion coefficients. These can be solved using either an initial-value approach, as in Section 6.3, or solutions asymptotic in some parameter, such as $T - T_c$, can be found. While formal convergence theorems have not been established, DiPrima and Habetler (1969) proved completeness of the eigenfunctions to the linear Taylor problem [time-dependent form of (11.125)], convective instability (Section 6.2), and Orr–Sommerfeld equation (Section 6.6).

THEOREM 11.29. The eigenfunctions of the operator

$$(L - \lambda M)\phi = 0 \tag{11.128}$$

are complete when either

$$\begin{aligned} (1) \quad & L = (D^2 - a^2)^2 - ia\mathrm{Re}[U(D^2 - a^2) - U''], \\ & M = -(D^2 - a^2), \\ & \phi = \phi' = 0 \quad \text{at} \quad x = 0, 1, \end{aligned} \tag{11.129}$$

or

$$\begin{aligned} (2) \quad & L = \begin{pmatrix} (DD^* - a^2)^2 & a\sqrt{T}f(x) \\ a\sqrt{T}g(x) & -(DD^* - a^2) \end{pmatrix}, \\ & M = \begin{pmatrix} -(DD^* - a^2)^2 & 0 \\ 0 & v \end{pmatrix}, \\ & \phi_1 = D\phi_1 = \phi_2 = 0 \quad \text{at } x = 0, 1. \end{aligned} \tag{11.130}$$

Eisenfeld (1968) has also proved completeness of the normal modes for (11.125).

11.5 Error Bounds Using the Maximum Principle

Many elliptic and parabolic differential equations obey a maximum principle: the maximum of the solution occurs on the boundary. Another form of the maximum principle transforms an inequality on the differential operator to an inequality on the function itself. We outline here a few of the results discussed in the excellent book on maximum principles by Protter and Weinberger (1967). The theorems are followed by applications to problems of physical interest.

THEOREM 11.30. Let $u(x)$ be a solution of the boundary-value problem

$$\begin{aligned} &u'' + H(x, u, u') = 0, \qquad a < x < b, \\ &B_1(u) \equiv -u'(a)\cos\theta + u(a)\sin\theta = \gamma_1, \\ &B_2(u) \equiv u'(b)\cos\phi + u(b)\sin\phi = \gamma_2, \end{aligned} \tag{11.131}$$

where $0 \le \theta, \phi \le \pi/2$, and θ and ϕ are not both zero. Suppose that H, $\partial H/\partial u$, and $\partial H/\partial u'$ are continuous and that $\partial H/\partial u \le 0$. If Z and z satisfy

$$\begin{aligned} Z'' + H(x, Z, Z') \le 0, \qquad & z'' + H(x, z, z') \ge 0, \\ B_1(Z) \ge \gamma_1 \ge B_1(z), \qquad & B_2(Z) \ge \gamma_2 \ge B_2(z), \end{aligned} \tag{11.132}$$

then the solution u obeys the inequality

$$z \le u \le Z. \tag{11.133}$$

This theorem gives error bounds for the solution, provided functions Z and z can be found with the requisite properties. Notice that the main condition on Z and z is that the differential operators have a particular sign. We take a function and compute the residual. If it is everywhere positive (or negative) the function is an upper (or lower) bound. Approximate solutions are of course candidates for Z or z. Usually, however, approximate solutions do not satisfy the inequalities on the residual unless they are especially constructed using some of the methods outlined below.

We next consider the problem

$$Lu + hu = f(\mathbf{x}) \quad \text{in } V, \tag{11.134a}$$

$$Bu \equiv \partial u/\partial n + \alpha(x)u = g_1(\mathbf{x}) \quad \text{on } S_1, \tag{11.134b}$$

$$u = g_2(\mathbf{x}) \quad \text{on } S_2, \tag{11.134c}$$

$$Lu \equiv \sum_{i,j=1}^{n} a_{ij}(x)\frac{\partial^2 u}{\partial x_i\,\partial x_j} + \sum_{i=1}^{n} b_i(x)\frac{\partial u}{\partial x_i}, \qquad h = h(\mathbf{x}). \tag{11.134d}$$

We assume L is uniformly elliptic [a_{ij} satisfies the inequality (11.47)], that $S = S_1 \cup S_2$, $\partial u/\partial n$ is an outward directional derivative, and $\alpha > 0$. Each point of S_1 lies on the boundary of a ball contained in V, and V is bounded. Protter and Weinberger (1967, p. 78) prove the following theorem.

THEOREM 11.31. Suppose there exists a function $w > 0$ on $V \cup S$ such that

$$Lw + hw \le 0 \quad \text{in } V, \qquad (\partial w/\partial n) + \alpha w \ge 0 \quad \text{on } S_1, \tag{11.135}$$

under the stated assumptions (if $h \leq 0$ then $w = 1$ has the desired properties). We assume that the following three conditions do not all hold: (1) $\partial w/\partial n + \alpha w \equiv 0$ on S_1, (2) $Lw + hw \equiv 0$ in D, (3) S_2 is vacuous. If Z and z satisfy

$$\begin{gathered}(L + h)Z \leq f \leq (L + h)z \quad \text{in } V, \\ BZ \geq g_1 \geq Bz \quad \text{on } S_1 \qquad Z \geq g_2 \geq z \quad \text{on } S_2,\end{gathered} \tag{11.136}$$

then the solution u to Eqs. (11.134) satisfies

$$z \leq u \leq Z \quad \text{in } V. \tag{11.137}$$

An immediate consequence is that the solution is unique. Suppose there were two solutions, u and u_1, satisfying Eqs. (11.134). Then u_1 is a candidate for both Z and z, so that $u_1 \leq u \leq u_1$, or $u = u_1$. The theorem also can be used to prove the following theorem.

THEOREM 11.32. The solution to Eqs. (11.134) with h, f, α, and g_1 equal to zero takes its minimum and maximum value on S_2.

Define the constants $a = \min_s u$, $b = \max_s u$. Then $u \geq a$, $b \geq u$ on the boundary. But a and b satisfy the differential equation and (11.136) is satisfied. Then (11.137) follows so that $a \leq u \leq b$ in V. A similar result exists for nonlinear elliptic operators, but it is obtainable from the theorem for parabolic operators which we discuss next:

$$\begin{gathered}L(u) \equiv F(\mathbf{x}, t, u, \partial u/\partial x_i, \partial^2 u/\partial x_i\, \partial x_j) - \partial u/\partial t = f(\mathbf{x}, t), \\ u(\mathbf{x}, 0) = g_1(\mathbf{x}) \quad \text{in } V, \qquad u(\mathbf{x}, t) = g_2(\mathbf{x}, t) \quad \text{on } S_T.\end{gathered} \tag{11.138}$$

We assume that F is elliptic in the domain Q, which means that

$$\sum_{i,j=1}^{n} \frac{\partial F(x, t, u, p_q, r_{pq})}{\partial r_{ij}} \xi_i \xi_j > 0, \qquad \xi \neq 0, \tag{11.139}$$

holds for each point in Q and for any u. In particular we need (11.139) for all functions of the form $\theta u + (1 - \theta)w$, with $0 \leq \theta \leq 1$. The operator L is parabolic whenever F is elliptic and assume that V is bounded. Protter and Weinberger (1967, p. 187) prove the following theorem.

THEOREM 11.33. Suppose u is the solution to (11.138). Let Z and z be functions satisfying the inequalities

$$\begin{gathered}L(Z) \leq f \leq L(z) \quad \text{in } V, \qquad z(x, 0) \leq g_1 \leq Z(x, 0) \quad \text{in } V, \\ z \leq g_2 \leq Z \quad \text{on } S_T,\end{gathered} \tag{11.140}$$

and assume that L is parabolic and $\partial F/\partial u$ is bounded for the functions $\theta u + (1-\theta)z$ and $\theta u + (1-\theta)Z$, $0 \le \theta \le 1$. Then

$$z \le u \le Z \quad \text{in } Q. \tag{11.141}$$

For the nonlinear elliptic case we must also assume $\partial F/\partial u \le 0$. This theorem says that the approximate solution is a lower bound on the exact solution if it satisfies the initial and boundary conditions and the residual is everywhere positive in Q. The theorem for infinite domains and linear parabolic operators has been proved by Il'in *et al.* (1962) under the added restriction that the coefficients in the operator satisfy certain growth conditions. For the system of linear equations

$$L_i(u_i) + \sum_{j=1}^{k} h_{ij} u_j = f_i, \tag{11.142}$$

Protter and Weinberger (1967, p. 188) prove a maximum principle under the assumption

$$h_{ij} \ge 0 \qquad \text{for} \quad i \ne j. \tag{11.143}$$

Kastenberg (1970) proves a theorem for a system of nonlinear equations.

Finlayson and Scriven (1966) use the maximum principle to obtain systematically improvable pointwise bounds for solutions to the transport equation. Consider a linear problem (11.133) with F given by Eq. (11.134d). The trial solution and residual are

$$u = u_0 + \sum_{i=1}^{N} c_i u_i, \qquad L(u) - f = R_0 + \sum_{i=1}^{N} c_i R_i. \tag{11.144}$$

We wish to choose the constants such that the residual is negative everywhere, and formulate the problem,

$$\text{minimize} \sum_{i=1}^{N} c_i u_i(x_l, t_l), \tag{11.145}$$

subject to the conditions

$$R_0(\mathbf{x}_k, t_k) + \sum_{i=1}^{N} c_i R_i(\mathbf{x}_k, t_k) \le -\varepsilon, \qquad k = 1, \ldots, M. \tag{11.146}$$

This represents a linear programming problem which can be solved with known techniques. For $\varepsilon = 0$ the inequality (11.146) is violated for points between the collocation points, while for ε large it should be possible to have (11.146) valid for all $\mathbf{x}$ and t. The result is an upper bound. Naturally we want to choose ε as small as possible in order to obtain the best bound (it can be shown that the bound is monotone increasing with ε).

For a sample application, Finlayson and Scriven (1966) consider

$$\nabla^2 T - 2\,\mathrm{Pe}\,\mathbf{u} \cdot \nabla T - \frac{\partial T}{\partial t} = 0,$$
$$\mathbf{u} = (x\mathbf{i} - z\mathbf{k}), \qquad T(\mathbf{x}, 0) = 0, \qquad T(z = 0, t) = 1. \tag{11.147}$$

The solution is independent of x (see Chan and Scriven, 1970). The problem can be transformed using a similarity transformation:

$$T(z, t) = \phi(\eta), \qquad \eta = z/\delta(t),$$

$$\frac{d\delta^2}{dt} + 4 \text{ Pe } \delta^2 = \alpha, \tag{11.148}$$

$$\phi'' + \frac{\alpha}{2}\eta\phi' = 0, \qquad \phi(0) = 1, \qquad \phi(\infty) = 0.$$

Trial functions are taken in the form

$$\phi = \begin{cases} 1 + \sum_{i=1}^{N} c_i \eta^i, & \eta < \kappa, \\ 0, & \eta > \kappa, \end{cases} \tag{11.149}$$

for the lower bound, and

$$\phi = \sum_{i=1}^{N} c_i e^{-i\eta} \tag{11.150a}$$

or

$$\phi = \sum_{i=1}^{N} c_i e^{-(i+1)\eta}. \tag{11.150b}$$

for the upper bound. Appropriate restrictions on the c_i are necessary to insure that the function and all derivatives appearing in the differential equation are continuous in the domain (particularly at $\eta = \kappa$) and that Eqs. (11.150a) and (11.150b) give a residual which is negative as $\eta \to \infty$. The flux at the boundary is also bounded, since there the upper and lower bounds agree and are continuous. The flux has the form

$$\text{Nu} = A \text{ Pe}^{1/2}[1 - \exp(-4 \text{ Pe } t)]^{-1/2} \tag{11.151}$$

with $0.864 \leq A \leq 1.145$. The exact value is 1.128 (Chan and Scriven, 1970). The upper and lower bounds for temperature are compared with the exact solution in Table 11.2. The lower bound is obtained using Eq. (11.149) with $N = 9$ and the upper bound comes from Eq. (11.150b) with $N = 2$. Other solutions are tabulated elsewhere (Finlayson, 1965).

The lower bound ($\varepsilon = 0.0078$) is very close to the exact solution. The upper bound is further away because of the severe restrictions necessary to ensure that the residual remain negative as $\eta \to \infty$. The approach is more useful in finite domains if close bounds are desired.

TABLE 11.2

COMPARISON OF UPPER AND LOWER BOUNDS FOR TEMPERATURE

η	ϕ lower	ϕ exact	ϕ upper
0.0	1.000	1.000	1.000
0.25	0.719	0.724	0.778
0.50	0.472	0.480	0.552
0.75	0.279	0.289	0.373
1.00	0.146	0.157	0.245
1.50	0.022	0.034	0.099
2.00	0.000	0.005	0.039

We next consider the nonlinear heat transfer problem. Boley (1964) obtained bounds for a similar problem, except that he did not require his trial functions to have continuous second derivatives as is required by the theorem:

$$\frac{\partial}{\partial x}\left[k(T)\frac{\partial T}{\partial x}\right] - \rho C\frac{\partial T}{\partial t} = 0, \tag{11.152}$$

$$k = k_0\left(1 + \frac{aT}{T_0}\right), \qquad \rho C = \text{constant}, \tag{11.153}$$

$$T(x, 0) = 0, \qquad T(0, t) = T_0. \tag{11.154}$$

We first use the solution to a linear problem and show that it is a lower bound. Take $k = k_0$ and call T_1 the solution to

$$k_0\frac{\partial^2 T_1}{\partial x^2} = \rho C\frac{\partial T_1}{\partial t} \tag{11.155}$$

under the same initial and boundary conditions (11.154). The solution is known:

$$T_1 = T_0 \operatorname{erfc}\left[\frac{x}{(4k_0 t/\rho C)^{1/2}}\right]. \tag{11.156}$$

We check the conditions of the theorem

$$NT_1 = \frac{dk}{dT_1}\left(\frac{\partial T_1}{\partial x}\right)^2 + [k(T_1) - k_0]\frac{\partial^2 T_1}{\partial x^2}. \tag{11.157}$$

For heating, $\partial T_1/\partial t \geq 0$ so that $\partial^2 T_1/\partial x^2 \geq 0$. Suppose a is positive, then $k(T_1) \geq k_0$ and $dk/dT_1 \geq 0$ so that $NT_1 \geq 0$. Thus T_1 is a lower bound by the maximum principle.

We next find an approximate solution by the integral method. Assume a trial function of the following form, which is one at the boundary, reduces to zero further into the domain, and has continuous second derivatives everywhere:

$$T_3 = \begin{cases} \left[1 - \dfrac{x}{q(t)}\right]^3, & x \le q, \\ 0, & x \ge q, \end{cases} \qquad \eta = \frac{x}{q}. \tag{11.158}$$

Take

$$qq' = \alpha k_0/\rho C \tag{11.159}$$

and determine α by the integral method

$$\int_0^1 N(T_3)\, d\eta = 0. \tag{11.160}$$

For $a = 2.5$ the result is $\alpha = 70$. The solution is then (11.158) with

$$q = (70 k_0 t/\rho C)^{1/2}. \tag{11.161}$$

If the residual is evaluated we find that it is both positive and negative, so that it is not a candidate for an upper or lower bound.

We next find the best bound of the form (11.158). Choose q as the solution to Eq. (11.159) with α unspecified, and the residual becomes

$$N = \frac{k_0}{q^2}\{15a(1-\eta)^4 + 6(1-\eta) - 3\alpha\eta(1-\eta)^3\}. \tag{11.162}$$

For $a = 2.5$, $N(u)$ is positive for $0 \le \eta \le 1$ if $\alpha \le 22$. Thus a lower bound is

$$T_2 = \begin{cases} \left(1 - \dfrac{x}{q}\right)^3, & x \le q, \\ 0, & x \ge q, \end{cases} \qquad q = \left(22 \frac{k_0}{\rho C} t\right)^{1/2}. \tag{11.163}$$

This is the best lower bound and is compared to the exact solution (derived using a numerical integration, Crank, 1956) in Table 11.3. Notice that T_3, which is not an upper or lower bound, is both above and below the exact solution.

TABLE 11.3

TEMPERATURE UPPER AND LOWER BOUNDS FOR $a = 2.5$

η	T_1	T_2	T_{exact}	T_3
0.0	1.00	1.00	1.00	1.00
0.25	0.72	0.71	0.87	0.83
0.5	0.48	0.48	0.73	0.68
1.0	0.15	0.18	0.43	0.44
1.5	0.03	0.04	0.16	0.26

More accurate results are possible using a computer. Collatz (1966a, b) gives upper and lower bounds for several differential equations, including

$$-y'' = 3x^2 - e^{-y}, \qquad y(0) = y(1) = 0. \tag{11.164}$$

The upper and lower bounds agree to five decimal places. Schröder (1966) solves the two-dimensional problem:

$$\nabla^2 u = e^u \quad \text{in } V, \qquad u = 0 \quad \text{on } S, \tag{11.165}$$

and the bounds are very close. For example,

$$-1.127690 \le u(0, 0) \le -1.127663. \tag{11.166}$$

Appl and Hung (1964) and Hung and Appl (1967) study an ordinary differential equation describing heat transfer from thin fins with temperature-dependent thermal properties and internal heat generation. For a five-term solution the upper and lower bounds agree to four or five significant figures. Mangasarian (1963) derives bounds for the solution to $\nabla^4 u = 0$, whereas Duncan (1932) provides bounds for the torsion of cylinders of symmetric cross section. See also Wetterling (1968). Rabinowitz (1968) surveys the applications of linear programming to derive bounds using the maximum principle and equations such as (11.146).

Nickel (1962) has proved a maximum principle for the boundary layer equations:

$$\begin{gathered} u\, u_x + v\, u_y - UU' - \nu\, u_{yy} = 0, \qquad v = v_0(x) - \int_0^y u_x(x, t)\, dt, \\ u(0, y) = \tilde{u}(y) \quad 0 \le y < \infty, \\ \left.\begin{aligned} &u(x, 0) = 0, \\ &u(x, y) \to U(x) \quad \text{as } y \to \infty \end{aligned}\right\} 0 \le x \le x_1, \end{gathered} \tag{11.167}$$

and constructs a lower bound for velocity.

11.6 Error Bounds Using the Mean-Square Residual

Some results above indicated the error was bounded in terms of the mean-square residual Eqs. (11.68), (11.72), and (11.78). Additional results are given below. For these problems, the mean-square residual provides a criterion for testing an approximate solution.

Sigillito (1966, 1967a, b) proves inequalities which give pointwise and mean-square error bounds in terms of the mean-square residual.

THEOREM 11.34. Let u and v be functions which are continuous and have continuous first derivatives in x and t and continuous second derivatives in x. The function u satisfies (11.168) and f satisfies a Lipschitz condition in u:

$$J(u) \equiv (a_{ij}u,_i),_j - \frac{\partial u}{\partial t} = f(x, t, u). \tag{11.168}$$

Then $w \equiv u - v$ and $R \equiv J(v) - f(x, t, v)$ satisfy

$$\int_0^T \int_V w^2 \, dV \, dt \le \alpha_1 \int_V w^2(x, 0) \, dV + \alpha_2 \int_0^T \int_S w^2 \, ds \, dt + \alpha_3 \int_0^T \int_V R^2 \, dV \, dt, \tag{11.169}$$

$$|w(x, t)| \le K \int_0^T \int_V w^2 \, dV \, dt + \mathscr{F}(x, t, R^2).$$

The constants α_i can be determined from the bounded domain; K and $\mathscr{F}$ depend on position in the domain and become infinite as the boundary is approached.

The theorem is valid for boundary conditions of the first and third kind. An example problem is solved:

$$\begin{gathered} Lu = u_{xx} - u_t = 0, \\ u(x, 0) = \cos x, \qquad u(0, t) = u(\pi, t) = e^{-t}. \end{gathered} \tag{11.170}$$

The error bounds given in Table 11.4 are clearly conservative.

TABLE 11.4

ERROR BOUNDS FOR HEAT CONDUCTION EQUATION[a]

x	t	Actual error $\times 10^7$	Error bound $\times 10^7$
$0, 0.8\pi$	0.2	<1	580
$0, 0.8\pi$	1.0	<1	72
0	2.0	<1	127
0.8π	2.0	1	620
0	4.0	1	4000
0.8π	4.0	23	30,000

[a] Sigilloto (1967b). Used by permission of the copyright owner, copyright © 1967, Association for Computing Machinery, Inc.

For many chemical reaction problems closer bounds can be obtained. Consider the semilinear problem

$$Lu = f(u, x), \tag{11.171}$$

where $f(u, x)$ is a nonlinear function of the variable u but not of its derivatives. We formally find an inverse to L:

$$u = L^{-1} f(u, x). \tag{11.172}$$

The approximate solution defines a residual, which can be manipulated to give

$$Lu_n - f(u_n, x) = R_n, \qquad u_n = L^{-1} f(u_n, x) - L^{-1} R_n. \tag{11.173}$$

Combination of Eqs. (11.172) and (11.173) gives the relation

$$\|u - u_n\| \le \frac{\|R_n\| \, \|L^{-1}\|}{1 - K\|L^{-1}\|}, \tag{11.174}$$

where K is the Lipschitz constant

$$\|f(u, x) - f(u_n, x)\| \le K\|u - u_n\|. \tag{11.175}$$

Equation (11.174) then gives the error in the solution in terms of the mean-square residual. The error applies to the Galerkin and other methods of MWR, although the least-squares method gives the minimim estimation of $\|R_n\|$.

Similar bounds are valid for systems of equations. Ferguson (1971) considers the diffusion and reaction problem in a spherical catalyst pellet:

$$\begin{gathered} \nabla^2 u_i = f_i(u_1, \ldots, u_N), \qquad i = 1, \ldots, N, \\ \frac{2}{\mathrm{Sh}} \frac{du_i}{dx} + u_i(1) = g_i \quad \text{at} \quad x = 1, \qquad \frac{du_i}{dx} = 0 \quad \text{at} \quad x = 0 \end{gathered} \tag{11.176}$$

The error bound is

$$\|\tilde{u}_i - u_i\| \le \|L^{-1}\| \left[\|R_i\| + \frac{K_i^{1/2} \|L^{-1}\| (\sum_k \|R_k\|^2)^{1/2}}{1 - \|L^{-1}\| (\sum_k K_k)^{1/2}} \right], \tag{11.177}$$

where

$$|f_i(u) - f_i(v)| \le \sum_{j=1}^{N} K_{ij} |u_j - v_j|, \qquad K_i = \sum_{j=1}^{N} K_{ij}^2 \tag{11.178}$$

and the mean-square Green's function is (for a spherical region)

$$\|L^{-1}\| = \left[\frac{1}{90} + \frac{2}{\mathrm{Sh}} \left(\frac{1}{9} \frac{2}{\mathrm{Sh}} + \frac{2}{45} \right) \right]^{1/2}. \tag{11.179}$$

The bounds are valid provided $\|L^{-1}\| (\sum_k K_k)^{1/2} < 1$. For a single equation bounds are valid for the reaction rate expressions listed in **Table** 11.5.

TABLE 11.5

CONDITIONS FOR ERROR BOUNDS IN TERMS OF MEAN-SQUARE RESIDUAL[a]

Reaction rate expression	Eq. (11.177) holds for restrictions
$f = \phi^2 c$	$\phi^2 < (270)^{1/2}$
$f = \phi^2 c^n$	$\phi^2 < (270)^{1/2}/n(1+\varepsilon)^{n-1}$ $n \geq 1, \quad -\varepsilon \leq \tilde{c}(\xi) \leq 1+\varepsilon$
$f = \dfrac{\phi^2 c}{1+\alpha c}$	$\phi^2 \leq 270^{1/2}(1-\alpha\varepsilon)^2$ $\alpha\varepsilon < 1$

[a] Ferguson (1971).

Bounds are calculated for the problems,

$$C \rightleftarrows 3A \rightleftarrows B,$$

$$\frac{1}{x^2}\frac{d}{dx}\left(x^2\frac{dc_B}{dx}\right) = k_1 c_B - k_2[1 - 3(c_B + c_C)]^3,$$

$$\frac{1}{x^2}\frac{d}{dx}\left(x^2\frac{dc_C}{dx}\right) = k_3 c_C - k_4[1 - 3(c_B + c_C)]^3, \tag{11.180}$$

$$\frac{dc_B}{dx} = \frac{dc_C}{dx} = 0, \quad \text{at} \quad x = 0; \qquad c_B = c_C = 0.05 \quad \text{at} \quad x = 1,$$

$$k_1 = 0.1, \qquad k_2 = 1.0, \qquad k_3 = 0.2, \qquad k_4 = 1.2.$$

The problem was solved using the orthogonal collocation method with two sets of polynomials: those corresponding to weighting functions $w = 1$ and $1 - x^2$. The residual was evaluated to determine the bounds shown in Table 11.6. Evident is the convergence as n increases and the residual becomes

TABLE 11.6

ERROR BOUNDS FOR CHEMICAL REACTION SYSTEM, POLYNOMIALS WITH $w = 1$[a]

N	$\|R_B\|$	Bounds for $\|\tilde{c}_B - c_B\|$	Bounds for $\|\tilde{c}_B - c_B\|_\infty$
1	3.2 (−3)[b]	5.1 (−4)	2.8 (−3)
2	1.8 (−4)	3.0 (−5)	1.6 (−4)
3	8.9 (−6)	1.5 (−6)	7.9 (−6)
6	6.6 (−10)	1.1 (−10)	5.8 (−10)
8	9.3 (−13)	1.6 (−13)	8.2 (−13)

[a] Ferguson (1971).
[b] For example, $\|R_B\| = 3.2 \times 10^{-3}$.

smaller. The mean-square residuals for polynomials using $w = 1$ are from 12 to 140% less than those obtained using $w = 1 - x^2$. The same problem was also solved for a boundary condition of the third kind (Ferguson, 1971). Error bounds such as these can be used to monitor an approximate solution, although their calculation increases the computer time (70 and 30% increase for $n = 6$ and 8, respectively). Ferguson (1971) also considers the nonisothermal problem with an irreversible first-order reaction:

$$\frac{1}{x^2}\frac{d}{dx}\left(x^2\frac{dT}{dx}\right) = -\delta[1 + \beta - T]\exp\left[-\gamma\left(\frac{1}{T} - 1\right)\right],$$
$$T'(0) = 0, \qquad T(1) = 1. \tag{11.181}$$

Calculations are done for $\delta = 0.25$, $\beta = +0.3$, $\gamma = 20$, corresponding to a unique steady state, and the convergence of the solution is shown in Table 11.7. These results indicate that the mean-square residual provides an error

TABLE 11.7

ERROR BOUNDS FOR NONISOTHERMAL CHEMICAL REACTION, POLYNOMIALS WITH $w = 1$[a]

N	$\|R\|$	$\|\tilde{T} - T\|$ bound	$\|\tilde{T} - T\|_\infty$ bound
1	2.9 (−3)[b]	3.1 (−4)	1.9 (−3)
2	1.3 (−4)	1.5 (−5)	8.6 (−5)
3	4.7 (−6)	5.1 (−7)	3.2 (−6)
6	1.1 (−10)	1.2 (−11)	7.0 (−11)
8	3.2 (−14)	3.6 (−15)	2.2 (−14)
10	4.0 (−17)	4.3 (−18)	—
12	6.2 (−19)	6.8 (−20)	—

[a] Ferguson (1971).
[b] For example, $\|R\| = 2.9 \times 10^{-3}$.

criterion, and the error decreases to zero as higher approximations are calculated. To calculate the results in Table 11.7 it was necessary to use double precision arithmetic for $N \geq 8$ on a CDC 6400 computer.

Ferguson (1971) has also converted these mean-square error bounds to pointwise error bounds. We define the maximum error as

$$\|\tilde{T} - T\|_\infty = \max_{0 \leq x \leq 1} |\tilde{T}(x) - T(x)|. \tag{11.182}$$

The mean square residuals in Tables 11.6 and 11.7 can then be converted into pointwise errors [see Ferguson (1971) for the proof]. We emphasize that the exact solution need not be known to calculate these bounds. The pointwise errors are proportional to the norm of the mean-square residual, so that as this is decreased the pointwise error decreases. The pointwise errors in Tables 11.6 and 11.7 are so accurate that the orthogonal collocation method, combined with the error bounds, can be said to give the exact solution. Exact, analytic solutions to Eqs. (11.180) and (11.181) are not known, so that the results on pointwise error bounds represent a significant advance.

REFERENCES

Anderssen, R. S. (1969). The Numerical Solution of Parabolic Differential Equations Using Variational Methods, *Num. Math.* **13**, 129–145.

Appl, F. C., and Hung, H. M. (1964). A Principle for Convergent Upper and Lower Bounds, *Int. J. Mech. Sci.* **6**, 381–389.

Babusko, I., Prager, M., and Vitasek, E. (1966). "Numerical Processes in Differential Equations." Wiley (Interscience), New York.

Birkhoff, G., DeBoor, C., Swartz, B., and Wendroff, B. (1966). Rayleigh–Ritz Approximation by Piecewise Cubic Polynomials, *SIAM J. Numer. Anal.* **3**, 188–203.

Birkhoff, G., Schultz, M. H., and Varga, R. S., (1968). Piecewise Hermite Interpolation in One and Two Variables with Applications to Partial Differential Equations, *Num. Math.* **11**, 232–256.

Boley, B. A. (1964). The Analysis of Problems of Heat Conduction and Melting, *in* "High Temperature Structures and Materials," *Proc. Symp. Naval Structural Mech. 3rd.*, pp. 260–315. Columbia Univ., New York, Pergamon Press, Oxford.

Ciarlet, P. G., Schultz, M. H., and Varga, R. S. (1967). Nonlinear Boundary-Value Problems I. One Dimensional Problem, *Num. Math.* **9**, 394–430.

Chan, W. C., and Scriven, L. E. (1970). Absorption into Irrotational Stagnation Flow, *Ind. Eng. Chem. Fund.* **9**, 114–120.

Cheney, E. W. (1966). "Introduction to Approximation Theory." McGraw-Hill, New York.

Collatz, L. (1966a). "Functional Analysis and Numerical Mathematics." Academic Press, New York.

Collatz, L. (1966b). Monotonicity and Related Methods in Nonlinear Differential Equations Problems, *in* "Numerical Solutions of Nonlinear Differential Equations" (D. Greenspan, ed.), pp. 65–87. Wiley, New York.

Courant, R. (1943). Variational Methods for the Solution of Problems of Equilibrium and Vibrations, *Bull. Amer. Math. Soc.* **49**, 1–23.

Courant, R., and Hilbert, D. (1953). "Methods of Mathematical Physics, " Vol I. Wiley (Interscience), New York.

Crank, J. (1956). "The Mathematics of Diffusion." Oxford Univ. Press (Clarendon), London and New York.

DiPrima, R. C. (1967). Vector Eigenfunction Expansions for the Growth of Taylor Vortices in the Flow between Rotating Cylinders, *in* "Nonlinear Partial Differential Equations: A Symposium on Methods of Solution," pp. 19–42. Academic Press, New York.

DiPrima, R. C., and Habetler, G. J. (1969). A Completeness Theorem for Non-Self-Adjoint Eigenvalue Problems in Hydrodynamic Stability, *Arch. Rat. Mech. Anal.* **34**, 218–227.

DiPrima, R. C., and Sani, R. L. (1965). The Convergence of the Galerkin Method for the Taylor–Dean Stability Problem, *Quart. Appl. Math.* **23**, 183–187.

Douglas, J., Jr., and DuPont, T. (1970). Galerkin Method for Parabolic Equations, *SIAM J. Num. Anal.* **7**, 575–626.

Dubinskii, J. A. (1967a). Weak Convergence in Nonlinear Elliptic and Parabolic Equations, *Amer. Math. Soc. Transl.* (2) **67**, 226–258.

Dubinskii, J. A. (1967b). An Operator Scheme and the Solvability of Some Quasilinear Equations in Mechanics, *Sov. Math. Dokl.* **8**, 1118–1121.

Duncan, W. J. (1932). On the Torsion of Cylinders of Symmetrical Section, *Proc. Roy. Soc.* (*London*) **A136,** 95–113.

Eckhaus, W. (1965). "Studies in Non-Linear Stability Theory." Springer-Verlag, Berlin,

Eisenfeld, J. (1968). Normal Mode Expansion and Stability of Couette Flow, *Quart. Appl. Math.* **26,** 433–440.

Ferguson, N. B. (1971). Orthogonal Collocation as a Method of Analysis in Chemical Reaction Engineering, Ph.D. Thesis, Univ. of Washington, Seattle, Washington.

Finlayson, B. A. (1965). Approximate Solutions of Equations of Change; Convective Instability by Active Stress. Ph.D. Thesis, Univ. of Minnesota, Minneapolis, Minnesota.

Finlayson, B. A. (1971). Convergence of the Galerkin Method for Nonlinear Problems Involving Chemical Reaction, *SIAM J. Num. Anal.* **8**, 316–324.

Finlayson, B. A., and Scriven, L. E. (1966). Upper and Lower Bounds for Solutions to the Transport Equation, *AIChE. J.* **12**, 1151–1157.

Finn, R., and Smith, D. R. (1967). On the Linearized Hydrodynamical Equations in Two Dimensions, *Arch. Rat. Mech. Anal.* **25**, 1–25.

Gallagher, A. P. (1969). On a Proof by Petrov of the Stability of Plane Couette Flow and Plane Poiseuille Flow, *SIAM J. Appl. Math.* **17**, 765–768.

Gelfand, I. M., and Fomin, S. V. (1963). "Calculus of Variations." Prentice-Hall, Englewood Cliffs, New Jersey.

Green, J. W. (1953). An Expansion Method for Parabolic Partial Differential Equations, *J. Res. Natl. Bur. Std.* (*U.S.*) **51**, 127–132.

Heywood, J. G. (1970). On Stationary Solutions of the Navier–Stokes Equations as Limits of Nonstationary Solutions, *Arch. Rat. Mech. Anal.* **37**, 48–60.

Hopf, E. (1950/51). Über die Anfangswertaufgabe für die Hydrodynamischen Grundgleichungen, *Math. Nachr.* **4**, 213–223.

Hung, H. M., and Appl. F. C. (1967). Heat Transfer of Thin Fins with Temperature-Dependent Thermal Properties and Internal Heat Generation, *J. Heat Transfer, Trans. ASME, Ser. C* **89**, 155–162.

Il'in, A. M., Kalashnikov, A. S., and Oleinik, O. A. (1962). Second-Order Linear Equations of Parabolic Type, *Russ. Math. Surveys* **17**, 1–144.

Johnson, O. G. (1969). Error Bounds for Sturm–Liouville Eigenvalue Approximations by Several Piecewise Cubic Rayleigh–Ritz Methods, *SIAM J. Numer. Anal.* **6**, 317–333.

Kadner, H. (1960). Untersuchungen zur Kollokationsmethode, *Z. Angew. Math. Mech.* **40**, 99–113.

Kantorovich, L. V., and Krylov, V. I. (1958). "Approximate Methods in Higher Analysis." Wiley (Interscience), New York.

Karpilovskaya, E. B. (1963). Convergence of the Collocation Method for Certain Boundary-Value Problems of Mathematical Physics, *Sib. Mat. Zh.* **4**, 632–640.

Kastenberg, W. E. (1970). Comparison Theorems for Nonlinear Multicomponent Diffusion Systems, *J. Math. Anal. Appl.* **29**, 299–304.

Keldysh, M. V. (1942). On B. G. Galerkin's Method for the Solution of Boundary-Value Problems, *Izv. Akad. Nauk USSR* **6**, 309–330; *English trans.* NASA TT F-195.

Kiselev, A. A., and Ladyženskaya, O. A. (1963). On the Existence and Uniqueness of Solutions of the Non-Stationary Problems for Flows of Non-Compressible Fluids, *Amer. Math. Soc. Transl.* (2) **24**, 79–106.
Ladyženskaya, O. A., Solonnikov, V. A,. and Ural'ceva, N. N. (1968). Linear and Quasi-Linear Equations of Parabolic Type, *Transl. of Math. Monogr.* **23**, Amer. Math. Soc., Providence, Rhode Island.
Leipholz, H. (1965). The Convergence of the Galerkin Procedure for Quasi-Linear Boundary-Value Problems, *Acta Mech.* **1**, 339–353.
Leipholz, H. (1967a) On the Choice of the Trial Functions in the Application of the Method of Galerkin, *Acta Mech.* **3**, 295–317.
Leipholz, H. (1967b). Über die Befreiung der Ansatzfunktionen des Ritzchen und Galerkinsehen Verfahrens von den Randbedingungen, *Ing. Arch.* **36**, 251–261.
Mangasarian, O. L. (1963). Numerical Solution of the First Biharmonic Problem by Linear Programming, *Int. J. Eng. Sci.* **1**, 231–240.
Mikhlin, S. G. (1964). "Variational Methods in Mathematical Physics." Pergamon, Oxford.
Mikhlin, S. G., and Smolitskiy, K. L. (1967). "Approximate Methods for Solution of Differential and Integral Equations." American Elsevier, New York.
Nickel, K. (1961). Parabolic Equations with Applications to Boundary Layer Theory, *in* "Partial Differential Equations and Continuim Mechanics" (E. Langer, ed.), pp. 319–330, Univ. of Wisconsin Press, Madison, Wisconsin.
Nickel, K. (1962). A Simple Estimation of Boundary Layer, *Ing. Arch.* **31**, 85–100.
Nitsche, J. (1969). Vergleich der Konvergenzgeschwindigheit des Ritzschen Verfahrens und der Fehlerquadratmethode, *Z. Angew. Math. Mech.* **49**, 591–596.
Petrov, G. I. (1940). Application of the Method of Galerkin to a Problem Involving the Stationary Flow of a Viscous Fluid, *Prik. Matem. Mekh.* **4**, No. 3, 3–12.
Petryshyn, W. V. (1965). On a Class of K-p.d. and Non-K-p.d. Operators and Operator Equations, *J. Math. Anal. Appl.* **10**, 1–24.
Petryshyn, W. V. (1968). On the Eigenvalue Problem $Tu - \lambda Su = 0$ with Unbounded and Nonsymmetric Operators T and S, *Phil. Trans. Roy. Soc.* (*London*) **262**, 413–458.
Protter, M. H., and Weinberger, H. F. (1967). "Maximum Principles in Differential Equations." Prentice-Hall, Englewood Cliffs, New Jersey.
Rabinowitz, P. (1968). Applications of Linear Programming to Numerical Analysis, *SIAM Rev.* **10**, 121–159.
Sani, R. (1968). An Extension and Convergence Proof of an Approximate Method due to Pellew and Southwell, *Z. Angew. Math. Mech.* **48**, 65–66.
Schröder, J. (1966). Ungleichungen und Fehlerabschätzungen, *Proc. Int. Cong. Math.* (*Moscow*) pp. 101–129.
Schultz, M. H. (1969a). Error Bounds for the Rayleigh–Ritz–Galerkin Method, *J. Math. Anal. Appl.* **27**, 524–533.
Schultz, M. H. (1969b). The Galerkin Method for Non-Self-Adjoint Differential Equations, *J. Math. Anal. Appl.* **28**, 647–651.
Schultz, M. H. (1969c). Rayleigh–Ritz–Galerkin Methods for Multidimensional Problems, *SIAM J. Numer. Anal.* **6**, 523–538.
Serrin, J. B. (1963). The Initial-Value Problem for the Navier–Stokes Equations, *in* "Nonlinear Problems" (R. E. Langer, ed.), pp. 69–98. Univ. of Wisconsin Press, Madison, Wisconsin.
Sigillito, V. G. (1966). Pointwise Bounds for Solutions of the First Initial Boundary-Value Problem for Parabolic Operators, *SIAM J. Appl. Math.* **14**, 1038–1056.
Sigillito, V. G. (1967a). Pointwise Bounds for Solutions of Semilinear Parabolic Equations, *SIAM Rev.* **9**, 581–585.

Sigillito, V. G. (1967b). On a Continuous Method of Approximating Solutions of the Heat Equation, *J. Assoc. Comp. Mech.* **14**, 732–741.

Shampine, L. F. (1968). Error Bounds and Variational Methods for Nonlinear Boundary-Value Problem, *Num. Math.* **12**, 410–415.

Szalek, M. (1970). Error Estimates for Some Variational Methods Applicable to Scattering and Radiation Problems, *Quart. Appl. Math.* **27**, 473–479.

Vainikko, G. M. (1965). On the Stability and Convergence of the Collocation Method, *Diff. Eqn.* **1**, 186–194.

Veliev, M. A. (1964). On Investigation of the Stability of the Bubnov–Galerkin Method for Nonstationary Problems, *Sov. Math.* **5**, 856–858.

Visik, M. I. (1961a). Boundary-Value Problems for Quasilinear Strongly Elliptic Systems in Divergence Form, *Sov. Math.* **2**, 643–647.

Visik, M. I. (1961b). Boundary Value Problems for Quasilinear Parabolic Systems of Equations and Cauchy's Problem for Hyperbolic Equations, *Sov. Math.* **2**, 1292–1295.

Visik, M. I. (1962). Simultaneous Quasilinear Elliptic Equations with Lower-Order Terms, *Sov. Math.* **3**, 629–633.

Visik, M. I. (1963). The Cauchy Problem for Equations with Operator Coefficients; Mixed Boundary Value Problem for Systems of Differential Equations and Approximation Methods for Their Solution, *Amer. Math. Soc. Transl.* (2) **24**, 173–278.

Visik, M. I., and Ladyženskaya, O. A. (1958). Boundary Value Problems for Partial Differential Equations and Certain Classes of Operator Equations, *Amer. Math. Soc Transl.* (2) **10**, 223–282.

Weinberger, H. F. (1962). "Variational Methods for Eigenvalue Problems." Univ. of Minnesota Book Store, Minneapolis, Minnesota.

Wetterling, W. (1968). Lokal Optimale Schranken bei Randwertaufgaben, *Computing* **3**, 125–130.

Yasinsky, J. B., and Kaplan, S. (1968). On the Use of Dual Variational Principles for the Estimation of Error in Approximate Solutions of Diffusion Problems, *Nucl. Sci. Eng.* **31**, 80–90.

Author Index

Numbers in italics refer to the pages on which the complete references are listed.

Subject Index

V

W